LIVE FORM

L F

I O

V R

E M

Women, Ceramics, and Community

JENNI SORKIN

THE UNIVERSITY OF CHICAGO PRESS *Chicago & London*

JENNI SORKIN is assistant professor of art history at
the University of California, Santa Barbara.

The University of Chicago Press, Chicago 60637
The University of Chicago Press, Ltd., London

Printed in the United States of America

25 24 23 22 21 20 19 18 17 16 1 2 3 4 5

ISBN-13: 978-0-226-30311-6 (cloth)
ISBN-13: 978-0-226-30325-3 (e-book)
DOI: 10.7208/chicago/9780226303253.001.0001

Library of Congress Cataloging-in-Publication Data
Names: Sorkin, Jenni, author.
Title: Live form : women, ceramics, and community / Jenni Sorkin.
Description: Chicago ; London : The University of Chicago Press,
2016. | Includes bibliographical references and index.
Identifiers: LCCN 2015040757 | ISBN 9780226303116 (cloth :
alk. paper) | ISBN 9780226303253 (e-book)
Subjects: LCSH: Women potters—United States. | Ceramics—United States.
Classification: LCC NK4008 .S67 2016 | DDC 738.092/2—
dc23 LC record available at http://lccn.loc.gov/2015040757

∞ This paper meets the requirements of ANSI/
NISO Z39.48-1992 (Permanence of Paper).

CONTENTS

ACKNOWLEDGMENTS

Toril Moi once wrote that philosophy proceeds from everyday language. I am borrowing the sensibility: scholarship also proceeds from the everyday, the lived experience of writing, thinking, talking, but in particular, the blessing of having people who listen and respond.

This book has its roots in my own formative years as an art student in the Fiber and Material Studies Department of the School of the Art Institute of Chicago in the late 1990s. It is this program's unique pedagogy and ongoing engagement to alternative modernisms, processes, women's labor, history, and materials that has profoundly shaped my own trajectory and scholarly pursuit of material culture and gender. I am proud and grateful to continue my relationship with SAIC and its community of artists and intellectuals, including David Getsy, Diana Guerrero-Maciá, Ellen Rothenberg, Shannon Stratton, Christine Tarkowski, and Lisa Vinebaum. For their ongoing support, generosity and friendship, I am especially indebted to Anne Wilson and Joan Livingstone. Also Barbara DeGenevieve, a formative mentor whose memory looms large in my own teaching, and I want to honor the memory of Kathryn Hixson, who took a chance on me and added my first stabs at criticism to the cacophony of voices that was *The New Art Examiner* before it folded.

The second point of contact for my development as a feminist writer and thinker came from the four years I spent working on *WACK! Art and the Feminist Revolution*, culminating in an exhibition at MOCA in 2007. The intensity of the research process, including hundreds of studio visits, and the ongoing recovery and archival work necessary to the preservation of feminist culture revealed to me the complexities of gender identity and artistic practice. It is what ultimately compelled me to reach back to an earlier generation of women artists. But I continue to be inspired by present and past dialogues with Jerri Allyn, Sheila de Bretteville, Norma Broude, Connie Butler, Judy Chicago, Mary Garrard, Susan Jenkins, Carole Ann Klonarides, Catherine Lord, Lisa Mark, Helen Molesworth, Griselda Pollock, Elaine Reichek, Martha Rosler, Paul Schimmel, Carolee Schneemann, Joan Snyder, and Faith Wilding, as well as Tee Corinne and Arlene Raven, in memoriam. I would also like to thank Michael Brenson for being a champion of my writing at an early moment in my career and a thoughtful interlocutor.

I would like to thank the Graduate School and History of Art Department at Yale University for sustained fellowship and research support and, in particular, Carol Armstrong, Ned Cooke, David Joselit, Mary Miller, and Sally Promey for invaluable feedback and numerous constructive discussions over many years. So much of what I do in the classroom and on the page is informed by my time at Yale. Thank you also to my extended cohort in New Haven, now dispersed: Bradley Bailey, Shira Brisman, Philip Ekhart, Freyja Hartzell, Jennifer Josten, Donny Meyer, Suzy Newbury, Jen Stob, and Bernie Zirnheld, for friendship and encouragement.

This project has benefited from the generosity of the following organizations and grant programs: Edith and Richard French Fellowship, Beinecke Rare Book and Manuscript Library, Yale, 2006; a Graduate Research Grant from the Center for Craft, Creativity and Design in Asheville, North Carolina in 2007; a Henry Luce Foundation/ACLS Pre-Doctoral Fellowship in 2008–9, at the Getty Research Institute; a library grant in 2009 and a Pacific Standard Time Post-Doctoral Fellowship from 2010–11; and the American Council of Learned Societies Fellowship in 2014–15 that permitted the completion of this project. I am deeply grateful to all the agencies and fellowship programs that have funded the research and writing of this book.

For images and the permission to reproduce objects, for sharing thoughts and memories, and for tracking down hard-to-source archival materials, I would like to thank the following people and institutions: Ara and Seth Cardew; Mary Beth Edelson; Erik Martz; Martha Rosler; Casey Allen and Andrew McAllister of the Nora Eccles Harrison Museum of Art; Mary-Beth Buegson and Garth Johnson at Ceramics Research Center, ASU Museum; Jane Kemp and Dr. Kate McAllister at Special Collections, Luther College; Karen Convertino of the Everson Museum; Anna Fariello and Liz Skene at Hunter Library, Western Carolina University; Heather South at the Western Regional Archives, State Archives of North Carolina; Steven Lee and Rachel Hicks at the Archie Bray Foundation; Jessica Shaykett, American Craft Council; Alice Sebrell of BMC Museum+Arts Center; Flora Brooks and Marty Stein, Museum of Fine Arts, Houston; Stephanie Soldner Sullivan of the Soldner Descendants' Trust; Pier Voulkos of the Voulkos

Family Trust; and Adam Welch of Greenwich House Pottery. A heartfelt thank you to Garth Clark and Margie Hughto for their enormous contribution to the history of ceramics and a genuine openness and willingness to engage in extended conversations, and to Billie Sessions for her prior scholarship on Marguerite Wildenhain, which has enabled my own.

A sincere thank you to Julia Connor/Estate of M. C. Richards; Jill Peterson Hoddick, Jan Peterson, and Taag Peterson/The Estate of Susan Peterson; and Jeffrey Spahn of Jeffrey Spahn Gallery for spirited commitments to the collective legacy of Susan Peterson, M. C. Richards, and Marguerite Wildenhain. A special thank you to Forrest L. Merrill, who opened up his collection, archives, and contacts to me. I greatly valued our discussions, visits, and the help of his assistant Dane Coultier. I was lucky enough to meet and interview Susan Peterson at her home in Scottsdale in 2007, before her passing, and I treasure this encounter.

The following friends and colleagues have been extraordinarily generous, extending invitations to write, think, and talk about this project or my scholarship publicly over the last ten years: Glenn Adamson, Elisabeth Agro, Rhea Anastas, Sarah Archer, Elissa Auther, Nick Bell, Lucy Bradnock, Kelly Cannole, Andrea Keys Connell, Vivien Green-Fryd, Ross Elfine, Geraldine Gourbe, Jennifer Harris, Del Harrow, Jessica Hemmings, Peter Held, Suzanne Hudson, Janis Jeffries, Alex Juhasz, Toby Kamps, Alex Klein, Jennie Klein, Katie Lee Koven, Elizabeth Kozlowski, Laura Kuhn, Heather Sealy Lineberry, Meg Linton, Joan Livingstone, Tom Loeser, Sue Maberry, Lisa Melandri, Sequoia Miller, Helen Molesworth, Mark Newport, Andrew Perchuk, Gabriela Rangel, Ezra Shales, Susie Silbert, Joshua Stein, Robert Storr, Cindi Strauss, Gloria Sutton, Anna Walker, Namita Gupta Wiggers, Anne Wilson, and Karin Higa, in memoriam.

At the University of Chicago Press, I am deeply indebted to my editor Susan Bielstein for her sustained interest and support, to my anonymous peer reviewers, and to editorial associates James Whitman Toftness and, previously, Anthony Burton, for seeing this book through.

I am thrilled to have wonderfully supportive, and intellectually engaged colleagues at the University of California, Santa Barbara, in particular Swati Chattapadhay, Laurie Monahan, Bruce Robertson, and Jerry White. Thank you to interim dean John Majewski for a publication grant and to Christine Fritsch-Hammes of the VRC for all her help in imaging. Thank you also to my previous colleagues at the University of Houston: Merriann Bidgood, Chris Conway, Sarah Costello, Rex Koontz, and Sandra Zalman.

Suzanne Hudson and Gloria Sutton have been key supporters and cheerleaders. Thank you to Katherine Litwin Swan for twenty-plus year of friendship and for the initial introduction to *Centering*. Special thanks are in order to my family: my siblings Amy Smith and Adam Sorkin; my uncles and aunts Sandi and Alex

Dubin, Steve and Sandy Solomon, and David Sorkin and Shifra Sharlin; and my stepparents Ira Moltz and Debbie Sorkin. I am deeply thankful for the time, encouragement, worry and kindness bestowed on me by my parents, Robbe Moltz and Sam Sorkin. They have been wonderfully, wildly supportive and generous to the point of being grantmakers without the application process. Their sustained pride in my work and career has been a true joy to me.

Finally, I am most grateful for my partner and spouse of twelve years, Cheri Owen, who has lived with this project for a very long time. She has good-naturedly exchanged vacations for countless research trips, kept me organized and task oriented, and remained a crucial sounding board and dedicated source of wisdom, warmth, and, most necessarily, humor. Her love, support, and hilarity have sustained me through my most inchoate moments of writing and revising. Thank you also to our furry and feathered brood: Riley, Pat, and Zima in the present and Prince, Macchiato, Oney, and Summit in the past, all cherished pets, whose wide-eyed, silent companionship and tail swishing I came to depend upon during innumerable late nights at my computer.

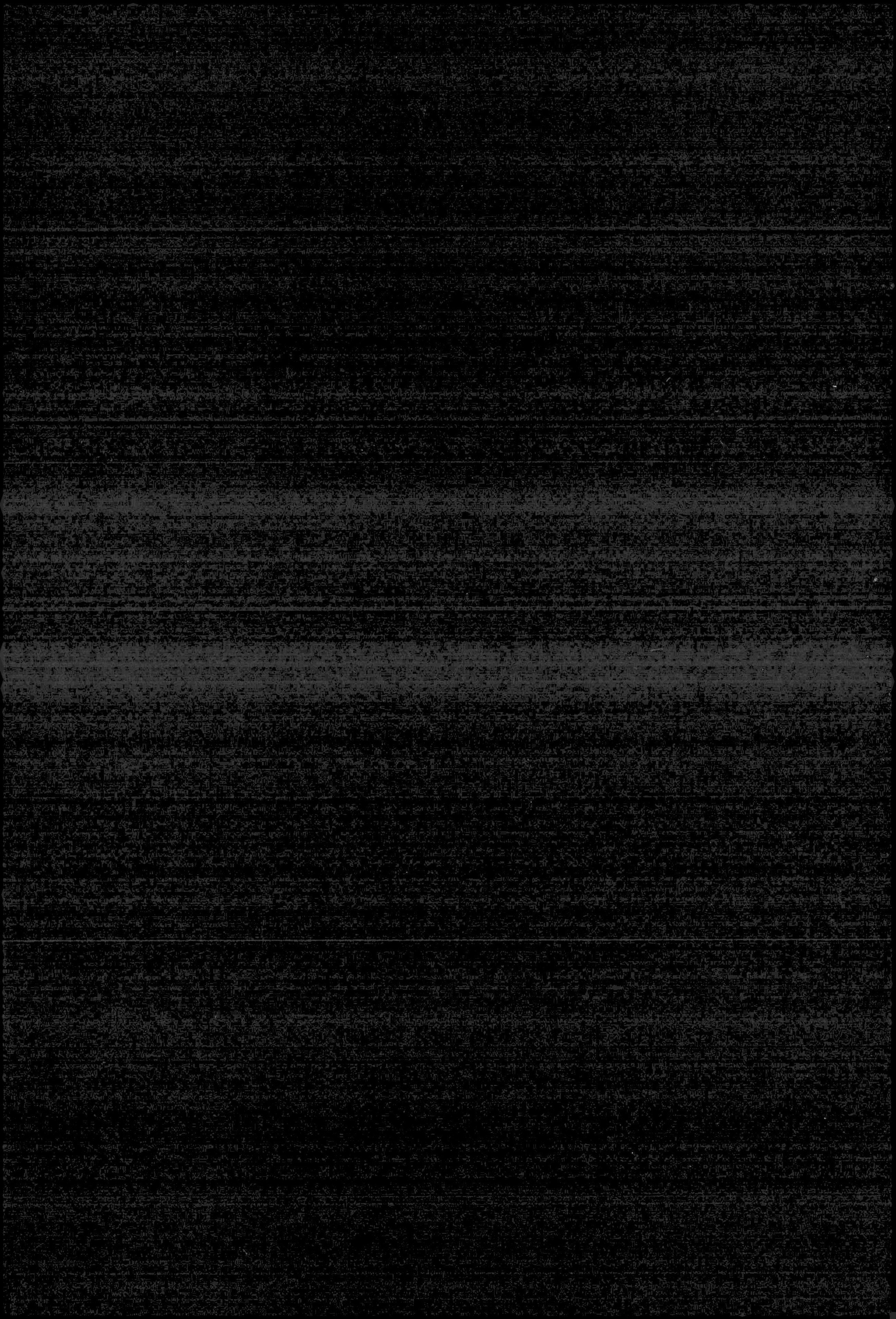

CRAFT AS COLLECTIVE PRACTICE

INTRODUCTION

One way of characterizing the social turn in contemporary artistic practice is to foreground its history in the pedagogical practices of previous generations, in this case, women ceramists whose careers throughout the mid-twentieth century expand and enrich our current understanding of what socially engaged artistic practice is today. This book argues that it was modern craft and not modern art that spearheaded nonhierarchical and participatory experiences, through the experiential properties endemic to craft practices and, in particular, ceramics. This runs counter to the existing genealogies of participatory art charted by Claire Bishop and Boris Groys, which are wholly tied to a European model of performance and non-object avant-garde practice.[1] Today, many artistic practices focus heavily on "socially engaged art," "institutional transformations," and "knowledge-exchange" between artist and audience.[2] Mid-century craft is an important but unacknowledged antecedent to the activist principles that service such contemporary ideologies. Moreover, it was women artists, many of whom were affiliated with social reform movements and spearheaded radical educational initiatives, who performed the teaching and transmission of craft skills and ideologies at midcentury.

This study is a thematic and gendered history of postwar American ceramics, which resituates a presently isolated and self-contained field as a malleable medium that could be manipulated to suit simultaneously avant-garde, nationalist, and regionalist constituencies. The tensions between these competing interests are compelling not just for their historical significance but for the prominent role craft has played in the political economy throughout the twentieth century.

Accordingly, the perpetuity of the art/craft divide is better understood politically, rather than aesthetically: craft is most often linked to fiscal policy, a redistribution of labor, production, and skill resulting in improved economic conditions, while avant-garde art is linked to social policy and its ensuing debates of morality. The circulation of handmade goods, no matter how expensive or inefficient their production, have markedly more potential for bestowing pride upon a nation through mass ownership and collection. In both formats of production, it is public efficacy that is at stake, turning on questions not of usefulness but of *use*, as political propaganda, the inherent attributes of craft's materialism, its sustainability and durability, is potent as a metaphor for community and nation-building. In this way, craft becomes the ultimate service discipline, its utopian and communal values both politically alluring and easily appropriated by ethnic, religious, diasporic, or cultural nationalisms.

In this book, I focus on three American women ceramists, each whom utilized the art form as a conduit for social contact: Marguerite Wildenhain (1896–1985), a Bauhaus-trained potter who taught form as process without product at her summer craft school Pond Farm; Mary Caroline (M. C.) Richards (1916–1999), who renounced formalism at Black Mountain College in favor of a therapeutic model she pursued outside academia; and Susan Peterson (1925–2009), who popularized ceramics through live throwing demonstrations on public television in 1964. These three artists were chosen in part for their direct relationship to teaching and writing—all left behind written legacies—and for their heavy involvement in the creation of alternative circuits and new forms of pedagogy.

At a time when women were virtually excluded from both the teaching and making of painting and sculpture, craft provided a vital arena as teachers, thinkers, and makers. It became a viable alternative to the mainstream, urban art worlds of New York City and Los Angeles, a space in which women could innovate, teach, and create lasting pedagogical structures. Ceramics in particular, with its emphasis on self-sufficient rural living, offered women unprecedented social freedoms, with the opportunity to live and teach in nontraditional settings, such as cooperative, experimental, or self-initiated communities. This unorthodox, largely rural livelihood was beholden to the formal requirements of their craft: the making, storage, and, most importantly, firing of ceramics. Due to the strict fire codes and zoning laws that made kilns illegal in most cities, the medium itself was ill suited to an urban setting. Infinitely more private, these off-the-grid situations were more conducive to alternative lifestyles and sexualities, minimizing the social pressure, judgment, and community policing endemic to the sexist and repressive 1950s. Able to barter their unique wares and skill sets, women, too, found varying degrees of financial autonomy in the informal economies of exchange that existed through pottery's social and pedagogical networks.

But to have chosen craft at that moment required a heightened awareness and rejection of conventional artistic structures, institutions, and hierarchies. As women foregrounded in both pedagogical and theoretical constructions of their shared discipline, Richards, Wildenhain, and Peterson were craftswomen engaged not just in highly skilled labor but, moreover, in the *language* of that labor. Their stories prefigure, or parallel, the masterful male pedagogical narratives now so well known—the ABCs, in effect—to scholars and students of the late twentieth and early twenty-first centuries: Josef Albers, Joseph Beuys, John Cage. In studio ceramics, as well, a version of this narrative has also been propagated, centered firmly on Peter Voulkos and his cadre of all-male students in the mid-1950s. As a parallel medium, ceramics offered a tantalizing proximity to the avant-garde movements of the era, including, but not limited to, abstract expressionism, Happenings, experimental music, minimalism, and, as I will argue, even early video art.

Indeed, its artists engaged in similar midcentury tactics: the expression of form through a suppression of narrative content, the embattled and sometimes exalted status of the artist as a maverick, the ambivalent relationship between artist and industry, and the increasing disregard for traditional artistic skills such as draftsmanship. In my project, the ceramic vessel becomes, in one sense, a foil, the shell object loaded with metaphorical and symbolic values of female labor, and, on the other hand, dematerialized, in tandem with the larger conceptual practices of the 1960s and 1970s, in which artists saw fit to engage with different forms of aesthetic experience not limited to traditional object making.

The similarities between the formation of craft and media discourses during the postwar era are striking. As a hand-based, ancient technology, ceramics became a powerful metaphor for encouraging critical awareness and adaptation in the face of new technologies such as radio and television, and its performativity developed alongside, not in spite of, this sensoria. If one of the criteria for newer media is remediation, that is, the ability to generate an adaptation, translating from one format to another, ceramics itself is marked by variability and transformation, no more consigned to its basic materiality than is a computer chip, originally derived from metals and sand (silica). Beyond art and aesthetics, clay consistently transforms into media formats that seem unimaginable, at the forefront of advanced technological innovation in fields such as dentistry, automotive, oil and gas, solar, industrial, electronics, and defense industries.

~~Outside~~ Beside

When it has been considered at all, ceramics has been a marginal practice within the history of postwar art. It is a field that is remote and maybe even a little bit exotic, particularly in settings where it is not taught with any regularity. It usually excites contemporary artists who revel in its wildly formless materiality and is regarded with some skepticism by modern and contemporary art historians. Within university art departments it has been consistently relegated to the basement (given the heavy equipment and toxic clay dust) or just as likely, it has been sequestered in design, applied, or manual arts departments. It is a medium that has been mired in formal discussions of technique, instead of its uniquely intertwined qualities of performance and pedagogical engagement.

Throwing functional pottery on a wheel is less a singular act than an integrated system of transference, a reciprocity between teaching/learning and social/affective imperatives. Eve Kosofsky Sedgwick used the phrase "texture and affect" to theorize discourses of narrative perception and dissonance in acts of touching and feeling, or, as she wrote, "Attending to psychology and materiality at the level of affect and texture is also to enter a conceptual realm that is not shaped by lack nor by commonsensical duality of subject versus object or of means versus ends."[3]

Using Sedgwick's terminology, "texture and affect" is a way of establishing the argument that in postwar ceramics, subject and object ultimately became interchangeable, and that throwing on the wheel was not a means to an object only. I am pursuing the discernable estrangement between the decorative arts histories where postwar ceramics has historically lived and the space that each functional vessel inhabits as an incomplete, process-based object that lives alongside, or *beside*, the rich dimensionality of collective practices of the 1960s and 1970s: dematerialized conceptual art, the Happening, and body-based feminist performance. Sedgwick used the term "beside" to encompass the generative ambivalence of ". . . a wide range of desiring, identifying, representing, repelling, paralleling, differentiating, rivaling . . ."[4] I've consistently framed pottery "beside" these other practices as a means of resisting the connoisseurship and hagiography that has passed for museum-based ceramics scholarship. The narrow purview of the encyclopedic decorative arts department or stand-alone craft museum has been to describe and analyze the formal attributes of significant objects, framed within the narrative biography of the maker and his or her influences.

To understand ceramics as a performative gesture made in real time and space is to witness an object-event. "Beside" also permits an ontology of *live form*, a term I employ to describe the performative gesture of collective skill building through the pottery demonstration. Throwing pottery on the wheel became a central tenet of craft curricula during the 1950s and 1960s, in which the process of making and the form itself were merged together as a seamless action. Ceramics, then, became a model for reintroducing authenticity, social engagement, and camaraderie into the creative process, overturning the traditional ideologies of craft as either an object-based commodity or a fetishization of labor.

In this book I argue that ceramics is central to a feminist history of this period: women ceramists of the 1950s pioneered the poststudio artistic production that characterized the decade of the 1960s—a decade Caroline Jones has marked as hypermasculine. But this has more extensive repercussions: 1950s women potters also set a precedent for 1970s feminist collectivism and, in particular, its legacy as a decade steeped in alternative-institution building.

Yet the status of Wildenhain's, Richards's, and Peterson's feminism is crucial: theirs was a protofeminist generation, invested in professional identities dependent upon what Anne Wagner calls "a gender-blind modernism," that is, the insistent separation between artistic identity and social identity.[5] If she upheld this distinction, a woman artist could not concede that the trajectory of her career was challenged or stymied by her gender. It is in the interstices of these dual identities, artist *and* wife, that Wagner performs her breathtaking analysis of Eva Hesse, Lee Krasner, and Georgia O'Keeffe.

But Wildenhain, Richards, and Peterson faced different predicaments. They were not artists *and* they were artists *beside*: their chosen medium was not even necessarily recognized as art. Certainly this limited their spheres of influence in the world beyond midcentury craft. The main inequity with which craftswomen lived lay within ceramics itself: demoted to the role of teacher—a conventionally acceptable female role—rather than regarded as an influential artist.

But functional pottery was especially well suited to willful genderblindness, as its primary concerns were formal even when ideological: embedded in a discourse of form and shape (traditional or not); materials (the composition of the clay body) and materialities (texture, tacility); and the infinite decorative effects of surface, made possible by original and inventive glaze chemistries.

In the 1989 reissue of *Centering*, Richards touched on the point of her own generational unawareness, writing:

> If I were to write the book now . . . I would not use the masculine pronoun to designate persons of both sexes. I would not write "man" when I mean "person" or "human being." . . . This would be changed even as our consciousness has changed, has become more discriminating, more subtle, more conscientious . . . Simply, we have outgrown the masculine pronounce as the referent for both sexes . . . We did not challenge the assumptions behind the grammar (the politics of language). We learned the rules and practiced them quite unconsciously, docile until meaning began to awaken in us and we began to hear what we were saying.[6]

For these reasons, I will use the term "craftsman" to preserve the language and feel of this now-defunct era. This was also how the craftswomen in question referred to themselves and their peers of both sexes. As writers of philosophical texts that lay claim to the spiritual impact of pottery, they also use "craftsman" and "artist" interchangeably in their writings. Consequently, I have done the same. In the 1950s, the studio potter represented a genderless ideal, a craftsperson who controlled the process of making from start to finish and successfully executed commissions without compromising quality or the integrity of workmanship—another gendered expression from the same era that grates on the ear when changed to "workpersonship."

Biography and Self-Reflexivity

Of all the crafts, ceramics is the medium with the largest body of self-reflexive literature: Bernard Leach, an English potter, is the wellspring, followed by a steady stream of English and American potters who produced books: F. Carlton Ball, Tony Birks, Michael Cardew, Michael Casson, Clary Illian, Susan Peterson, Philip Rawson, Daniel Rhodes, M. C. Richards, and Marguerite Wildenhain.[7] The majority of these volumes married the technical to the philosophical, sharing information and offering expertise to fellow potters alongside occasional spiritual or metaphysical reflections on the craft. Peterson, Richards, and Wildenhain all departed from this model in that their contributions cultivated the collective, rather than the intensely individual strains of the discipline.

Leach's *A Potter's Book* (1940) was heavily circulated in both Britain and America and considered the earliest and most influential text of the studio pottery movement.[8] The book celebrates Leach's conception of the potter: an individual driven by the vitality of human expression, guiding hand-based processes from beginning to end, independent of the conjoined systems of design and industry. This definition of the "studio potter" was embraced by craftsmen everywhere in the United States and England, as was the wealth of aesthetic and technical knowledge the book offered in its wide-ranging discussions of technique, kiln technologies, ancient Chinese and Korean ceramics, and instructions for building one's own pottery. This last section, "The Workshop," was widely retained as a methodology, the first portrayal of what it meant to enact, in Leach's own words, the "actual life of a potter."[9]

But for the three women ceramists of this study, such an "actual life" was differently constructed. Protofeminists all, the women of this particular narrative, Marguerite Wildenhain, M. C. Richards, and Susan Peterson, were as unconventional as their male counterparts but with more longevity—characteristic of nearly all women artists—and less opportunity for comparable recognition. Certainly there was never the money nor the prestige attached to studio craft of the 1950s, in any medium. But functional pottery was more than just an art form;

it was a lifestyle, offering women extraordinary autonomy, both economically and socially, through experimental artistic communities that were collective in nature.

The terms of engagement were amplified for women ceramists, since the stakes were higher both personally and professionally. I argue that Wildenhain, Richards, and Peterson were, in their own lifetimes, cognizant of both their own difference and their own neglect: they knew they had already been left out of the narrative of postwar art, even as they continued to circulate in its networks. My project examines the paradox of their trajectories in that their contributions are not tangible as artistic *oeuvres* but rather as protofeminist epistemologies that worked sometimes quietly, but in concert, with the larger post-1960s turn toward process. As Briony Fer observes, in her book on Eva Hesse's studio practice, "There is something threatening in the prospect of there being no product to speak of that could conceivably be classified as complete."[10] One of the main issues is that their collective contributions were provisional: their only index is in the form of personal letters, fliers, and numerous out-of-print books and essays. Or in long-ago summer workshops and weekend symposia, at institutions or programs now defunct or that exist in other iterations altogether. But however fragile or forgotten, their entwined personal and professional histories have lessons to offer. Ceramics offers a compelling site for examining the sexism and media hierarchies embedded in modernist art histories. Generationally distinct, it is entirely possible that the three potters never stood in the same room together, yet there is a core group of artists, critics, writers—students, colleagues, friends, peers—who knew each other and circulate throughout this text.

These participants were not just individual players in social and pedagogical networks but also people with vested interests, biases, and ideas. I cast these characters purposefully, as I have made extensive use of personal and institutional papers, using biography as a means of enlivening a particular era and what was at stake for its women artists: what they hoped to gain, what we might learn from their struggles, and how anti-institutional forms over sixty years old continue to serve as a blueprint for sustaining an artistic career over a lifetime, outside the traditional avenues of recognition: galleries, museums, and the art market.

In chapter 1, I survey the historical debates of the ceramics field regarding vessel making versus sculpture and the gendered tensions that played out in its exhibition record as the field slowly professionalized over a period of nearly forty years. In chapter 2, I map the early presence of non-object ceramics production through the forgotten site of Pond Farm, a Bauhaus-inflected summer workshop established by Marguerite Wildenhain in Guerneville, California, which ran concurrent to Black Mountain College. Chapter 3 is the partial biography of an institution, Black Mountain College, as seen through the lens of its little-known

1952 pottery seminar. While the workshop itself featured male potters like Leach, all three women potters are embedded in this historical moment: Wildenhain as the host potter for the event, Richards as its uncredited curator, and Peterson as the organizer of a similar event held in Los Angeles with the same male potters just two months later. The Black Mountain Pottery Seminar and its West Coast iteration set the stage for two decades of continuous development and expansion of American ceramics practice in and beyond the university setting.

Chapter 4 is a case study of M. C. Richards's career trajectory and the depth of her influence as a connective force in the 1950s New York avant-garde beginning at Black Mountain College and continuing through her presence at Gate Hill Cooperative in Stony Point, New York. Finally, chapter 5 returns to the West Coast and provides an account of Susan Peterson's foray into broadcasting as ceramics elided with television through her fifty-four-part show, *Wheels, Kilns, and Clay*, that ran throughout 1964.

Chapters 2, 4, and 5 are biographically driven case studies that seek to destabilize the master narratives of postwar modernist history and offer the little-known world of ceramics as evidence of a powerful and productive protofeminist sphere of influence. As Anne Wagner writes, "for a woman to be educated as an artist is not the same thing as her becoming one."[11] Wildenhain, Richards, and Peterson all received exemplary educations but were still subject to various exclusions. These gendered circumstances compelled each artist to redirect and proceed in interesting and unexpected ways, as I have detailed at length in these chapters.

The communities examined here in depth include not only Black Mountain College but also the lesser-known enclaves of Pond Farm and Stony Point. They represent a rural avant-garde, a set of utopian communities with spiritual, egalitarian, and educational philosophies as their core mission, offering a particular set of epistemologies that evoke an ethics in artmaking that is now virtually nonexistent, despite the intense contemporary interest in relational art.

Together, the depth and texture of the case studies initiate the idea that *live form* itself constituted a form of risk that was not tied to reward but to the embodiment of an ethical consciousness achieved only through the participation in community and, moreover, a community focused on the acquisition of skill. The overlapping histories of Wildenhain, Richards, and Peterson constitute a 1950s bohemia that anticipates the 1960s commune, the 1970s feminist cooperative, and the 1980s alternative space movement. The players and the details might change, but the terms of engagement remained largely the same: women artists who led the integration of social commitment with aesthetic intent. This structure raises the stakes of production exponentially, introducing, much earlier than previously known, a form of sophisticated, female-led social engineering.

It is only very recently in contemporary art circles that communities engaged in active collaboration have been admitted to the canon as valued partners in the production of art. In the work of collaboratives like Claire Fontaine, Fierce Pussy, Guerilla Girls, and LTTR, the art object produced is often considered secondary to the collaboration. Previous to this trajectory, however, such a practice conjures the late-nineteenth-century urban settlement house, which, as the penultimate amalgam of social, pedagogical, and artistic goals, is arguably the first-ever alternative art space. I discuss this antecedent at length in the first chapter.

The Proverbial Wheel

The cliché "reinventing the wheel" would not be out of place in describing the postwar status of ceramics: studio ceramics was still inventing itself professionally during the period of 1945 through 1965, establishing numerous footholds throughout the country in art schools, community centers, evening classes, and summer programs. Even as late as 1969, there was still a mandate in some places to ensure that ceramics served the community, not through providing objects for use but as therapy. For example, at the University of Southern California, ceramics was still a featured component of the occupational therapy curriculum, its beginning courses a requirement for all therapy majors, even though it was housed in the art department. The mandate for one humanistic discipline to be of service to another, rather than a stand-alone discipline invested in its own aesthetic concerns, consigns ceramics to a unique position within the academy and the culture at large: ceramics became valued for its process rather than its product. This is a stunning reversal of its traditional object-based orientation.

Thus, at midcentury, ceramics became valued not as a commodity but as an experience, as well as a helpmate discipline that sometimes served the greater good. Within art schools and university art programs, ceramics became the locus for experiencing and building community beyond the confines of master-student lineages and, rather, as a structuring device for nonhierarchical and participatory experiences. The period recognition of ceramics as an experiential medium that encouraged social transformation has been, on the whole, lost to history.

Clay is a plastic medium that is always in process, offering a pliability and potential for self-exploration within the framework of a group dynamic—the flexibility of an open ceramics studio—without the traditional expectations of either an audience or a finished object. The process of throwing is itself an accumulation of forms that might never make it out of the studio. Classroom instruction is often predicated on encouraging the formation of vessels that are themselves ephemeral: smashed down, balled up, and turned back into an unrefined mass of raw material.

For artists steeped in the solitude of life drawing, painting, or sculpture, ceramics at midcentury represented an aesthetic migration toward the collective enterprise of trial-and-error making, firing, and glazing. Purposefully (un)structured by play, ceramics fully inculcated participation, rooted in the primacy of experimentation and performance. At the dawning of a revivalist endeavor, ceramics was new territory, filled with endless permutations of techniques and materials.

In the earliest years after World War I, craft was widely introduced in Great Britain as a therapeutic method with which to rehabilitate returning war veterans.[12] The early twentieth-century Arts and Crafts romance with hand labor had come to an abrupt end; neurasthenia, that malady of modernity, was supplanted by a new world order: injury, death, and despair. Never mind the nervousness and fatigue, craft had real work to do, pressed into service as a treatment for battle shock.

Despite its elite status in the Arts and Crafts movement half a century earlier, art pottery, as it was once known, became democratized, appealing to an entirely new demographic: young, working-class men, who had neither the patience to copy vaguely Oriental paintings of flowers and trees onto the surface of preformed vessels, nor the slender fingers to mold ornate bottleneck vases and delicate tea services.

For a brief moment in the early 1940s, this was to be repeated in the United States, when craft was offered by diverse cultural constituencies such as the Museum of Modern Art, Goodwill Industries, the Red Cross, and the newly formed American Craftsmen's Cooperative Council. This proliferation was hardly spontaneous; rather, it was a carefully conceived response to the war by well-meaning women, arts organizations, and social service agencies. Taught by women but aimed at men, the history of postwar ceramics is particularly conducive to an examination of gender inequity.

In her recent revisionist history of the GI Bill, historian Kathleen Frydl notes that one-third of all GI Bill money spent on university training was put toward postsecondary education.[13] Women seldom received GI Bill benefits, though they made significant domestically based contributions during the war years, often anticipating the needs of the war wounded. As therapeutic craft drew to a close at the end of the 1940s, the influx of male GI Bill student-veterans into newly formed MFA-granting university art programs firmly divided women ceramists—many of whom already held advanced degrees—from their male counterparts who received advanced degrees during this celebrated era. Rudy Autio, Paul Soldner, and Peter Voulkos are among the most well-known names in postwar American ceramics, but, as we will see, they are also representative of this skew: all were veterans whose graduate education was financed by GI Bill monies. Talent aside, it is no surprise that all three went on to tenured university positions and, in time, esteemed careers.

Live Form

In her book *Pottery: Form and Expression* (1959), the Bauhaus-trained California potter Marguerite Wildenhain coined the term "live form" to describe a wheel-thrown vessel, in which the body of the craftsman, through his or her physical manipulation of the clay, determines the size and shape of the most intimate spaces of the vessel itself: its girth and weight, the delicacy of the rim, the strength and placement of a handle, and so on.[14] Seen here, Wildenhain's practiced fingers coax forth a spout, making a physical decision that earmarks the vessel as pourable: for use as a creamer, or a pitcher (figure 0.1). Fascinated by the tactility and careful choreography of her process, the *LIFE* magazine documentarian Otto Hagel (1909–1973) produced a series of photographs on Wildenhain's ceramic process. One of the most compelling images is a montage of her hands in motion, captured as progressive movements frozen in time (figure 0.2). Read as a language of signs, her hands offer a series of visible gestures similar to the signaling techniques utilized by orchestra conductors, underscoring the inherent musicality and performative nature of her art form. In both instances, the hands are a conduit for the utter absorption of technique in search of perfect form: the precision of an exact tempo at a particular moment, striking notes in unison, the potter opening and lifting the vessel form from a raw mass on the potter's wheel, making instantaneous decisions as she creates width, depth, slope, and height.

For Wildenhain, the vessel is "live" in that the actions of the artist are permanently registered within the very form of the vessel itself, such as the thumbprint-sized curvatures on the external sides of the pouring spout in Wildenhain's stoneware *Batterbowl* (ca. 1960) (plate 1). Her term goes beyond mere maker's mark, conveying the artist's embodiment of form itself, through an indexical presence that becomes ever present and unceasing. Another emigré potter, Otto Natzler, called this same phenomenon "immediacy," writing in 1968:

> The hand-thrown pot is something that happens at the spur of the moment, and as it happens in a matter of minutes, it reflects the mood of that moment. The pot is formed without tools, entirely by hand, and the soft clay retains the fingermarks of its maker. In no other art is there such an immediacy, such personal close contact.[15]

The unique combination of the artist's body producing an immediate form in real time is what makes the work *live*. This sort of live demonstration became a

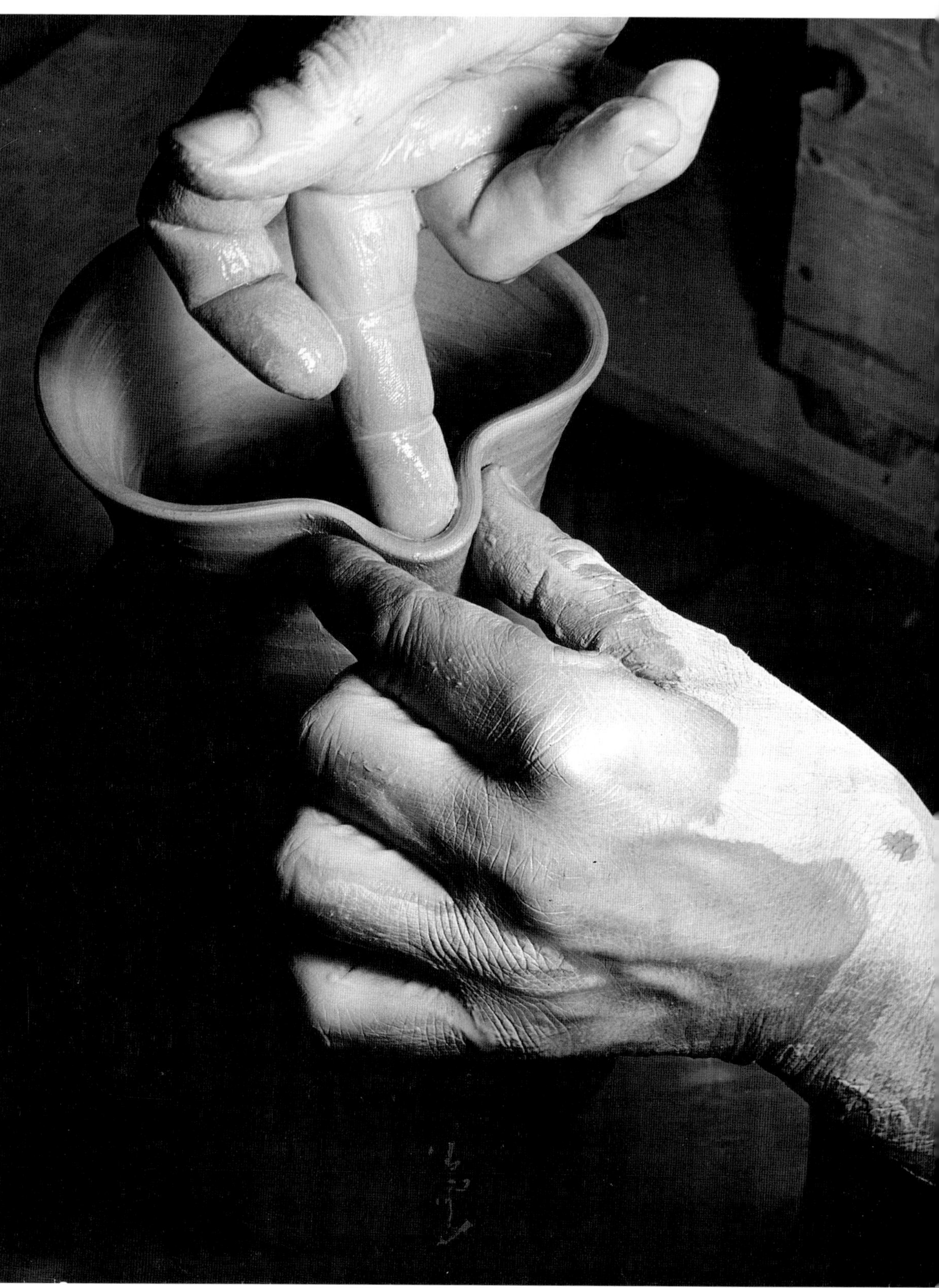

Figure 0.1 *Feeling the possibilities of their material, the fingers invent the form of the spout,* 1956. Photo by Otto Hagel. Collection Center for Creative Photography. ©1998 Center for Creative Photography, The University of Arizona Foundation.

performative gesture of skill and virtuosity enacting mastery through the size and difficulty of the forms, endurance, and the persona of the maker. Ceramics is one of the unacknowledged models of the object-action in which the live presence of the pedagogical process was the gesture of its maker.

The indeterminacy of *live form* offered the unity of experience and expression predicated entirely on the presence of the artist-performer. For instance, during the same span of four months in the fall of 1952, *live form* came to three seemingly disparate postwar mediums of American music, pottery, and painting: John Cage's *Silent Piece* (August), the Black Mountain Pottery Seminar (October), and Harold Rosenberg's oft-quoted *Art News* article, "The American Action Painters" (December). As Rosenberg wrote in his essay, "In this gesturing with materials, the aesthetic, too, has been subordinated. . . . What matters always is the revelation contained in the act."[16]

In this book, I will argue for the complexities of *live form* as a heuristic tool that has implications beyond individual artistic production, in that the "revelation contained in the act" is a process not just of artmaking but also of learning. This book examines the ways in which women ceramists improved upon formalism's use value, transforming the standard hegemonic discourse into a necessary life skill.

The primary method by which ceramists teach their students is by example. In the classroom, the live demonstration confers the potter's performance upon a group of student-bodies expected to perform in turn, on their own individual pottery wheels. The participatory spectatorship enforces the mimesis inherent to midcentury pottery: both learned and imitated, throwing on the wheel is heightened in its complexity by the evolution of a highly personal style, while adhering to certain standard conventions, such as the use of the thumb—the strongest digit—to penetrate a ball of clay. Such a mimetic gesture presupposes not only skill but also the ability to grow form—the alchemical transformation of the lump of clay into a vessel, accruing height and depth through the speed of the wheel and the deftness of the hand.

While such a dramatic transformation appears to be a highly skilled endeavor, it is just as easily an elementary and imitative gesture, akin to the child in Little League throwing a ball in the manner of his favorite professional pitcher, or the

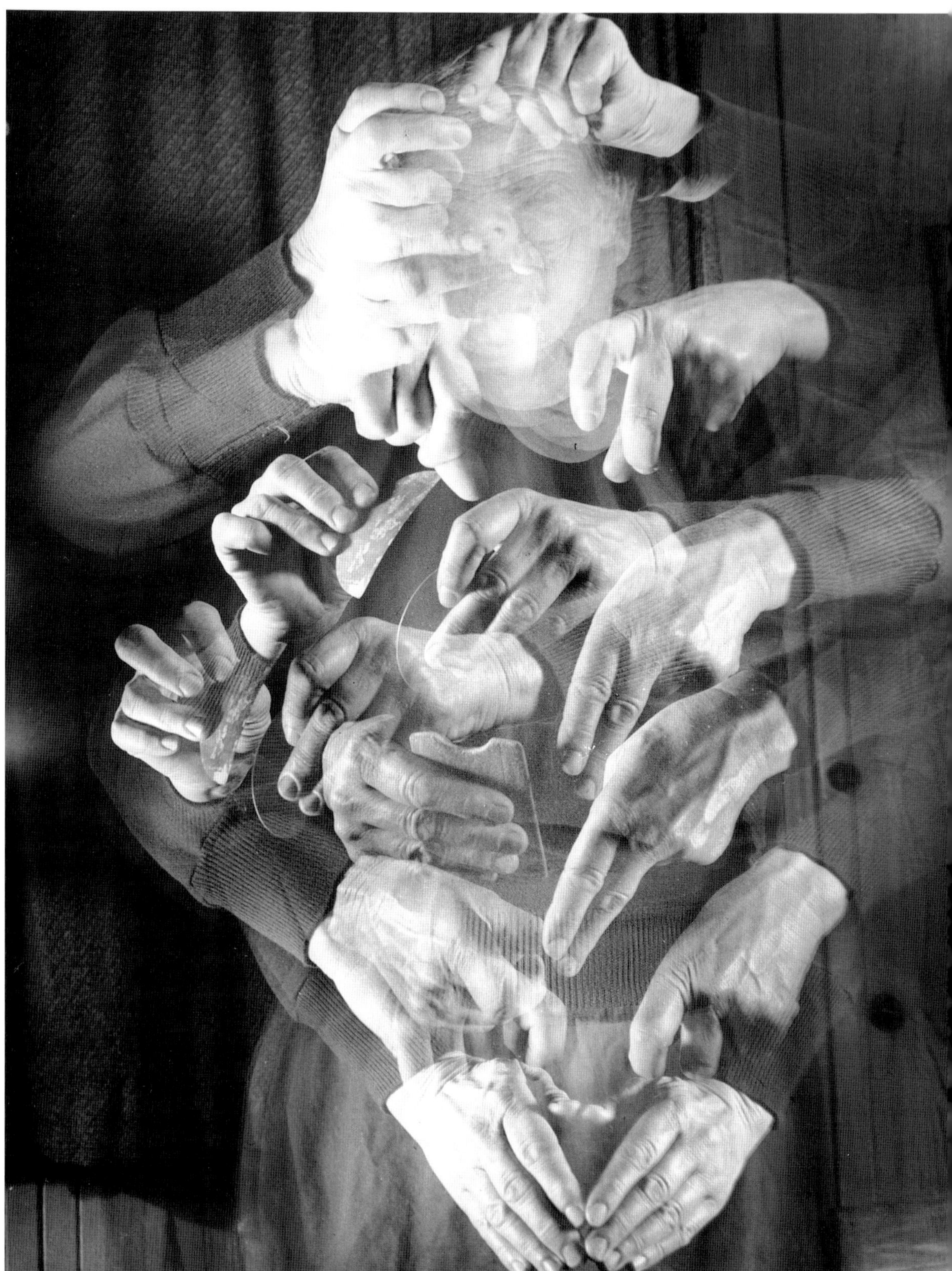

Figure 0.2 *Marguerite Wildenhain's Hands*, 1956. Photo by Otto Hagel. Collection Center for Creative Photography. ©1998 Center for Creative Photography, The University of Arizona Foundation.

piano student imitating the stylistic flourishes of the professional concert pianist. For this reason, ceramics instruction most closely parallels methods of other performance-based classrooms, such as music and dance, in which the rigor of constant practice is the well-trodden path toward the skill-agility of artistry.

Yet ceramics deviates. Though its production is far from the realm of traditional visual art objects, the finished vessel is similar in appearance to the sculptural object. Not living sculpture, like dance, nor the "frozen music" of architecture but the heavy, hulking thud of the object, demoted to its utilitarian roots: a vessel with an undeniably domestic nomenclature, a creamer, a pitcher, a vase, a teapot.

For this reason, ceramics is a fascinating and entirely confounding medium, in that its resultant object is a work that entirely conceals the performativity of its process and instead extols the virtues of its materiality.

CERAMICS

The Medium and Its Discontents

CHAPTER ONE

In 2009, a contemporary museum outside the craft world organized the first major American exhibition on clay in four decades. Co-curated by Jenelle Porter and Ingrid Schaffner with Glenn Adamson, *Dirt on Delight: Impulses that Form Clay* originated at the Institute of Contemporary Art, Philadelphia, and traveled to the Walker Art Center, Minneapolis (figure 1.1).[1] The exhibit was notable as a medium-specific exhibition that relished the exploration of a so-called minor art form. It combined disparate modern artists—for instance, the Italian modernist Lucio Fontana and the American folk potter George Ohr—in concert with a selection of contemporary practitioners who utilized clay as an experimental sculptural medium. But these disjunctive examples did not fully make sense either together or apart: materiality became the fallback premise of the show when generalists like Porter and Schaffner capitalized on the momentary trendiness of ceramics without a fully comprehensive understanding of the longstanding historical debates of the field itself.

Their sudden attentiveness is what the queer Canadian potter Paul Mathieu declared to be "tokenism," an empty gesture in which pottery was reduced to its most basic formal attribute: its viscous sensuality, also considered to be a kind of modernist-inflected *informe*, or formlessness. This is particularly resonant in relation to Mathieu's vantage point as a functional potter and, further, as a queer artist concerned with issues of gender in his own sculptural production. In his article, Mathieu wrote this sarcastic diatribe:

> Form is all it takes; formlessness even better. Base transformation is what this is all about. No need for intent, no need for content, no need for context. Even less for concepts.

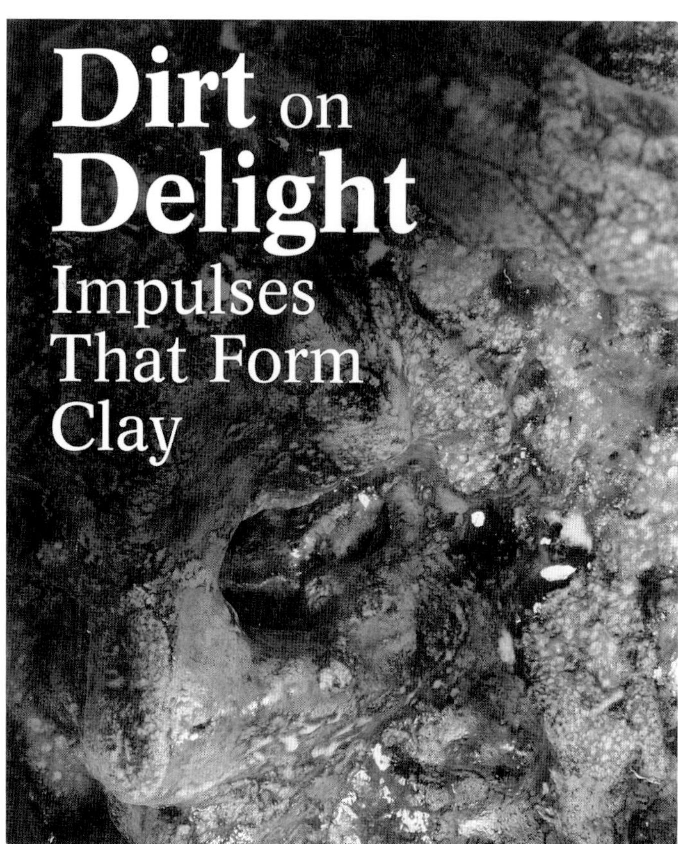

Figure 1.1 Exhibition catalog cover, *Dirt on Delight: Impulses That Form Clay*. January 16–June 21, 2009, Institute of Contemporary Art, University of Pennsylvania. Designer: Purtill Family Business. Courtesy of the Institute of Contemporary Art, Philadelphia.

> Just act, push the stuff around, make something, anything. Express yourself. Form is all that counts.[2]

In his view, *Dirt on Delight* is, in actuality, not a ceramics show but an exhibition of sculptors who luxuriate in the messy formalism of raw clay, or what the artist Anne Wilson has termed "sloppy craft."[3] Within *Dirt on Delight*, this purposeful disorder is facile in its classification as a feminist contribution, rooted in the amateur explorations of 1970s-era feminist artists, for example Judy Chicago's appropriation of china painting in her famed work *The Dinner Party* (1975–1979).

Porter and Schaffner routinely misappropriate the language of feminism to champion contemporary women artists like Nicole Cherubini while simultaneously overlooking the many fine examples of women ceramists who pioneered the ceramics medium and its critical and pedagogical trajectories throughout the twentieth century: for example, Adelaide Alsop-Robineau, the first woman employed as a professor of ceramics at Syracuse University from 1920 until 1929 and the founding editor of *Keramic Studio*, the first journal established for art pottery in the United States.

These examples illuminate the complicated terminologies and histories associated with ceramics production and the routine misunderstandings and missteps in its legacy of twentieth and twenty-first-century exhibition history and display. The medium's problem of reception, in which it was soundly dismissed as an applied art rather than a conceptually rigorous practice, intersects with a fundamental resistance to recognizing this marginalization as gendered.

But exhibition practice was only one of the public stages for American ceramics. Certainly, it was the most limited. This chapter explores the unique social conditions of the ceramics field as it developed during the twentieth century: the golden era of its artistic development and display occurred in tandem to important therapeutic endeavors pioneered by its women artists. Ceramics was therefore uniquely positioned as a socially-engaged art form. This distinction continues to mark the field methodologically and linguistically. *Dirt on Delight*'s most glaring errors were twofold: its failure to acknowledge the complex, longstanding tensions between ceramics-as-pottery versus ceramics-as-sculpture and its lack of attention toward the medium's curative properties and gendered social conditions.

Potter in particular has remained a loaded moniker, a term that remains firmly attached to the tradition of producing pottery by hand, in which the studio craftsperson oversees all aspects of production, from beginning to end. Midcentury studio potters were required to hone their craft artistically, learning to throw on the potter's wheel; master a vast repertory of vessel forms, such as bowls, cups, jugs, pitchers, plates, teapots and vases; and then innovate stylistically, conceptualizing new and fresh forms from these basic structures and shapes. Beyond the potter's wheel, studio ceramists also engaged in hand building, creating dimensional form using ancient hand construction techniques such as pinch or coil pots. In order to produce uniform wares for sale (also called production lines), many used molding and casting techniques that employed slip, or clay liquefied with water.

But this was not the studio potter's only job: ceramists also functioned as amateur materials scientists and chemical engineers, building home laboratories and adapting and altering industrial equipment for the purpose of making backyard kilns in which to fire their wares. This educated tinkering resulted in the invention of specialty ceramics equipment, which became largely the domain of male ceramists like Paul Soldner, who patented and sold his eponymous clay mixers and pottery wheels, a vast array of gear that billed itself as familiar, familial, and easily interchangeable, transferring seamlessly between his workshop and those of his peers, from one studio setting to another (figure 1.2).

Studio potters prepared the clay body via a laborious process of mixing and kneading mineral components like silica, ash, quartz, and feldspar. Part of this process engaged in constant reuse: crushing a form and rolling it back into a ball,

LIKE ALL SIBLINGS,
THEY COME IN DIFFERENT SHAPES AND SIZES

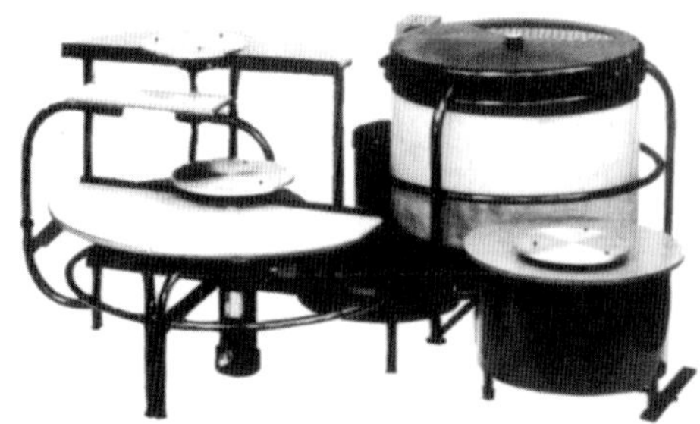

BUT THE FAMILY RESEMBLANCE IS THERE

AND ALL THOSE TRAITS THAT HAVE MADE THEM SO WELL KNOWN AND SOUGHT AFTER

They are stable, have a wide stance and hold up well under stress
They are ready for rough use; no skimping on bearings, shaft or materials
They are overbuilt, powerful, and long lived
They don't take up much space, considering the duties they perform
Their controls are outstanding
They all have built-in safety features
Their motors are fan cooled and enclosed against dust
They can each be personalized because there are many options to choose from
Their designs have been simplified, are direct, effective and efficient
They are aesthetically pleasing to the eye
They are used and appreciated in the best studios, schools, colleges and universities

THEY'LL DO WHATEVER JOB YOU HAVE IN MIND
THEY BELONG TO A FAMILY THAT HAS SET THE PACE IN POTTERY EQUIPMENT SINCE 1955
SOLDNER EQUIPMENT CAN BE DEPENDED UPON

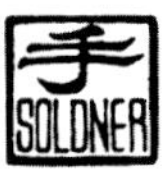

Send for a free brochure showing all of the family members, including those you can thriftily put together yourself

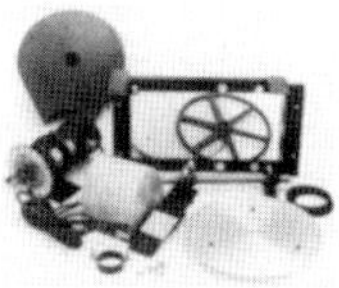

SOLDNER POTTERY EQUIPMENT, INC., Box 427, Silt, Colo., 81652 or call 303-876-2935

12 CERAMICS MONTHLY

Figure 1.2 *Like All Siblings,* ca. 1960. Advertisement for Soldner Pottery Equipment, Inc. Copyright © Soldner Descendants' Trust, Colorado. Image courtesy of the Soldner Descendants' Trust.

liquefying old clay into slip, and utilizing grog, crushed up bits of previously fired clay, to temper new clay. Potters refined their chemical knowledge to prepare a melted, glass-like substance, glaze, that fused permanently to the vessel during firing. Made from mixing combustible powdered materials, glaze chemistries were fiercely guarded and rarely shared at midcentury, amounting to secretly held formulas that were used to create signature colors, decorative motifs, and textures. American ceramics offers a microhistory of betrayal: master-student lineages that were ruined due to the perceived slight of copying or otherwise transgressing the mimetic boundaries of admiration.

Pottery was labor intensive: a multipartite process of throwing vessels on the potter's wheel, allowing pots to dry overnight, to achieve a "leather hard" surface, meaning clay that has dried to the point of losing its plasticity but is still soft enough to be carved or altered. At this stage, pots could then be wheel turned, a secondary process used to tidy up vessels before being placed in the kiln; stacking and balancing the kiln so that pots would not explode or otherwise damage each other; firing and controlling the kiln; unloading the kiln; glazing and decorating; and firing again to create a finished object.

The idea that a single person had all these necessary skills to execute this laborious process from beginning to end points to a supreme investment in authenticity but, moreover, a critical agenda that prioritized the complete mastery of a material, which, in its natural state, was formless. In turning out beautifully made, hand-thrown or mold-made wares, studio potters could conquer this formlessness and imbue it with a subtle spirituality. Such pots became known as the "ethical pot," instilled with moral superiority of traditional form.[4] As modern-day alchemists, potters could, in effect, make something from nothing. This specifically magical quality of watching a vessel form emerge from a ball of raw, wet clay is one key attraction of watching wheel-thrown pottery being made: giving volume and shape to a previously inert mass and the illusion of its ease, a form literally enlivened by the deft touch of human hand.

Such values were contrary to the Arts and Crafts movement of the late nineteenth and early twentieth centuries, in that many of the vessels produced during that era comprise a division of labor: the production potter, turning forms, and the decorator, who painted and glazed them. These objects were made for the mass marketplace and visually masterful, intended as part of an interior design schema in the form of figurative reliefs, decorative tiles, and ornamental vessels that were often displayed on raised pedestals, which emphasized their preciousness and fragility. Women played a crucial role in American Arts and Crafts; for instance, Maria Longworth Nichols was the owner of Rookwood (established 1885), the famed commercial pottery in Cincinnati. Or the example of Newcomb Pottery (New Orleans, 1895–1940), an education-based setting which was organized expressly to employ women. However, British design historian Cheryl Buckley points out that as the Arts and Crafts movement gave way to more modernist approaches in 1930s-era ceramics, "the term 'potter' was widely used to denote those who used 'clay' to make either sculpture or vessels by hand."[5]

This early freedom to move between "sculpture or vessels" solidified into a language about use value, or what is more succinctly described as functional versus non-functional pottery. Functional pottery is not merely decorative; it obligates the user, simultaneously seducing and educating the hands with its curves, rims, spouts, and edges. Known colloquially as the "hand-feel" of a vessel, this knowledge is instinctive, quick, and propels consumer decisions: how it rests just so between the palms, if its textures are sensuous and inviting, or how grainy or smooth the lip might be when brought to the mouth of the user.

Even in a contemporary Starbucks, "hand-feel" is still the reigning ethos of keeping a stockpile of mugs and drinking vessels for purchase. Though nothing is handmade, the instinct to touch has not dissipated. Customers handling crockery while waiting is a common occurrence, even—or especially—when distracted, feeling if the size and shape elide with an already established personal ritual, for instance, the preference for a single handle or for none at all.

At midcentury, it was this carelessly automatic tactility that compromised non-functional pottery. The 1950s became a decade of liberation: not only did many (male) ceramists have sculptural ambition, but also their work became intensely visual, invested in scale and surface. But such visuality could only be accomplished through a suppression of its (female) haptic properties: pottery made with the intent to be looked at and admired, rather than handled and used. Opposite a woman's touch sits the male gaze. Both are clichés, yet what they suggest is an easily negated, subconscious history: use eliding with usefulness, the object redirected, made over as an immaterial supposition, the clay thrown again and again, quieting the mind, offering up its healing touch. Because of the immediacy and tactility, women ceramists pressed formlessness into service, carefully cultivating the *live form* as a tool for social healing. As a performance of nurturing, teaching itself became a form of making.

Therapeutic Craft

Craft as an occupational therapy dates to World War I in Britain.[6] Likewise, the American Occupational Therapy Association was founded in 1917. In the famed children's book *The Velveteen Rabbit* (1922), one of the aloof nursery toys is described as: "Timothy, the jointed wooden lion, who was made by disabled soldiers, and should have had broader views, put on airs."[7] Its British author, Margery Williams Bianco, was married to a World War I veteran and familiar with the social programs cultivated on behalf of maimed and traumatized soldiers.

In Chicago, Goodwill Industries offered what was known as "sheltered work," a remedy for idleness among the disabled that functioned as both physical and psychic therapy. Because workers were considered patients rather than employees, "work-cures" were unpaid; they included millinery, pottery, cabinet making, weaving, cooking, and training in mechanical repair. In 1936, a female health professional described the situation:

> Picture a despondent grown man out of a job and leading an invalid's life for months with no idea he will ever work again. He is put at a table covered with broken toys. He starts fussing with them, trying to make them "go" again, gluing heads or guns onto broken soldiers. It seems like play, not work, and he "doesn't mind" it, is even willing to come back the next day and go on with the repair job. Gradually his interest revives and he "graduates" to another job, perhaps repairing an electrical toaster instead of

a mechanical toy. In time he will have learned to like work itself again, and if he has developed some skill at his new task, he may be fitted into a real job. He is a workshop "cure."[8]

Intentional or not, the author's usage of the term "broken soldiers" is apt. Instructed by women professionals, the "broken soldiers" were themselves repaired through their exposure to handicrafts, in time returning to more suitable—that is, paid—work.

Such situations feminized craft for most of the twentieth century. Historian Eileen Boris has thoroughly mapped out the impact of the British Arts and Crafts movement on American women social reformers and their dissemination of handicrafts and manual skills in settlement houses as a means of social and economic improvement for impoverished women, children, and immigrant families.[9] For instance, in Chicago, the antecedent for Goodwill Industries was Hull House, the pioneering settlement house established in 1889 by Jane Addams and her female companion Ellen Gates Starr. Having studied bookbinding in London, Starr was the conduit to Hull House's expansive craft programs, which aspired to train immigrant women as skilled, and therefore higher wage, laborers.

Similarly, Beth Linker has written of occupational therapy as a "female-dominated" health occupation and of its inception of female professionalism during the Progressive Era.[10] Together, these narratives offer a more complete picture of the era: occupational therapists, social activists, and craft teachers were women professionals, engaged in the care and instruction of men—many of whom were former soldiers. In 1921, Nelbert Chouinard (1879–1969), a former settlement house worker, established the Chouinard Art Institute in Los Angeles, the West Coast's first woman-run art school, as a way to offer vocational training to World War I veterans.[11]

Such an insistence on preparing men for their reentry into the professional and public spheres asserts the tenacity of pre-1940 gender values, reinscribing women in their traditional social roles as healers, caregivers, teachers, and reformers. Likewise, their task had strong moral dimensions: guiding men away from melancholia and leading them back to their rightful place in the world. Craft, then, was not a manual skill but a life skill, a technique used to evolve the male psyche past its internal dilemmas. Through its professionalization, rehabilitation evolved into a redemptive strategy, and craft, gendered female, became a service discipline.

Such female moral and spiritual guidance is compelling: a way of evolving nineteenth century constructions of womanhood and recoding them to fit within a twentieth-century context, offering women the autonomy of paid work rather than the volunteerism of the previous century found in settlement houses, benevolent societies, and charitable religious organizations.[12]

Craft worked to modernize these settings as well: Greenwich House, for instance, opened as a settlement house in 1902 in New York City's Greenwich Village, offering social and medical services to newly arrived immigrants. Yet it achieved notoriety for its arts programming. Greenwich House Pottery opened in 1909 and was entirely female-led for the whole of the twentieth century. Its first director, Maude Robinson, was a well-known New York potter who held the post for thirty years, from 1911 to 1941. Robinson also served as a technical consultant to the curatorial staff at Metropolitan Museum of Art, advising its Greek and Roman curators, and was credited on several occasions in the museum's bulletin for her connoisseurship.[13]

It wasn't until the 1950s and 1960s, however, that Greenwich House attained national recognition for its cultural contributions, under the leadership of Jane Hartsook (1916–2004) (figure 1.3). Hartsook was a professionally trained potter who led the organization from 1945 until 1982, reorienting the pottery's mission by founding an exhibition space, a permanent collection, and book and slide libraries and maintaining a teaching staff of well-regarded artists.[14] Hartsook invited star-quality men of international reputation: Bernard Leach, Kitaoji Rosanjin, Peter Voulkos—creating visibility for Greenwich House Pottery but perhaps at her own expense, reinforcing an unconscious distinction between the legendary male artist and herself, the female craft teacher, despite her own educational pedigree. New York State College of Ceramics, Alfred University, in Alfred, New York, was the premiere advanced ceramics program in the United States, matriculating its first Master of Fine Arts (MFA) student in 1940, Daniel Rhodes, who went on to an illustrious career as a ceramics professor and critic.[15]

While Hartsook also held an MFA degree from Alfred, when she began her career in 1946, few, if any, faculty positions would have been available to women. Instead, she flourished as the director of a community pottery, applying a serious and sophisticated pedagogy to a decidedly amateur environment. Hartsook's programming fulfilled Greenwich House's early, original mission, which described its neighborhood presence as "a popular university for all ages and all groups."[16] Today, Greenwich House Inc. is a strangely postmodern hybrid that evolved as the neighborhood around it did: split between medical services for at-risk populations, including AIDS counseling and an in-house methadone clinic, and its famed pottery, which still offers classes and workshops for children and adults, taught by professional ceramists. In 2011, Adam Welch became the pottery's first male director.

Because efforts were localized and often temporary, staffed primarily by civic-minded volunteers and charitable organizations, craft was continually reinvented anew. Following World War II, craft was reintroduced as a therapeutic method with which to rehabilitate returning war veterans, taught by women and offered

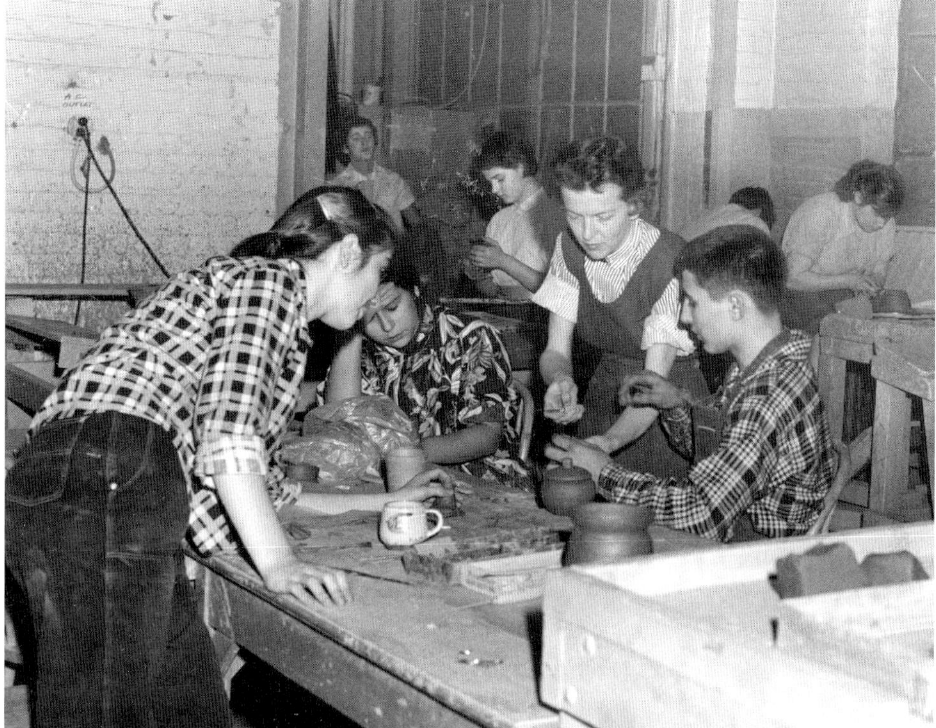

Figure 1.3 Jane Hartsook (1914–2002) with students. Greenwich House Pottery, ca. 1960. Courtesy of Greenwich House Pottery, New York.

by diverse cultural constituencies. It was only in 1942 that the first national craft organization, the American Craftsmen's Cooperative Council (ACCC), was established in New York, and it worked to coordinate regional activity.

The School of American Craftsmen was one such site for therapeutic craft endeavors, opened in 1944 by the ACCC, in connection with the Dartmouth College Student Workshop in Hanover, New Hampshire.[17] Wood, metals and ceramics were taught. That same year in New York, the Museum of Modern Art (MOMA) instituted a Department of Manual Labor, headed by the museum's future director Rene d'Harnoncourt. As a member of its educational board, d'Harnoncourt maintained close ties with the ACCC (later shortened to the American Craft Council, ACC). The Metropolitan Museum of Art also made advances to support the war effort, operating a foundry for veterans on its property during the early 1940s.[18]

While well intentioned, these gestures were ultimately short lived, lacking policy and direction as to what role the veteran would play. Was he meant to manufacture products that would advance the war effort, or was he to heal from his own wartime experiences? This reveals a tension between industry and industriousness: whether craftsmen could aid production, versus engage in a form of busywork, an Arts and Crafts-derived "joy in labor" that resulted in nothing of value to an industrialized society.

In May of 1942, Horace H. F. Jayne, then vice-director of the Metropolitan Museum, suggested that craftsmen be grateful for any sort of interaction with industry when he commented:

> Perhaps one workman in his own shop, with helpers under his guidance, can turn out something of singular usefulness or can bring his skills to bear at vital places in the assembly line of the defense program. Let such a one not bewail that in doing so he is sacrificing the individuality of his craft and becoming a mere cog in the machine; there should be a measure of thankfulness instead, that he can contribute so directly to the great work in hand.[19]

As therapeutic craft gravitated toward the domain of the museum, the ceramics field responded accordingly. Whether or not pottery had a role to play in American industrial society became a key point of debate, expressed between potters Daniel Rhodes, F. Carlton Ball, and Marguerite Wildenhain in 1957, when the American Craft Council convened the First Annual Conference of American Craftsmen in Asilomar, California. Rhodes believed pottery was a spiritual force, while Ball argued for its aesthetic value. Wildenhain, on the other hand, took a different tack, ignoring her colleagues' philosophical affectations and focusing on pottery as a bastion of social freedom.

All three were well regarded as teachers: Rhodes is deeply embedded in the history of Alfred University, as both its first graduate student in studio ceramics (1940–42) and, later, its most important postwar professor, from 1947 until 1973. A student of Grant Wood at the University of Iowa, Rhodes had started as a figurative painter, employed by the Works Progress Administration (WPA) as a mural painter during the 1930s. Ball, on the other hand, spent the entirety of his career on the West Coast, and was a talented and important innovator in the field, experimenting extensively with clays and glazes and building the first salt kiln west of the Mississippi.[20] He had a checkered academic existence, setting up an important program at Mills College, where he taught from 1939 until 1950, but changed jobs often, spending time at University of Southern California and Southern Illinois University before resettling in Washington State. Ball pioneered large-scale forms, constructing kilns as high as six feet tall to accommodate his pots, which were scaled as sculptural forms. His *Jar* (1958) is just over a foot in diameter, too large and heavy for domestic use but highly decorative, with carefully carved, iron-stained gridwork around the exterior. Playing with offbeat color combinations, the upper register of the vessel is turquoise, a striking and modern tone (figure 1.4). Wildenhain had a distinctly different trajectory; she was

Figure 1.4 *Jar,* 1958. F. Carlton Ball. Stoneware. Courtesy of the Nora Eccles Harrison Museum, Utah State University, Logan, Utah, Museum Permanent Collection. Photo by Andrew McAllister.

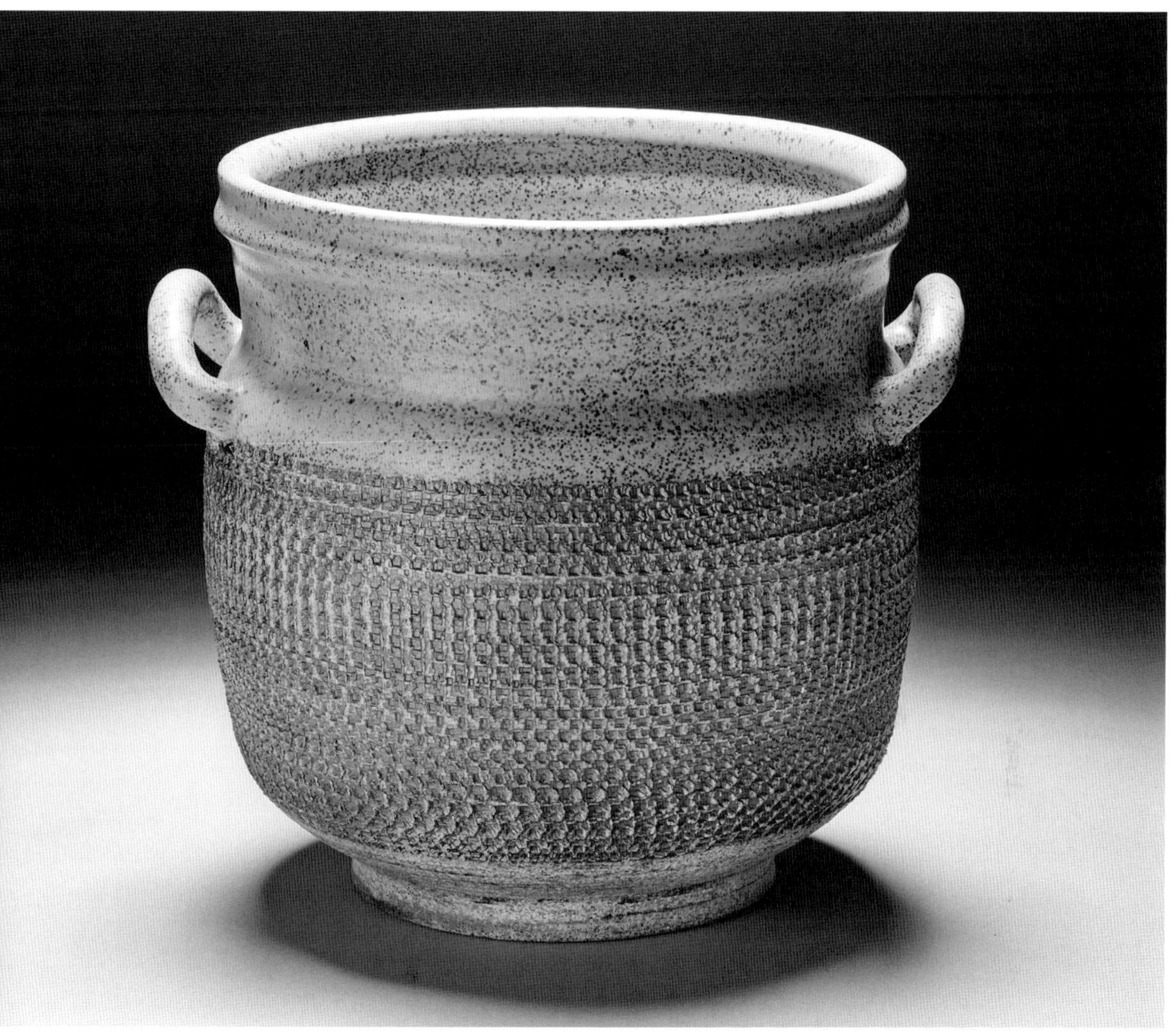

a Bauhaus-trained potter who immigrated to the United States and established her own independent pottery, Pond Farm, in Northern California. She produced highly textured functional wares with deep incising such as a stoneware *Vase* (ca. 1960s) (plate 1). Her career is examined at length in the next chapter.

At a panel called "Ceramics: The Socio-Economic Outlook" the three potters took opposing views:

RHODES: The potter is in a special position in the crafts. Industry has rightly taken on the task of furnishing inexpensive wares for daily use. New materials such as plastics and stainless steel now serve many of the functions formerly satisfied with ceramics. But it is still the potter's task to continue a great tradition; the tradition of pottery form which has been expressive of man's spiritual as well as practical needs from the first.

BALL: With the material items of pottery well supplied, the need of the people has changed to an emotional one, a need for the creative expressions of artist potters. The potter is now being recognized as an artist on a level with painters and sculptors . . . I'll stick my neck out here and say I think it's practically impossible for a craftsman to earn his living at doing his craft from the purist's point of view. That means just making pottery and earning his living at pottery. I think that it's quite necessary to have to do some other thing besides just making pottery.

WILDENHAIN: I've done it since 1925 and it's only for five years that I've done some teaching with it. Nevertheless, I have lived from the money that I got from my pottery. So it can be done, and I'm a woman myself. So, don't you ever, man, tell me it cannot be done. You can't do it when you come right out of school. Obviously, that is not possible. Then is the time for a young man to go to a good craftsman and work with him another five years, and then start on his own. . . . because I think a craftsman can only learn craft from a man who makes his living at it . . . I mean, I'd rather live for $100 a month, the way I want to do it, then have a $10,000 or $100,000 salary and have the chain around my neck and have to sit at meetings. Much rather. I feel my bare living of $100 makes me a thousand times richer than the one that has a half a million.[21]

In these excerpted remarks, both Rhodes and Ball are fully cognizant of being outmoded by industry. Looking to move the field forward, they are split on their answer: Rhodes believed in the spiritual efficacy of the potter and his/her distinction as the overlooked philosopher of society, the "mad potter" as the poet John Hollander envisioned, when he wrote "Meanwhile the slap and thump of palm and thumb/On wet mis-shapenness begins to hum/With meaning that was silent for so long."[22] As Hollander describes, the potter is the alchemist that coaxes forth meaning from an inchoate material.

For Rhodes, the potter's hand was a divining rod that draws upon a holy source of greatness through its proximity to tradition, that is, an adherence to traditional forms. Ball, on the hand, wasn't interested in the potter as a spiritual presence but, rather, as an artistic one. He desired the cultural recognition that established the potter on par with the painter and sculptor. He employed the term "artist potter," as a means of distinguishing common, functional pots from pots made with full artistic intent—pots meant to be taken seriously as art objects. Ball was concerned with pottery's reception and its cultural value beyond utilitarianism, while Rhodes saw the potter himself as exceptional: a role in which the purity of form itself is enough.

But there is a rupture in this philosophical musing on form. Wildenhain makes herself into an exception. The strength of her position lay in her own independence: unattached to an institution, her assertion about bare living illustrated a hand-to-mouth existence that was the price she was willing to pay for noncompromise. Her comments regarding the richness of her own poverty became an extravagant critique of her male colleagues' deeply entrenched positions in academia. She is particularly remarkable for calling Ball out for his male privilege, referring to him specifically as "man." In 1957, before a live audience, Wildenhain asserts the difference in their thinking as attributable to gender. Publicly savoring her own freedom, she makes clear that the stakes for such a sovereign position had an enormous impact on the livelihood of being a "woman myself," failing to extend outward beyond the claim of her own individual situation. She identifies with the category but cannot argue for the group known as "women." This deep-seated limitation is a failure of her time and place, the midcentury masculine world around her, which did not reward women for their ambition. For Wildenhain to have imparted even this, a gendered sense of self, seems a noteworthy victory.

Ceramics and Display

One of the key distinctions that has barred ceramics from being identified alongside other modernist practices has been its separatist history of collection and display. In 1932, two New York museums of American art decided to host concurrent annual exhibitions. The Whitney Annual was held at the Whitney Museum, in New York City, while the Ceramic National was held upstate, at the Syracuse Museum of Fine Arts, later the Everson Museum, in downtown Syracuse. The Ceramic National toured internationally in 1937—with a stop at the Whitney Museum itself before traveling to Denmark, Sweden, and Finland. In 1939, the show had a section of the Golden Gate International Exposition in San Francisco. In 1941, the exhibition embraced the Americas, including Canadian and Latin American ceramic work, as well as recent archeological finds from Mexico and Peru.[23] For prewar America, such experiments ran the gamut from internationally minded to genuinely radical.

So how was it the Ceramic National ran its course, closing its doors for good in 1972, while the Whitney Biennial became an important international event, eagerly anticipated as a biannual trendsetter in contemporary artistic practice? Rather than the old, tired paradigm of fine art versus craft, this question suggests a case study in the problem of medium specificity and its steady but eventual rejection not just within post-1960 artistic practice in the United States but within the ceramics field itself.

Buckley has already thoroughly examined the early years of the Ceramic National through the lens of 1930s era New Deal cultural initiatives and the role that women played in advancing aesthetic alliances. For example, the two museum directors, Anna Wetherill Olmsted of the Syracuse Museum and Juliana Force of the Whitney, were in constant dialogue during monthly meetings of the New York committee of the Public Works of Art Project, finding ways to support projects like murals that would appeal to the public good while aiding joblessness among artists. But one of Olmstead's key concerns was to put ceramics on the map as a stand-alone, artistic discipline.[24]

A juried, medium-specific show, with an open call for entries, rather than a tightly controlled, curated exhibition, the Ceramic National was democratic in its inception in that its format included a first round regional panel from areas throughout the United States that then sent on the best grouping of works to the yearly, national exhibition in Syracuse. At the national level, however, the show catered to the taste and sensibilities of not only the jurors—a mix of high profile ceramists, curators, and museum administrators—but also the exhibition's sponsors, which ranged from high-end department stores like Neiman Marcus to the industrial ceramics factories like Homer Laughlin and Syracuse China that once helped shape the field. Sponsorship allowed for purchase awards, which was a fairly inexpensive way of building up the Everson's own holdings in contemporary ceramic art. However, as Ezra Shales observes, "Corporate sponsorship for the Nationals had been vigorous in the 1940s and 1950s, but diminished thereafter in direct proportion to the clay-workers' desire to be granted the prestigious title of artist."[25]

Indeed, exhibiting and winning prizes at the Ceramic National was a mark of excellence and a career builder, helping artists to achieve national recognition and placement in invitational ceramic-only exhibitions, many of which were held at major museums throughout the 1940s and 1950s, such as at the Baltimore Museum of Art and the Art Institute of Chicago (figures 1.5 and 1.6). For instance, Ball exhibited at the National in 1947, 1948, 1949, 1952, 1954, 1956, 1958, 1960, 1961, and 1966. He was awarded an honorable mention in 1948 and first prize in 1958.[26]

In select cases, the Ceramic National was fully career anointing. The most innovative and influential potter of the postwar period, Peter Voulkos (1924–2002), had his work discovered at the National in 1949 while he was still an undergraduate, earning him a national reputation for his well-made functional forms, such as *Green and White Bottle* (figure 1.7) (1950) with its feathery, ikat-like glaze, as well as a directorship at the Archie Bray Foundation, a burgeoning residency program in Helena, Montana.[27] The Bray Foundation waited for him to complete graduate school before he assumed his position. While still under 30, between 1949 and 1955, he won a total of twenty-nine national and international prizes for his pottery, including the gold medal at the International Ceramic Exhibition in Cannes, France, and purchase prizes from the Ceramic National in 1949 and 1954.[28]

However, ceramics increasingly turned away from its functional-only orientation toward new sculptural formats. The National accounted for this by offering, beginning in 1952, two different categories of awards, making a distinction between objects called pottery and those called ceramic sculpture.[29] Voulkos became renowned for his non-utilitarian ceramic forms, built out of heavily worked slabs of clay. Ball's influence can be felt strongly in Voulkos's slab constructions, which towered similarly. Often gashed, slashed, or punctured, Voulkos's works were controversial both for their destructive qualities and for their size, which approximated sculpture in an era when craft and fine art were regarded as disparate disciplines. He sought to bridge that divide through his blatant disregard of scale and, most crucially, through his insistence on an artistic vocabulary within a medium that had come to revere technique and skill as proof of mastery. Yet Voulkos staged a showy rejection of functional pottery, as in *Untitled Plate* (1961) (figure 1.8), using an off-center slash and crudely drawn cross to counter the traditional symmetry of functional, perfectly formed wheel-thrown wares.[30]

Part of the waning influence of the Ceramic National came from the ACC's establishment of the Museum of Contemporary Crafts in New York City in 1953, which became the most prestigious national venue at which to exhibit. While the museum organized and took its share of traveling medium-specific shows, like the Ceramic National, its director, Paul J. Smith, focused his efforts on thematic exhibitions that showcased craft's larger relationship and relevance to modernity. Over time, then, the Ceramic National was demoted to the status of the outmoded and specialized, a regional exhibition in a regional city that had not faired well economically as American manufacturing died out. While the show was suspended for the duration of World War II, it settled into a biennial format beginning in 1954, touring nationally for the off year.[31]

In 1968, the 25th Ceramic National became the exhibition's peak, the longest-planned exhibition tour of two full years, and a grand celebration of its twenty-fifth anniversary planned to coincide with the opening of a new building by the international architect I. M. Pei. The streamlined modernism of the building, with its central, curving concrete staircase, embodied the museum's hopes for reemergence in the national spotlight and, with it, the Ceramic National's planned

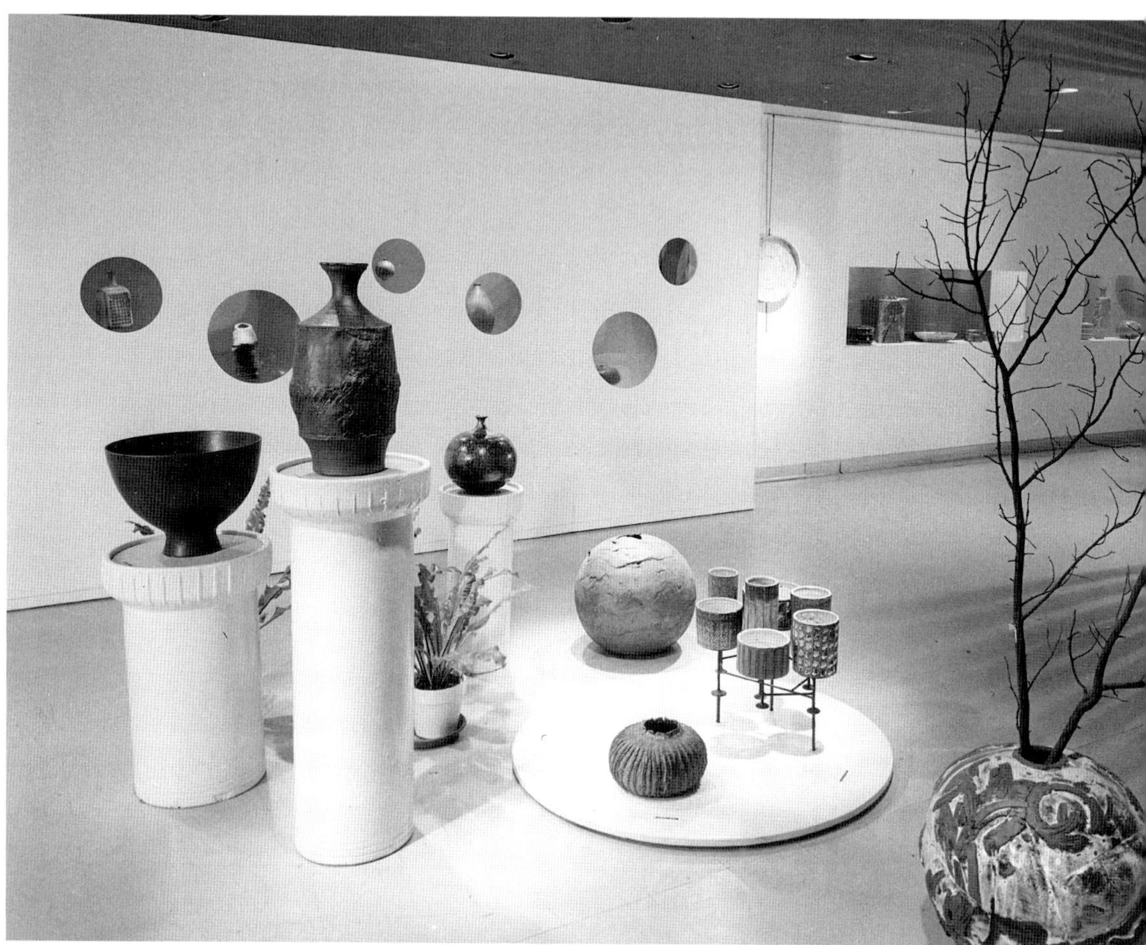

return to glory. It is no coincidence that the Everson's director, Max Sullivan, wrote in the catalog's introduction:

> One's initial reaction to the Twenty-fifth Ceramic National is that it is really a sculpture show; utilitarian pottery as such is barely represented. Individual entries exist as "one of a kind" works of art. The media, the fact that they are ceramic, seems incidental . . . The resulting form no longer has to be true to the material—to be "unclay-like" does not automatically eliminate an object as a work of art.[32]

As the introduction to a medium-specific exhibition celebrating its longevity, Sullivan's commentary seems anxious and ashamed of the ceramic field itself: the

Figures 1.5 and 1.6 19th Ceramic National, 1957, installation photographs. Museum of Contemporary Crafts, New York. Photo by Ferdinand Boesch. American Craft Council Archives, Minneapolis

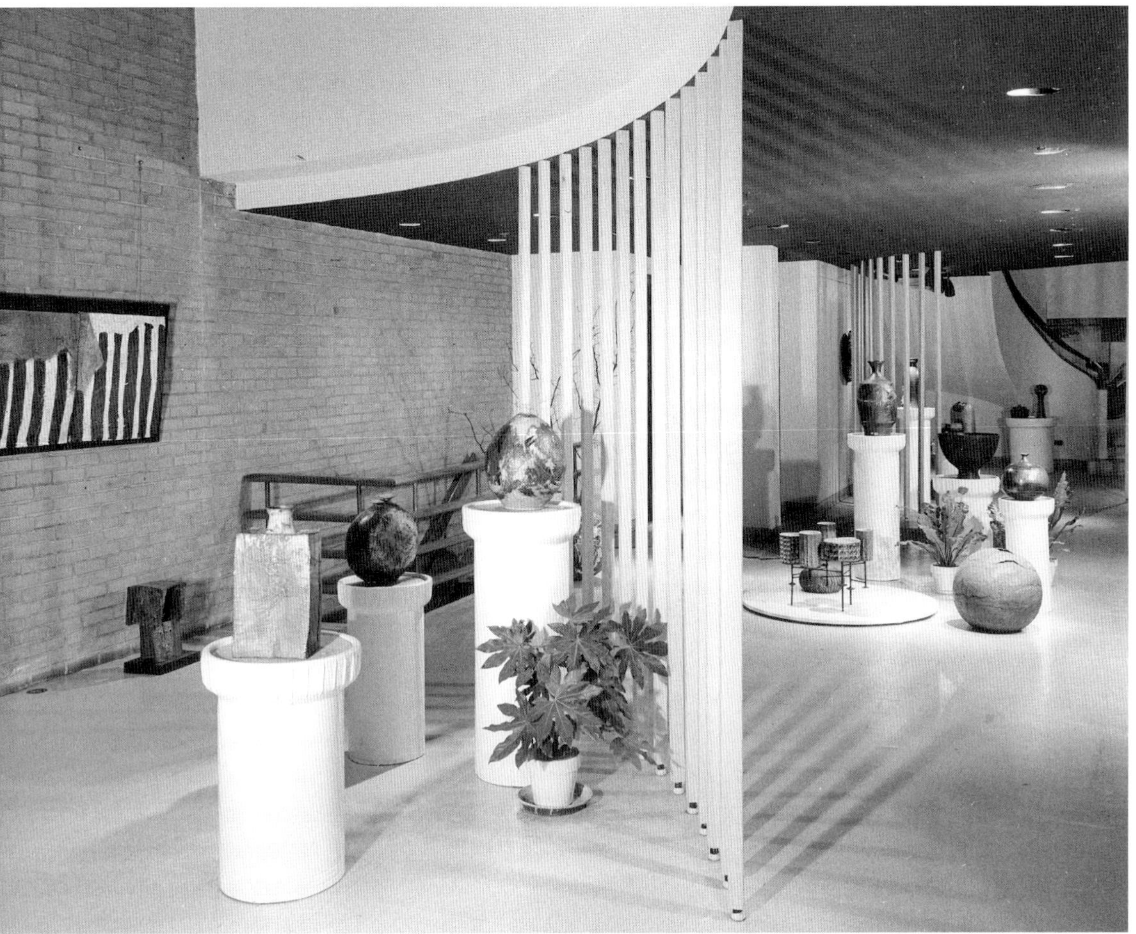

strong roots of functionalism and masterful vessels, long sensitized to register touch, texture, and the intimacies of use. Insistent on suppressing the Everson's own pioneering institutional history as the sole supporter of contemporary ceramics, Sullivan wants clay to sublimate its own viscous materiality, preferring for it to register visually as "unclay-like" as possible, a feat indicative of "art." Sullivan's ethos is eerily echoed forty years later, in Porter's approach to clay in *Dirt on Delight*, or, as she writes, ". . . seeing how far form can go before it reaches its material limits, or becomes just a bad sculpture."[33]

But the fact was, there was little consensus regarding vessel making versus sculpture. The prominence of a high profile new building and the excitement leading up to the 25th Ceramic National made what happened next all the more

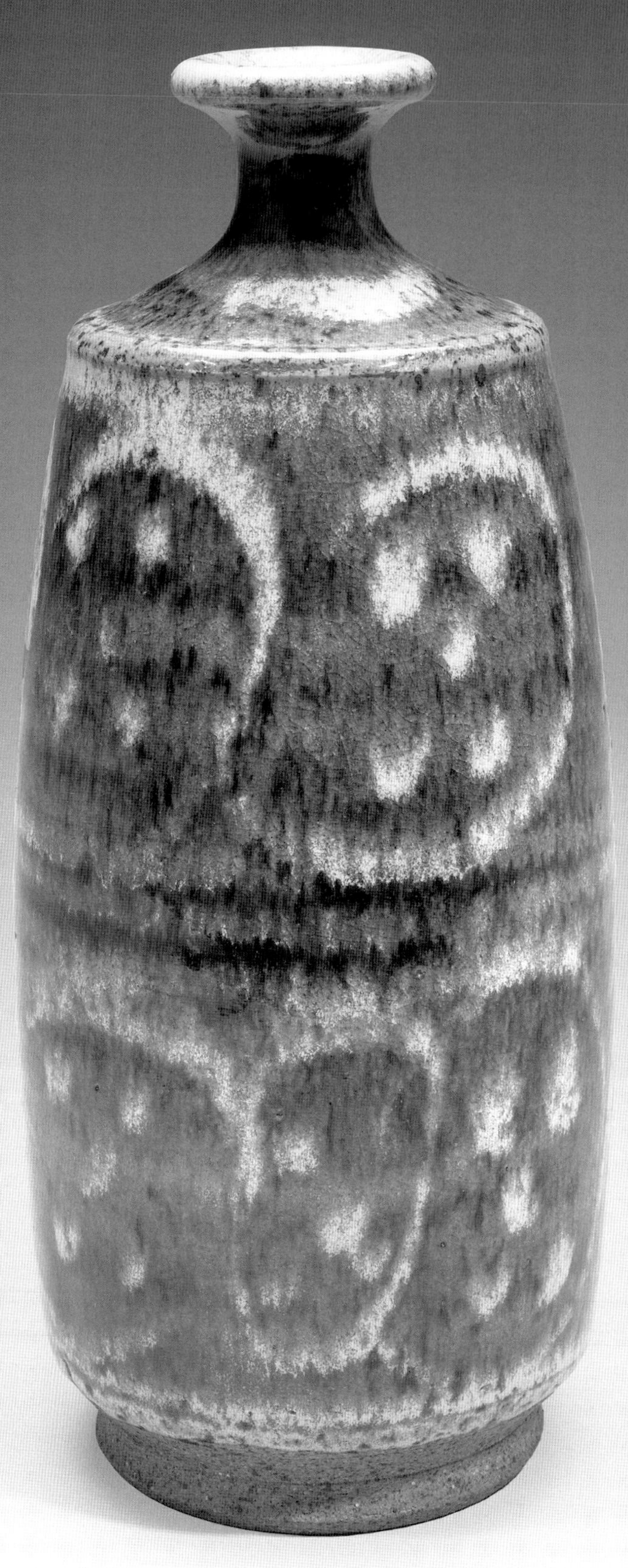

Figure 1.7 *Green and White Bottle*, ca. 1950. Stoneware, 9 1/2 x 4 5/16 x 4 1/2 inches. The Museum of Fine Arts, Houston, Garth Clark and Mark Del Vecchio Collection, gift of Garth Clark and Mark Del Vecchio. Courtesy of the Voulkos Family Trust.

disappointing. What was to be the 26th Ceramic National, the 1972 exhibition, became a banner year of public confusion and soul searching for the Everson, after a trio of highly respected ceramist-jurors Jeff Schlanger, Robert Turner, and Peter Voulkos determined they could not find quality works out of a pool of over 4,000 images. Their refusal to participate, either as jurors or as artists, forced the hand of then-director James Harithas—incidentally the Everson's most celebrated director—best known for establishing its video collection there and doing early solo shows of Yoko Ono and Nam June Paik.

The 1972 Ceramic National was cancelled and the show format terminated. The anxieties and aspirations of the Everson, which sought to raise the cultural capital of ceramics from an association with labor and industry and lift it into the realm of so-called high art, was a bruising lesson in the waning cultural authority of not only museums but, most of all, medium specificity. For two years, the Everson floundered in what the new, incoming director Ronald Kuchta later described as "the apparent identity crisis the ceramics world seemed to be undergoing at the time . . . [which] felt that something should be done to get the attention of the 'art-world,' and essentially break down the barrier caused by a certain disdain for the medium of clay."[34]

From this distance, such remarks are breathtaking for their self-loathing. The medium was not only being treated by its own makers as lesser; some of its top artists believed things were so bad that they collectively deserved to have their national show publicly shamed, then dissolved. The field seemed to be mired in a kind of full-fledged depression: its own materiality, its very center, so to speak, seemed irrelevant. Yet it is not unheard of for artists to get fed up with their own medium. Due in part to the hegemony of abstract expressionism, 1960s-era artists such as Donald Judd, Robert Morris, and Sol Lewitt all turned to sculpture, feeling that painting was no longer a viable or meaningful art form. Why was it that ceramists, blessed with an inherently plastic medium, could not recommit or at least address the failures of their medium with wit and aplomb?

Resuscitating the Ceramic National was out of the question: what ceramic artist, other than hobbyists and amateurs, would aspire to be in a show tainted with scandal and embarrassment? "The mood in 1974 was a moment for innovation," the ceramist Margie Hughto recalled.[35] Hughto taught locally at Syracuse

Figure 1.8 *Untitled Plate,* 1961. Peter Voulkos (American, 1924–2002). Stoneware, 2 1/2 x 12 1/4 inches. Collection of The Museum of Fine Arts Houston, The Leatrice S. and Melvin B. Eagle Collection, museum purchase funded by the Caroline Wiess Law Accessions Endowment Fund. Courtesy of the Voulkos Family Trust.

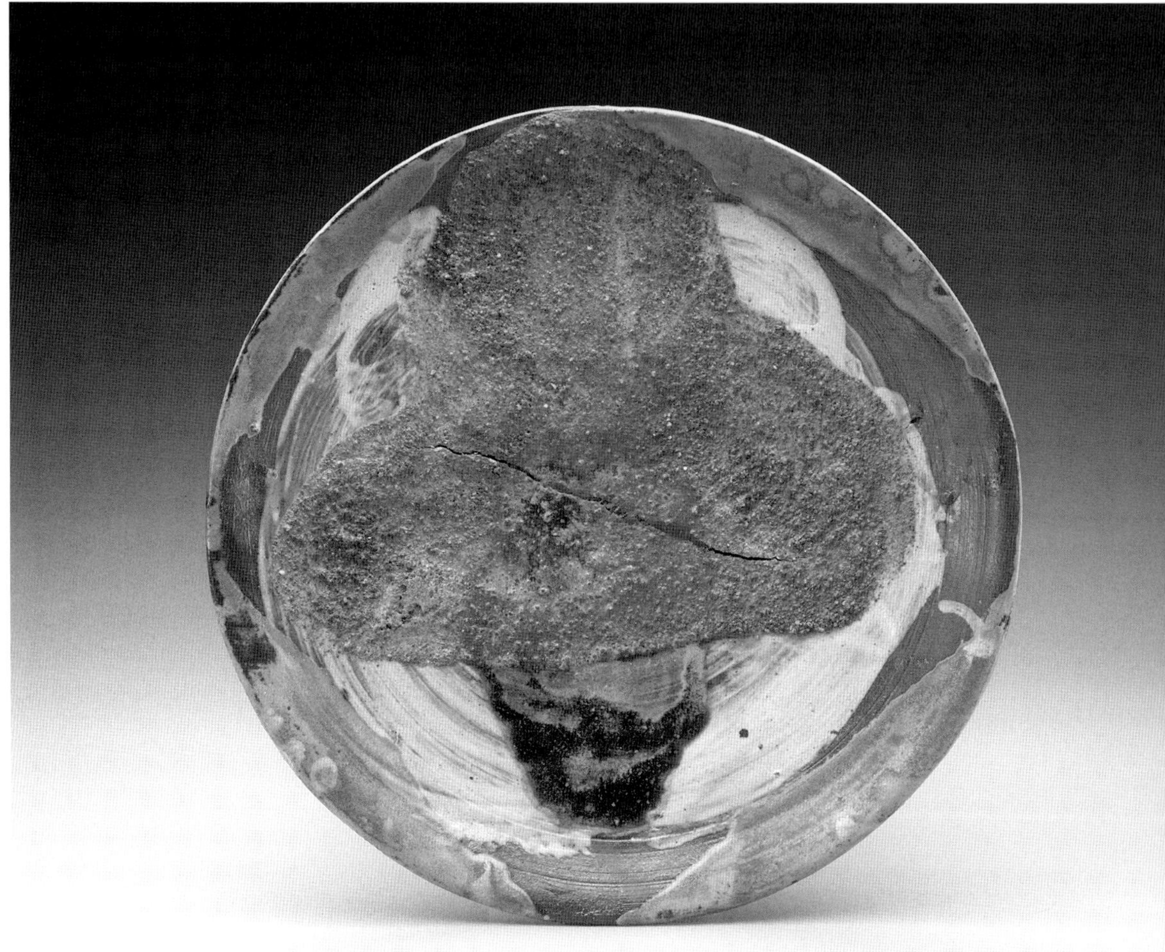

University and was brought onto the museum staff as curator of ceramics at the Everson. From 1974 until 1982, Hughto pioneered a series of exhibitions known as *New Works in Clay*, tied to the Syracuse Clay Institute, a residency program she established in tandem. The institute proselytized clay to established fine artists, offering leading painters and sculptors the means to produce experimental works in clay with the help of student assistants in a vast industrial space. Given such resources, artists usually produced large-scale sculptures. The painter Jules Olitski, for instance, created a minimal form of concentric circles, *Iron Cone 02–3*

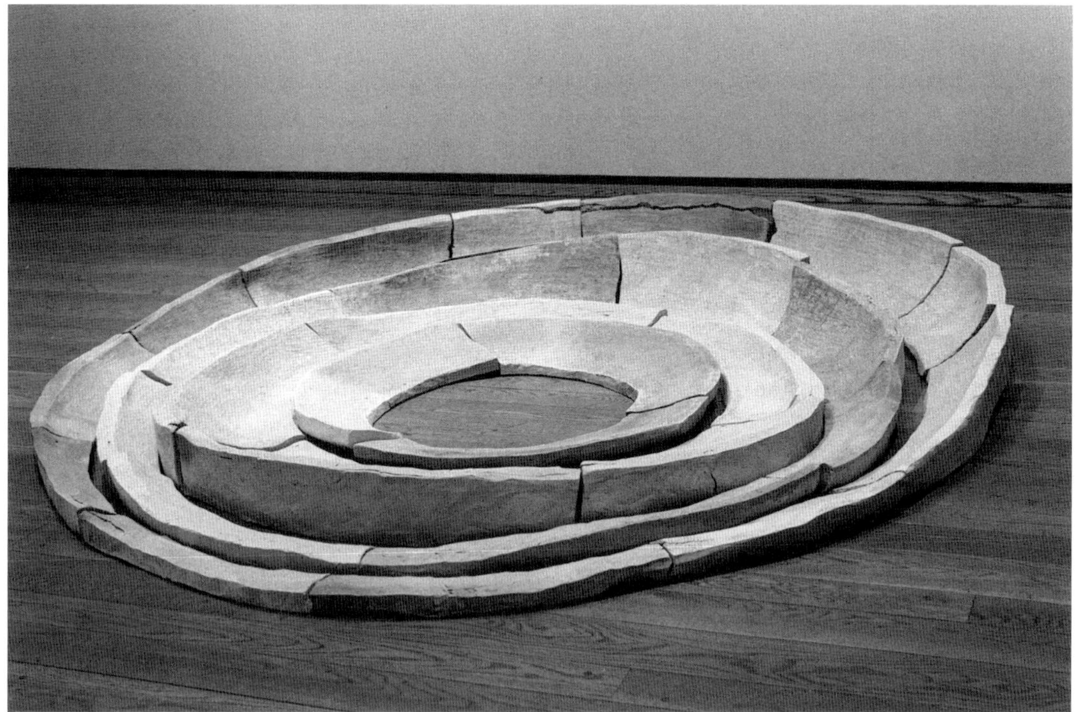

Figure 1.9 *Iron Cone 02–3*, 1975. Jules Olitski (American, b. Russia, 1922–2007). Stoneware, 84 x 120 x 12 inches. Collection of the Everson Museum of Art, gift of Joan C. Olitski and the Olitski family in memory of Jules Olitski, PC 2009.7. Photo credit: Stuart Lisson. Art © Estate of Jules Olitski. Licensed by VAGA, New York, NY.

(1975), made out of modular stoneware and named for the temperature at which it was fired (figure 1.9).

Hughto received funding for the venture through the National Endowment for the Arts, with the pledge to keep the project in Syracuse. As she explained:

> I compare it to drama: imagine if Robert Redford came to work on a play with you. The students got to have the experience of working with high-level people. We would run tests, fire the things, sometimes they liked them in their bisque state. Kenneth Noland came for quite a while, Anthony Caro was with us for six years. I call it a million dollar education. My students were the primary workers. In the long run, they tell me that it was the thing that made the heaviest impact [in their educations].[36]

The first edition, *New Works in Clay by Contemporary Painters and Sculptors* (1976), included work by artists including Billy Al Bengston, Anthony Caro, Helen Frankenthaler, Olitski, Larry Poons, and David Smith (figure 1.10). *New*

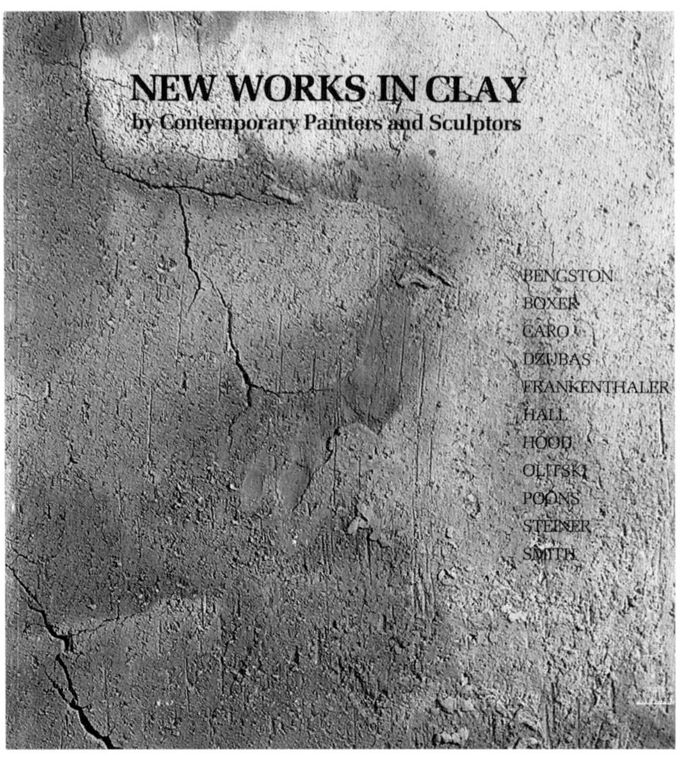

Figure 1.10 *New Works in Clay* (1976). Exhibition catalog cover. Curator: Margie Hughto. Courtesy of Margie Hughto, Jamesville, NY.

Works in Clay was Hughto's way of propelling the field forward, finding new forms of recognition by allowing so-called outsiders into the inner sanctum of the ceramics process. Or, as Kuchta put it in the catalog's foreword, "We applaud the artists' adventurous encounter with what to many was an unfamiliar medium."[37]

As a joint venture between Syracuse and the Everson, which, despite being a mile apart, were never affiliated, the program proved fruitful, bringing a host of visiting artists to campus. But part of its success was that the program functioned similarly to an alternative art space: a time-limited residency program Hughto organized, leading a team of talented technical assistants, mainly women, to perform the medium's laborious tasks, as well as to advise artists on their projects. As Hughto gently but wryly observed in the catalog, ". . . the artists often changed their ideas after spending time with the clay."[38]

In considering the photo of Hughto and her women students (figure 1.11) assembled before a kiln for a photo, the group could easily pass for a feminist cooperative. It is 1976, the right era for women's cooperatives, collective skill building, and consciousness-raising. Yet Hughto's group was not making and firing their own work but, rather, that of established, mainly male artists. They performed as helpmates, the traditional female role of cheerful and efficient

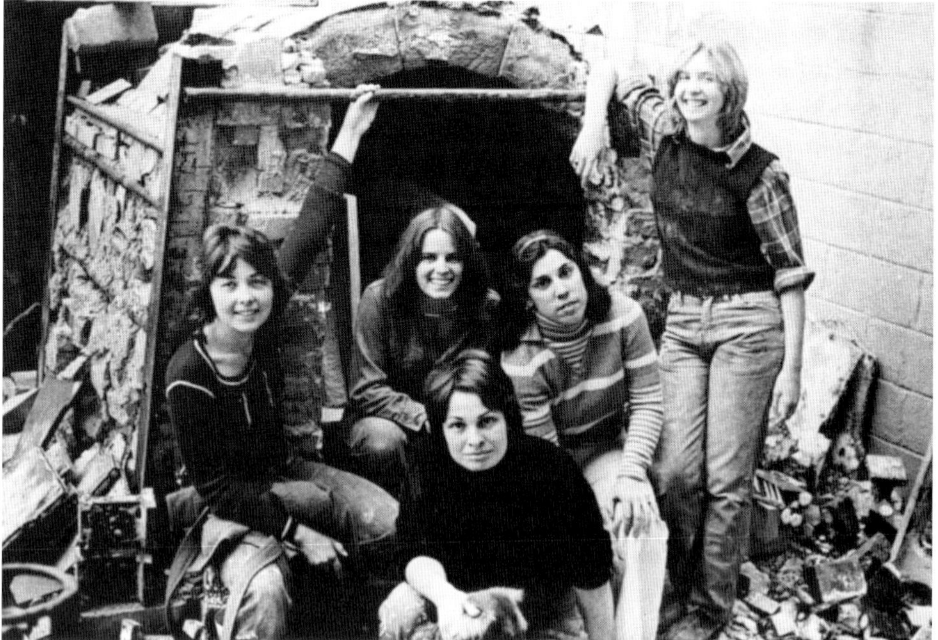

Figure 1.11 Margie Hughto and project staff in front of salt kiln, 1975. Front: Jo Buffalo. Rear (L to R): Margie Hughto, Colleen Lisson, Deborah Staats, Leslie Le Fevre. Courtesy of Margie Hughto, Jamesville, NY.

assistants, rather than as artists themselves. Hughto details the many trips taken in her four-wheel drive vehicle to the offsite industrial kiln and the scramble to solve and remedy various technical problems, which required multiple test materials, moving, packing, carrying, and the like.[39] Such a role seems the natural evolution of the therapeutic craft studio: women artists cheerfully managing crisis, creating an opportunity that benefited a particular, local community, in this case the students and artists in and around Syracuse, a city that had been demoralized by the existential crisis of faith in the ceramics medium itself.

The 1970s is marked by the rise of the alternative space and non-object-based practices like performance, video, land art, and conceptual art. This image is evidence of the slippage between ceramics and the rest of the art world, in that its women artists had not yet politicized but were still at the forefront of thinking about alternative spheres of production. Hughto's project was the first to mark this shift. She went on to produce two more editions of the *New Works* exhibitions through 1982. This span of five years coincides exactly with the alternative space movement, many of which were rooted in women's labor, cooperation, and energies.[40] Further, during a moment when internal divisions had shuttered the ceramics community's primary, national exhibition venue, the politics of gender became a consideration far secondary to the medium's fight for viability as an enduring sphere of contemporary artistic practice.

Hughto's exhibition series prefigured Clayworks Studio Workshop (1978–1981), founded by Susan Peterson on the Lower East Side of Manhattan, and also the Fabric Workshop (1977-present), a contemporary art space in Philadelphia, both with the somewhat suspect mission of bridging the divide between contemporary artists and contemporary craftspeople. Clayworks paired contemporary artists with their ceramist counterparts as a collaborative gesture. Some of its collaborations, such as the one between Joyce Kozloff and Betty Woodman, were incredibly fruitful. Others, like the one between Tony Hepburn and Robert Morris, ended early in bitter acrimony.[41]

Inherently flexible, and catering to the needs of artists, the ceramic-based program spaces were short lived but actively sought to move the field beyond the realm of the lone studio potter working in isolation and toward the collaborative situations that reflected women ceramists' own commitments to teaching. Hughto's program in particular, with its strong pedagogy, functions as a counternarrative to the museological historicism she pursued simultaneously at the Everson alongside her emerging colleague Garth Clark.

Garth Clark

In 1974, the Everson Museum began its own search for identity and purpose under the new direction of Kuchta. Two years later, a little known South African-born art historian by the name of Garth Clark (b. 1947), who had been recently awarded an MA in ceramic history from the Royal College of Art, was invited by Hughto to be an active collaborator. Clark was the author of two monographs, *Potters of Southern Africa* (1974) and *Michael Cardew* (1975), a formidable English studio potter who had a difficult history in Ghana as part of British colonization.

While the exhibition's catalog was authored jointly, Hughto always conceived of herself as an artist first, admiring Clark's articulate and erudite writing:

> Garth and I worked very closely for three years. We had a great working relationship. He gave over the material from his book [he'd been working on]. Garth, unlike me, has tremendous writing skills and he speaks well and he is very convincing. So I asked him to do the show with me. But it was my idea for the show.[42]

It was this winning combination of a ceramist and critic working closely together that proved to be a gift to the field: Clark and Hughto's collaboration culminated in a landmark blockbuster exhibition, *A Century of Ceramics in the United States*,

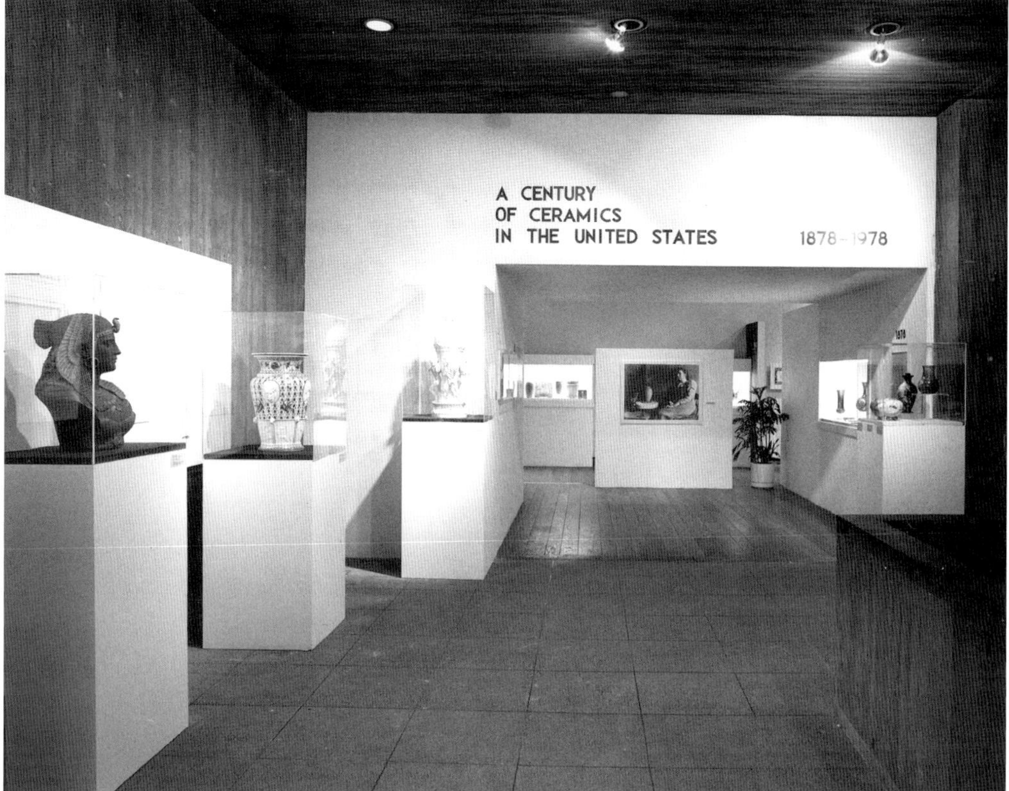

Figure 1.12 *A Century of Ceramics in the United States, 1878–1978* (1979). Exhibition entry, Everson Museum of Art, Syracuse, NY. Courtesy of the Garth Clark Archives, Museum of Fine Arts, Houston. Reproduced with permission of the Everson Museum of Art, Syracuse, NY.

1878–1978, which opened at the Everson in 1979 (figure 1.12). Comprised of 450 objects, Hughto and Clark's careful connoisseurship redirected postwar ceramics from its contemporary moment and encouraged a reconstruction of its heroic and storied past, an exciting new narrative that had not been previously told or seen. By returning the Everson to its roots, the duo guided the museum to find its footing through their joint exploration of its outstanding modern ceramics collection, which came from a combination of Robineau's estate and the heyday of collecting that took place during the era of the Ceramic National.

The resulting catalog became the go-to history of modern American ceramics. Last revised in 1990, it is now in its third printing, titled *American Ceramics: 1876 to the Present*. Clark was motivated by the problem that modernist discourse had eclipsed functional ceramics and saw an urgent need to compile an historical record of the artistic achievements of individual ceramists. Rather than seeking to restore the Ceramic National, it was history making that became the primary vehicle for his interests. This was a very effective strategy for combatting the field's depressive slump: if ceramists didn't know their own history, how could there be anything but a generalized malaise at an art history that ignored their own contributions? It stands to reason that if one doesn't know one's own history, it is hard to be inspired. One of its biggest successes was that *A Century* depicted failure as inevitable, citing specific historical examples such as Rookwood's near closure due to the rise of overseas manufacturing.

The volume is chronological and organized by decade, charting the evolution of ceramic styles, the medium's technical developments, and cultural history that begins with the Philadelphia Centennial Exposition in 1876 and continues up through the late 1970s, when ceramic sculpture dominated. The book's methodology reflects a decorative arts history of connoisseurship and collecting, offering a chronicle through the selection and elevation of masterful objects. This approach is no doubt problematic: it is straightforward and monolithic. Like a steamroller, it stays its course, flattening out everything along the way. But it is also an instructive model, unafraid to make unpopular assertions or to speak broadly and authoritatively, as a way of working *against* the set model of impermanence in the field; the Ceramic National had been defined by its task of documenting the "best of" in a given set of years, rather than setting a course or a comprehensive vision. As a means of recouping a lost history, nearly eighty of *A Century*'s 300 pages are artist biographies, and another thirty pages provide a chronology of the field, compiling exhibition, pedagogical, and institutional histories into one exhaustive timeline.

American Ceramics is something akin to a dissertation, albeit without an advisor, classmates, or a university structure. This is what makes Clark and Hughto's archival recoveries all the more admirable, that they committed to ceramics history in an era when it was patently unfashionable, she as a ceramist, he as a connoisseur who together spent time carefully, lovingly looking at early-twentieth-century American pottery as a means of trying to understand its contemporary iterations. The museum became their classroom.

When Clark moved to the United States in 1976, there must have been a certain culture shock in encountering the bizarre world of bicentennial fever, which swept the nation as a patriotic and sometimes kitsch campaign, jangling not only newly minted coins but also heady regionalisms as a tribute to the nation's own history. *A Century of Ceramics in the United States, 1878–1978* was, in some sense, a version of the museum's belated bicentennialism. But it was also its first attempt at introspection: the museum's own collection was virtually unknown outside a local or specialized audience, since it had never produced a permanent collection catalog, and there was not yet an articulated disciplinary strategy of either craft or design history within contemporary ceramics.

Reviewed in major media outlets, the show became a critical success and solidified Clark's reputation as a scholar. Hughto returned to her studio practice, gravitating toward public art and securing numerous architectural commissions. *A Century* toured for four years, underwritten by Phillip Morris, and was rebranded as a documentary shown on PBS titled *Earth, Fire, and Water: A Century of Ceramics in the United States*. According to Clark himself, the exhibition is also responsible for having launched "a vibrant marketplace."[43]

A prodigious writer, Clark became the most became the most prominent voice in shaping the field of ceramics history, producing numerous exhibition catalogs, books, edited volumes, and monographic essays over the last thirty-five years.[44] But he also became the field's primary dealer, debuting his eponymous gallery in Los Angeles in 1981, relocating to New York City in 1983, and combining efforts with his life partner Mark Del Vecchio.

From 1981 until 2008, Garth Clark Gallery was the commercial and intellectual cornerstone of American ceramics. This was accomplished through a dual role of critic and art dealer, a dichotomy that consigned the field to seeming trapped in the past, with a powerful and dominant figurehead, analogous, perhaps, to Daniel-Henry Kahnweiler, Cubism's first interpreter and dealer in early twentieth-century Paris. The couple partially gifted the bulk of their collection and archives to the Museum of Fine Arts, Houston, in 2010. In the exhibition catalog that was produced, Glenn Adamson skirts the issue in a tactful way, writing, "Clark has necessarily occupied a complex position in relation to the ceramics marketplace; after all, he did more than anyone else to shape it."[45]

In creating a discursive framework for the field of ceramics, Clark also helped legitimize his own tastes, which, early on, tended toward functional pottery that prioritized clean lines and geometric shapes, such as Michael Cardew's *Vume Bowl with Lily Design* (1945) (figure 1.13), in which the artist formulated an olive green glaze made of local materials, including crushed oyster shells, and used a wax resist technique to allow the stylized flower designs of the deep purple-ochre clay to glow against the complementary green background. Such a decorative technique results in a fusion of colors and a smooth and unbroken texture, reinforcing the roundness of the circular form itself.

This is quite a different aesthetic in opposition to Marguerite Wildenhain, whom Clark never personally collected or showed. Her incised surfaces can seen on her *Vase* (ca. 1960s) (figure 1.14), in which the potter squared a wheel-thrown vessel, keeping its tight circular rim and neck, which descends into boxy shoulders and a broad front and back surface and then returns to a vessel shape at the bottom. In such a pot, Wildenhain expends her energies on coaxing out a form, batting and shaping it with hand tools, and then intricately imprinting it with a tool, deeply marking the surface with tiny uniform abstract markings reminiscent of petals or feathers. The core of the vase, its inside, is glazed blue,

Figure 1.13 *Vume Bowl with Lily Design*, 1945. Michael Cardew (British, 1901–1983). Stoneware. The Museum of Fine Arts, Houston, Garth Clark and Mark Del Vecchio Collection, gift of Garth Clark and Mark Del Vecchio. Courtesy of the Estate of Michael Cardew.

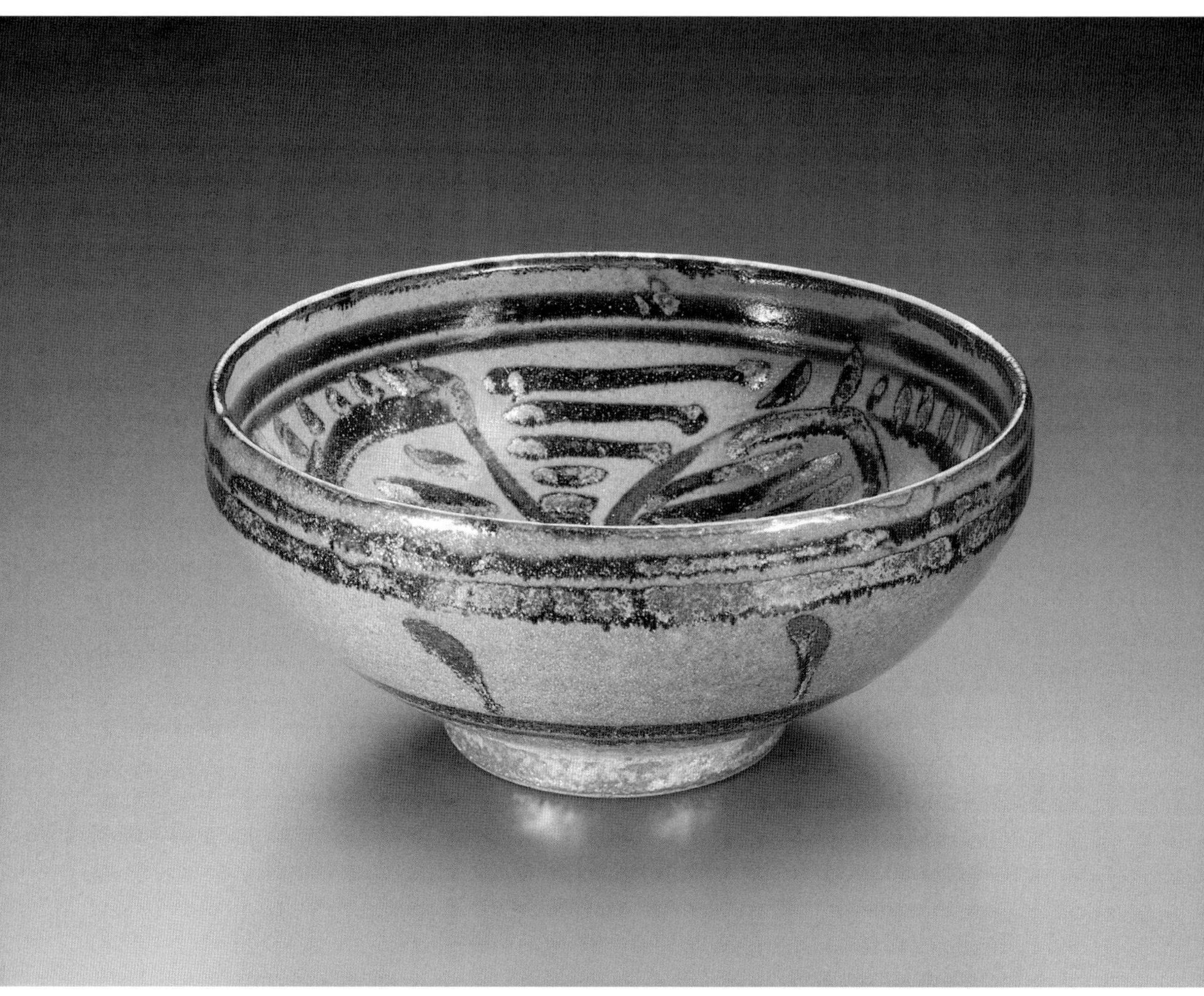

hinted by its blue rim and outer edge. Cardew's simple wax resist is painterly, resulting in an even visual surface, a bowl that is ultimately decorative, while Wildenhain focuses on creating a highly textured hand-feel, a haptic surface that would compete with a flower arrangement, if the vessel were indeed pressed into service. Clark described her as a potter who "... adhered to a strong functionalist view. This attitude, based on repetition and production ware, was anathema to Beatrice Wood, Maija Grotell, and others who worked with the same intent as painters and sculptors."[46] In such a statement, Clark makes his biases clear, disavowing Wildenhain's production as differently intentioned, not in the realm

Figure 1.14 *Jar*, 1958. Marguerite Wildenhain (American, b. France, 1896–1985). Stoneware, 6 1/2 x 5 1/2 inches. Courtesy of the Nora Eccles Harrison Museum, Utah State University, Logan, Utah, Museum Permanent Collection. Photo by Andrew McAllister.

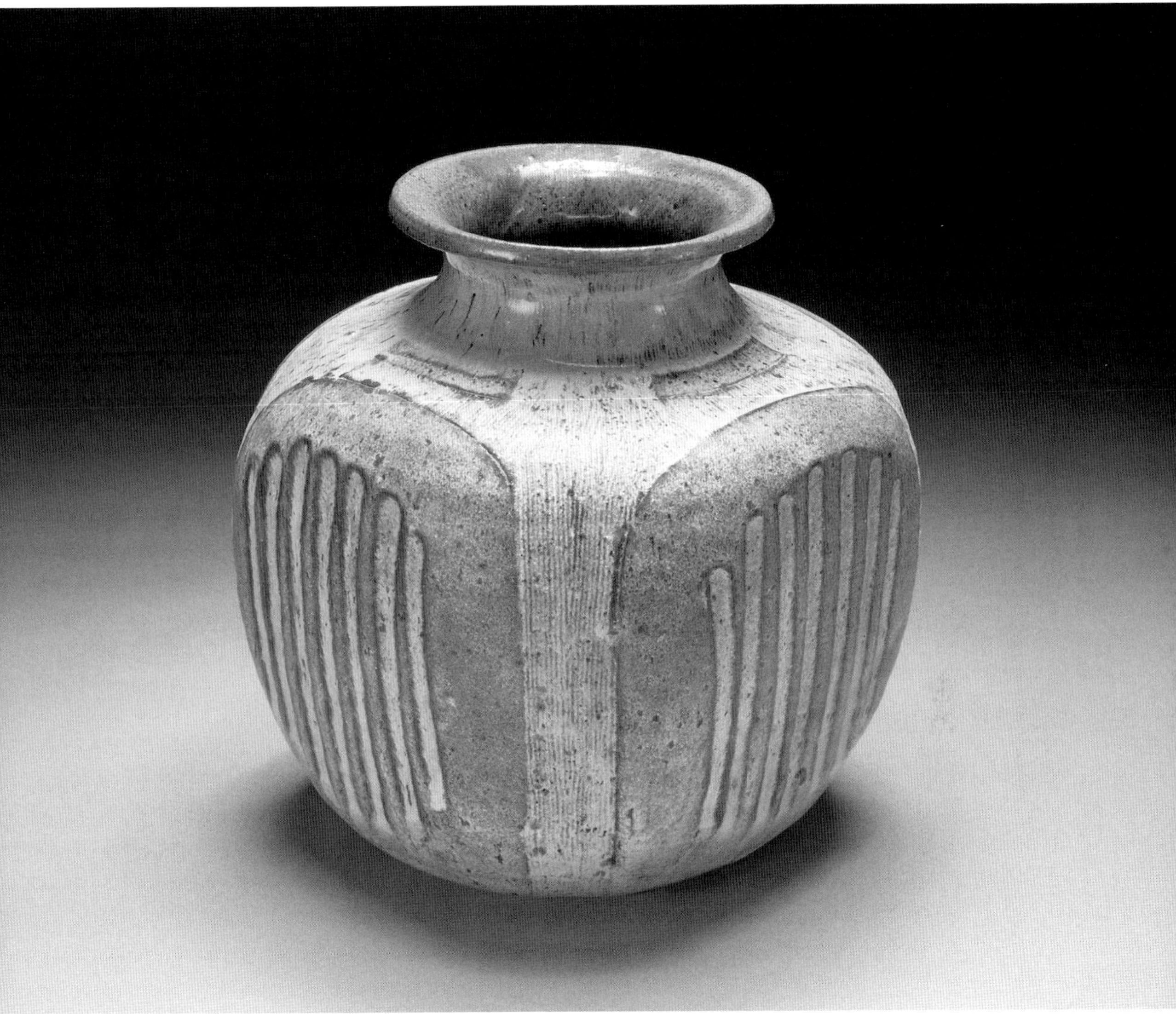

of the artistry he attributed to either the aforementioned women potters or to Cardew, whom he called "the best of Europe's ceramists."[47]

While Del Vecchio took over the day-to-day operations of the gallery, Clark continued to produce focused studies of ceramics in the form of exhibition catalog essays, including the first monographs on Beatrice Wood (1983) and Viola Frey (1984).[48] In 1983, Clark curated a small gem of a show called *Ceramics Echoes: Historical References in Contemporary Ceramics* in Kansas City, Missouri, which was based upon a comparative strategy between contemporary ceramists and the ancient pots to which they looked for influence and inspiration. In the introduction, Clark writes that the show was inspired by his work at the Everson: "I was working with Margie Hughto . . . we were unpacking the pots from Bob [Robert] Turner at the same time that several cases of Tang and Sung masterworks were being cleared. So the old and the new ended up freely mixed together on the vault floor."[49] Clark goes on to state that connoisseurship offers the viewer the possibility of making contact with the past, with clay as a symbolic, but ultimately universal, language.

Alongside his curatorial work and art criticism, Clark simultaneously staged symposia through the Ceramic Art Society, an organization he founded with the intent of creating a platform for scholarly conferences. The first was held in 1979 and featured Clement Greenberg as a keynote speaker. Clark regarded this as quite a coup, a sign of the changing fortunes of ceramics, in that an ageing modernist critic, synonymous with the high art of painting, would consent to engage critically with a field long classified as decorative.

Throughout the 1980s and 1990s, using the format of the solo show, Clark ushered in an appreciation for the mainly male ceramists who worked in a postmodern idiom making content-rich work driven by an interest in visual jokes, sexual innuendo, slick surfaces, and high-low dichotomies, such as the work of Ron Nagle, Ken Price, and Adrian Saxe. Saxe's *Untitled Ewer (Chou-ccr)*, 1992 (figure 1.15), for instance, appropriates a household vegetable, the humble cabbage, and remakes it in a delicate porcelain form, a golden and bejeweled teapot that refers to obliquely to India, as CCR is the airport code for Kolkata (Calcutta).

In 2008, the year of his gallery's closure, Clark wrote one of the most influential and important craft essays of the still-young twenty-first century, realized in a format for the new era. "How Envy Killed the Crafts Movement: An Autopsy in Two Parts" began as a lecture at the Museum of Contemporary Craft, Portland, that was released as an e-book that was widely downloaded and circulated.[50] The essay became a diagnosis for the craft field's self-conscious and beleaguered status in relation to either art or design. It offers an examination of the contentious split between the American Craft Council and its once-innovative museum, the Museum of Contemporary Crafts (now the Museum of Art and Design, MAD), in New York.

Taken together, Clark's numerous interventions offer a seductive, selective version of the field. They present a teleological history told not only chronologically but also through a history of like-objects, offering an overarching, dominant viewpoint of the medium. A careful and cogent writer, Clark's is a masterful narrative that begins to look something like a canon: it is a historicism that works against the ephemeral, the minor, and the secondary. It makes clear,

Figure 1.15 *Untitled Ewer (Chou-ccr)*, 1992. Adrian Saxe (American, b. 1943). Porcelain and mixed media. The Museum of Fine Arts, Houston, Garth Clark and Mark Del Vecchio Collection, gift of Garth Clark and Mark Del Vecchio. Courtesy of the artist.

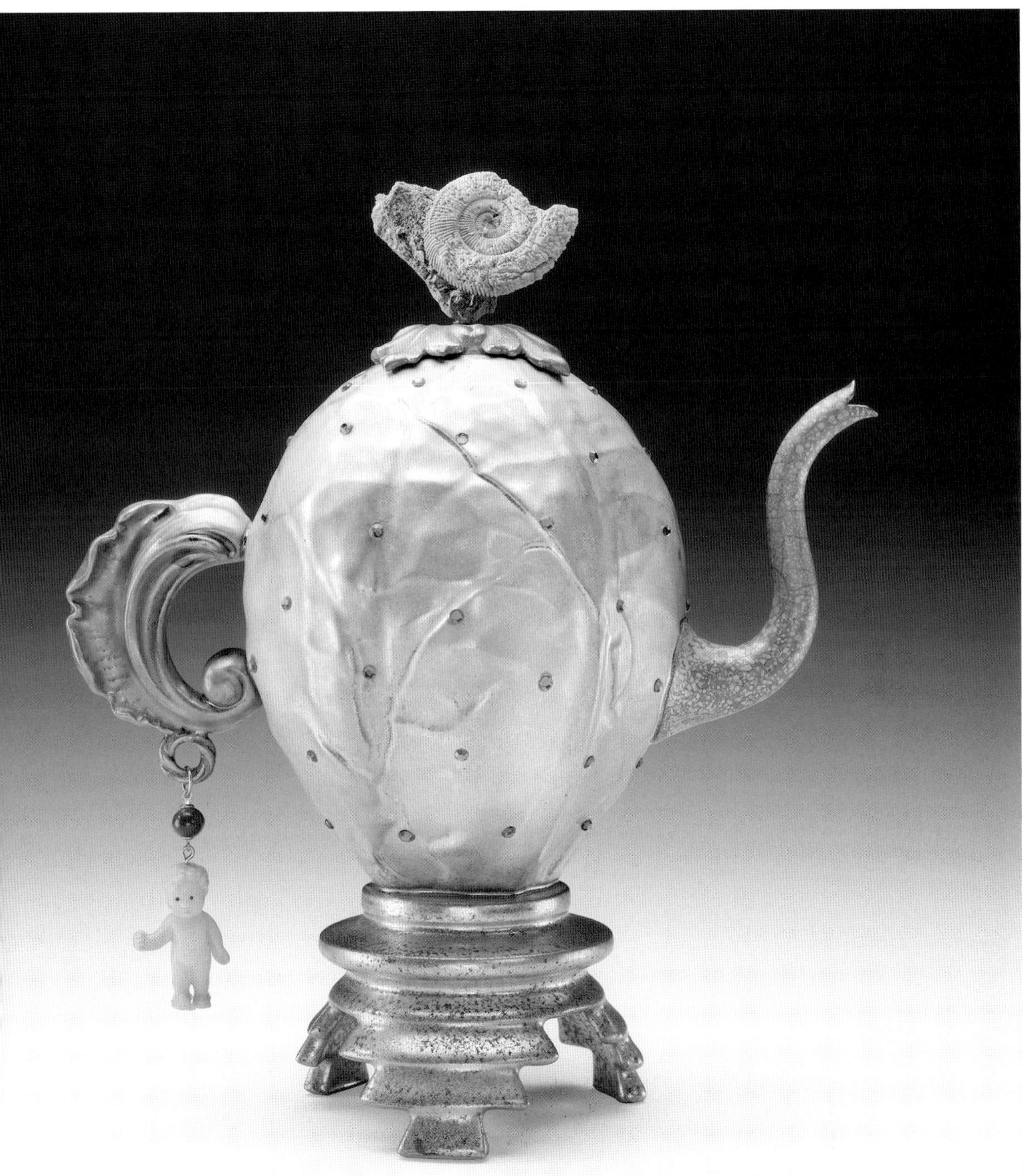

straightforward arguments for objects engaged in the principles Clark champions: pure form, authenticity, and, if non-functional, then conceptually sound. Clark casts off the majority of works that resemble, but do not rise, to the level of his favorite gold standards.

Over time, Clark's exhibitions and books shifted the taste of the field, moving it toward a visual vocabulary on par with sculpture in its theoretical underpinnings and methodologies. Meanwhile, the entwined supremacy of his profound intellect, business acumen, and an outsized personality has left any other critic on unequal footing: the quandary of dealing, and indeed profiting, from the artists one seeks to examine critically. Clark's writing, while thoughtful and lucid, has, at times, seemed promotional due to the inherent conflict of financial relationships with artists, even if it that was not the intent. Yet Clark singlehandedly brought rigor to the field of ceramics.

Further, as a combination of object analysis and social art history, *American Ceramics* predates much material culture scholarship as it has been retroactively applied to the twentieth century. In 1982, Jules Prown published his now-classic essay "Mind in Matter: An Introduction to Material Culture Theory and Method," which argues that objects of the past are primary cultural data—that is, texts—that we can read for clues about historical situations, that a man-made object of any kind—not just the handmade—transfers the conditions of its production, conveying, consciously or not, entire belief systems of a particular culture to which the maker belonged. Prown institutes a practice: description, content, deduction, and speculation.[51]

But by that time, Garth Clark had already begun his own version of Prown's material culture method, through his own meticulous inventorying and historicizing of the first American century of the studio pottery movement and elucidating its connections to present-day production. Clark's process—salvaging, then extracting, reclaiming, and building upon—parallels the process of working with clay. In effect, he uses its materiality as a methodology. Clay is itself a salvage material that embodies a cycle of recovery, and its sustainability is extremely familiar to artists of the medium. Part of why ceramists have put their faith in Clark is his immersive understanding of ceramics as a material history, its various formats, chemistries, eccentricities, and histories. This is the kind of hands-on, empirical understanding that few trained art historians have ever experienced firsthand.

Los Angeles and Postwar Ceramics History

It made perfect sense that Clark first opened Garth Clark Gallery on the West Coast in 1981. Los Angeles was the epicenter of American ceramics throughout the postwar period. The dominant narrative has centered on what is known as the Otis group. From 1954 to 1959, as the figurehead of the ceramics program at the Los Angeles County Art Institute (later known as Otis), Peter Voulkos taught and wielded influence over an entire generation of young male ceramists and artists such as Billy Al Bengston, Michael Frimkess, John Mason, Mac McClain, Jerry Rothman, Paul Soldner, and Henry Takimoto. The program attained widespread acclaim, what *Craft Horizons* editor Rose Slivka termed "the new ceramic presence" in 1961.[52] Voulkos was voracious and destructive in his experiments, training his students in, as Glenn Adamson writes, "a particular variant of their leader's 'how not to' manner of using clay."[53]

What has dropped out of this narrative is the extent to which Voulkos was affected by the many women potters who trained and influenced him, including Wildenhain, examined in the next chapter, and M. C. Richards, considered in chapter 4, whom Voulkos met during the 1953 summer he spent at Black Mountain College and followed to New York City, his first-ever trip. It was after this significant summer that Voulkos abandoned the potter's wheel, resorting to so-called primitive hand-building methods to achieve a uniquely raw and asymmetrical ceramic sculpture, in which the clay was pushed, like bread dough, to its very limits, past the point of plasticity—punctured, marked, and collapsed. In a 1966 article in *Artforum*, John Coplans called this "abstract expressionist ceramics."[54] In deconstructing this terminology, Andrew Perchuk has already demonstrated the strong magnetism that the work of John Cage and Robert Rauschenberg exerted on Voulkos.[55]

For Voulkos, clay became finite—used up, rather than renewable—and formless, rather than formed. A decade later, formless, or *informe*, became the most salient feature of such expanded sculptural practices as kineticism, installation, and land art, as Rosalind Krauss came to define them in her famed essay "Sculpture in the Expanded Field" (1978).[56] Thus, the absolutism of the two unifying principles of studio ceramics—namely, utility and traditional form—lost its allure in favor of unencumbered movement in the wider world of avant-garde sculpture and, arguably, access to its more lucrative market. The collector Fred Marer, a Los Angeles City College mathematics professor, bought the work of this generation right out of the studio. Voulkos's unorthodoxy got him fired at Otis, and in 1959 he began teaching at UC Berkeley, training a second generation of ceramists, including Robert Arneson and James Melchert, who came to be affiliated with Funk, a subsidiary of 1960s Pop that also embraced popular culture.

By showing his large-scale, impressionistic ceramic sculpture at the Ferus Gallery, Los Angeles's hip 1960s art spot, rather than at craft exhibitions, Voulkos forced the art world to confront its long-held assumptions about ceramics as a minor art. Ferus's stable of artists included peers such as Ed Ruscha and Andy Warhol. Likewise, Voulkos's colorful lifestyle—his motorcycles, womanizing, and all-night, caffeine- and amphetamine-fueled group work sessions—contributed to his legend and dazzled his posse of all-male students. As well, Voulkos received the critical support of Rose Slivka, the most important American critic of studio craft in the 1960s as the powerful editor of *Craft Horizons* magazine, who promoted him assiduously. In 1977, she published the first monograph on his work and curated his first retrospective, held at the Museum of Contemporary Crafts in New York.[57]

Voulkos's history is well worn: first documented by Slivka, followed by Clark's *American Ceramics* a year later and the numerous exhibition catalogues that championed his position and standing in the ensuing decades. In *20th Century Ceramics*, the potter and critic Edmund de Waal recounts a popular version of this same history using the previous source material.[58] Further, the two major exhibition narratives themselves are skewed to reflect the primacy of the Southern California permanent collections being represented. Mary Davis McNaughton's *Revolution in Clay* (1994) recounts the history of Los Angeles through Fred Marer's collection at Scripps College. This was followed by Jo Lauria's *Color and Fire: Defining Moments in Studio Ceramics, 1950–2000*, another medium-specific exhibition drawn from the Los Angeles County Museum of Art's permanent collection. The latest iteration of this history, *Common Ground: Ceramics in Southern California, 1945–1975* (2012), showcased the collection at the American Museum of Ceramic Art in Pomona. All three catalogs contain major essays by Clark.[59]

But there was a great deal of activity that led up to Voulkos's formation and beyond it. Los Angeles was an intensely energetic and innovative art scene, or, as the potter Vivika Heino described it, "a livewire place."[60] In October of 1956, *Craft Horizons* produced a special issue on California, reporting that sixteen teaching institutions throughout greater Los Angeles had "complete modern ceramic workshops and equipment."[61] Fully ten of the schools were community colleges, and still other institutions, not least of all public high schools, offered continuing studies in the evening, fully equipped with electric kilns.[62] Beatrice Wood, for instance, honed her skills in such a setting, at Hollywood High School, during the early 1960s.[63] Such sites were especially conducive to popularizing ceramics beyond a specialized community of craftspeople.

The earliest university-based ceramics programs on the West Coast were located in Los Angeles, both established in 1933. UCLA's program was run by Laura Andreson, who was a trained painter but self-taught potter who learned to throw from émigré potter Gertrud Natzler. The University of Southern California's was helmed by Glen Lukens, which he established at the School of Architecture. Lukens is responsible for training a crop of ceramist-teachers, including F. Carlton Ball, Vivika Heino, and Harrison McIntosh, as well as the architect Frank Gehry and jeweler Sam Kramer.[64] Yet within the wider world of the city's cultural production, Los Angeles's ceramics scene represented an avant-garde with which most of the city was patently unfamiliar.

The Chouinard Art Institute (1921–1972), the West Coast's first woman-run art school, was another of the medium's primary training sites, and the focus of its pedagogy and instruction was consciously, and unapologetically, directed toward returning war veterans. Nelbert Chouinard, a Pratt-trained artist and educator who had worked in settlement houses in New York and Minneapolis, returned to her hometown of Los Angeles to found an art school on a shoestring budget, or what she called "a war-widow's pension."[65] Her husband had died in 1918, at the end of the First World War. As such, her primary allegiance was toward returning war veterans of both wars. In 1952, fresh out of the MFA program at Alfred, Susan Peterson was hired to establish the school's renowned ceramics program. The ceramics program was only a two-year track, and it lost its accreditation for the 1954–1955 school year due to its insufficient liberal arts offerings.[66] Peterson's first student was John Mason, who went on to study and share a studio space with Voulkos.

Ceramics in Los Angeles was convivial: Peterson was one of the youngest members of a loosely affiliated group of ceramist-teachers, who gathered regularly. As Ralph Bacerra, a Chouinard student under Vivika Heino and, later, faculty at the school, described it:

> Everybody belonged, everybody that did ceramics or talked ceramics. And once a month there were meetings, and Susan [Peterson] was maybe president one year and Vivika the next or Bernie Kester from UCLA or Laura Andreson and Peter Voulkos. So they all got together, and there was a nice communal atmosphere. And that's how I met all these people, is through these meetings and visiting their campuses and their studios.[67]

The community eventually formalized, calling itself the Ceramics Education Council, a fledgling subcommittee of the American Ceramics Society, a rather stodgy turn-of-the-century industrial organization comprised mainly of ceramic engineers and industrial, commercial potteries. In 1966, the group eventually splintered off, forming its own professional organization, known today as the National Council on Education for the Ceramic Arts (NCECA).[68]

In 1955, Peterson took over the ceramics department at the University of Southern California. The first student she accepted who also became her graduate assistant was Ken Price, another student who defected to Voulkos's circle. Though they were a generation older, Vivika and Otto Heino became Peterson's eventual successors at Chouinard. Vivika was the recommended candidate; however, the couple was hired together. To her credit, Nelbert Chouinard properly compensated Vivika for her talent and experience, allowing Otto to teach alongside her for less than half her earned income.[69] Since neither potter earned anything close to the salaries of male professors in the animation and advertising departments, it cannot be said, however, that female solidarity was the sole reason for Chouinard's equality: more likely than not, the negotiated price was market driven. As Otto commented:

> It was the best time. Teaching was vital in California because there was so much competition. There were so many good students. We were there at the right time. Everybody was going full speed at the colleges: Chouinard, Otis, and the community colleges.[70]

Yet a 1959 Chouinard faculty meeting transcript belies another viewpoint, highlighting the lack of respect for disciplines outside commercial art. Harry S. Diamond, the chairman of advertising, refers sardonically to ceramics and its women teachers, specifically Vivika Heino as "a Pots & Pans Director."[71] Such a slur is telling, relegating the discipline, and its pedagogy, to the kitchen—the only fully acceptable, and universally recognized, sphere of women's production.

At midcentury, making a living as a full-time studio potter was a precarious existence. Therapeutic and community-based craft workshops offered women ceramists the possibility of paid, professional work in their chosen creative field. Yet these community-based roles kept them in relatively lesser professional situations than many of their male colleagues. Cognizant of this difference, women like Wildenain, Richards, and Peterson capitalized on their ability to produce both goods and skill sets as a form of social mobility, utilizing functional ceramics as a means of securing their own professional opportunities. In the best sense, they became directors of their own destinies, literally extending the reach of their own "pots and pans" to displace male authority.

FORMS-OF-LIFE

Marguerite Wildenhain's Pond Farm

CHAPTER TWO

Consider Marguerite Wildenhain's *Milk Pitcher* (1923), made in the long-standing tradition of sturdy German domestic vessels (plate 3). Made of stoneware, the open vessel is a mottled cobalt blue, made uneven by small pits on the surface, a trademark of salt glazing. Salt is thrown into the kiln during the firing process to react with the silica in the clay body in order to produce a dappled, pleasantly textured surface. This rougher exterior offers the hands a bit of abrasion, to prevent slippage when full.

Produced during her years at the Weimar Bauhaus, Wildenhain was the only woman at Dornburg, a provincial village eighteen miles (thirty kilometers) from the main Weimar campus, where the ceramics workshop and its handful of students were housed in the former stable building that once belonged to the Grand Duke of Saxony-Weimar. According to Wildenhain, there were five students in 1920, that first year of the workshop, and never more than seven. The group remained intact for five years, banding together to survive as Germany's economic situation rapidly deteriorated, resulting in severe inflation and an insecure food supply. They lived through near starvation, eating oatmeal, supplemented with homegrown Swiss chard, three times a day.[1]

Slightly potbellied, the form of *Milk Pitcher* suggests the humble abundance of raw, fresh milk, the crown jewel of the farmhouse breakfast table. But there was none to be had. This functional pouring vessel was made during a period of deficit, without access to dairy products. It likely represented hope and the future promise of basic sustenance.

Born Marguerite Friedlaender in Lyon, France, in 1896 to Jewish parents of English and German origin, Wildenhain was the first woman to be named a

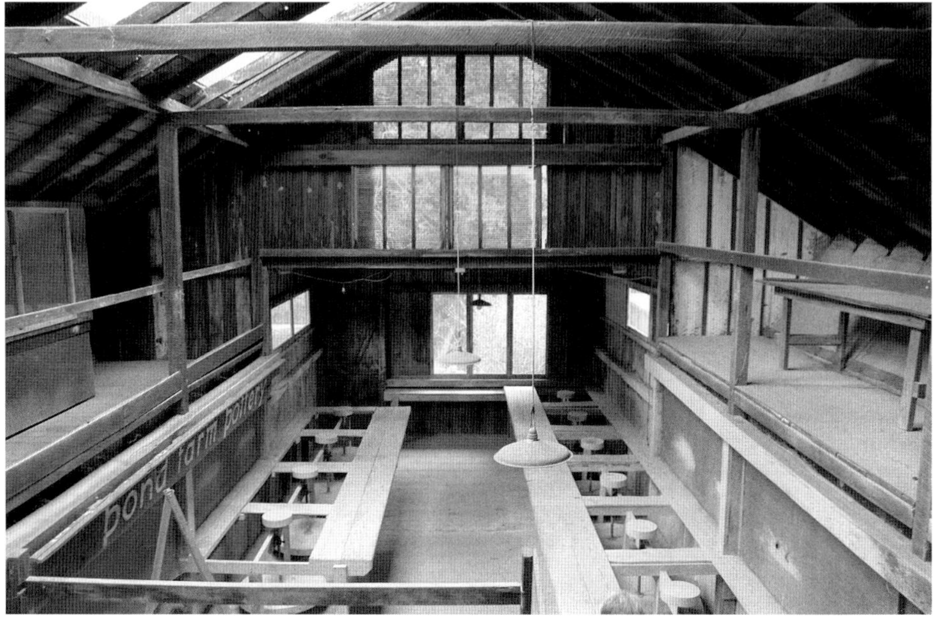

Figure 2.1 Interior view of Pond Farm work barn, 2015. Photo by Cheri Owen-Sorkin. Courtesy of the author.

master potter in Germany, earning her certification in 1925, the same year the workshop closed. The isolation and solidarity of the Dornburg workshop was a formative experience for Wildenhain. She would return to it spatially and psychically through her career, reconstructing facets of its particular culture over and over again. Through Pond Farm, the pottery she founded, Wildenhain implemented a Bauhaus-style training camp in Sonoma County, California, long after functional pottery had yielded to ceramic sculpture (figure 2.1). But Pond Farm had a particular mission: the dissemination of the Bauhaus's mythic pedagogy to unsuspecting American students, who would be inculcated with ceramics as a way of life and go on to imprint its rigor onto future generations.

Milk Pitcher must have nourished its maker spiritually as well: it is the only Bauhaus-era work that accompanied Wildenhain to the United States when she arrived as a refugee in 1940. In the intervening years, she became a prodigious functional potter, well regarded for her stoneware pots but notorious for her opinions, including a public rebuke of Bernard Leach in *Craft Horizons* in 1953, fiercely defending her adopted homeland from a perceived outsider, and her robust refusal of an award the American Craft Council tried to offer her for her career-long contributions to the field in 1977.[2] In rejecting the latter, she wrote: ". . . I detest what the ACC has been standing for these last 20 years or so, their point of view toward the crafts and art, their horrible magazine, their museum etc etc. So forget me as a prospective 'Fellow.'"[3]

From 1952 until 1980, Marguerite Wildenhain presided over a community entirely of her own making: a summer program devoted to the discipline of ceramics, admired for its uncompromising rigor in pursuit of the craftsman ideal. Yet her practice has been misconstrued as object oriented: Wildenhain's ideology was entirely process driven. Over eight or twelve weeks, students threw five or six hundred pots and left with nothing. By purging her students of the desire to produce finished pots, Wildenhain set her students on a course of non-object production. This is a radical reformulation of pottery, redirecting it away from its traditional object-based orientation in favor of process. Thus, at Pond Farm, ceramics became a *live form*, valued not as a commodity but as an experience.

Each summer, around twenty-five students would come to study pottery with her for about nine weeks at a time. Wildenhain did not accept beginners; students were routinely turned away or sent elsewhere to gain basic skills. Pond Farm, unlike Black Mountain College, willfully lacked institutional status. Without any kind of accreditation or degree structure, it had no credibility within the larger American educational system. As Wildenhain herself wrote, "Pond Farm is not a 'school'; it is actually a way of life."[4]

However, Wildenhain's history functions as an allegory of illegibility: she was misinterpreted again and again by her American colleagues, who rejected her Old World functionalism, even as they profited from her pedagogy. Wildenhain's career is a case study in the gender inequity and media hierarchies that plagued women artists, even those with the best training, at midcentury. She represents a series of exclusions: a Bauhaus-trained Jew, a French refugee/German émigré, a woman in the elite reaches of her male-dominated profession, and moreover, a pedagogical demagogue—a proclaimed, and self-proclaimed, master. But neither Wildenhain nor any other Bauhaus-affiliated woman (not even Anni Albers) ever attained an artistic stature equivalent to that of their male counterparts, in particular Josef Albers. Given that Josef Albers shunned her chosen medium as "not intellectual enough," no correspondence exists between Wildenhain and the Alberses, but surely they were well acquainted, as Wildenhain preceded Josef Albers at the Weimar Bauhaus by one year, training from its inception in 1919.[5] Yet Wildenhain's students were simultaneously endowed with, and differentiated from, her Bauhaus legacy. Her forgotten production is testimony to the limits of both assimilation and empathy in her new homeland.

Forms-of-Life

Wildenhain's pottery was functional, pragmatic, and domestically scaled: made with the intent to sell simultaneously in high-end department stores such as Gump's in San Francisco, in art galleries, and on-site in her showroom, the first

Figure 2.2 *Marguerite Wildenhain, Bauhaus potter, settled in Pond Farm, California*, ca. 1958. Photo by Otto Hagel. Collection Center for Creative Photography. ©1998 Center for Creative Photography, The University of Arizona Foundation.

space visitors saw in the main building at Pond Farm. This undated image (figure 2.2), likely from the mid-1950s, casts her at as a lone but proud figure, foreshortened by a perspectival frame. She is smiling slightly, her head cocked toward the fruits of her labor: a series of worktables laden with pots, vessels, vases, candleholders, candelabras, and pitchers.

In the foreground, on the right, stands one of the few narrative works she created, one of a series of four vases, based upon the four acts of T. S. Eliot's play *Murder in the Cathedral* (1935). Drawing in a purposefully naïve style, Wildenhain secures her affinity to modernism by flattening her figures and textures them with an incising technique known as sgraffito, made through scratching upon the surface to produce distinct surfaces and outlines. Yet the effect is not sculptural, offering the same flatness found in illuminated medieval manuscripts, in which the vase and its surface are one and the same. Raising their weapons, the grouping of knights depicted is one of the key moments of the play, which reenacts the assassination of Archbishop Thomas Becket of Cantebury, in 1170, but it is also an allegory for martyrdom and the problem of religious faith, both subjects that ostensibly plagued her as a Jew who had escaped the Holocaust. In her review of the play, the poet Marianne Moore wrote that Eliot's ". .austerity assumes the dignity of philosophy and the didacticism of the verities incorporated in the play . . ."[6] Austerity, dignity, and didacticism are a trio of descriptors that are applicable to Wildenhain, who performed her own austere philosophy of teaching through the strict methodology she applied to her students.

She never wrote about what the play represented, but it must have touched her deeply: she offered one of the four vases to the playwright himself and hoped to present it to him in person during the holiday season of 1952. T. S. Eliot was not in his office at Faber and Faber Publishers when she came calling, but upon her homecoming, she mailed it from California and received a warm response from him in return. As Eliot wrote: "It is more than kind of you to have given me such a magnificent present, and I am happy to say that the parcel arrived intact. The jar is at present standing on the mantelpiece of the Boardroom here, where it has been admired by my fellow directors."[7] Eliot's grateful acknowledgement and praise of her work must have meant a great deal to Wildenhain, whose career has gone largely unacknowledged in the literature on the Bauhaus tradition in America.

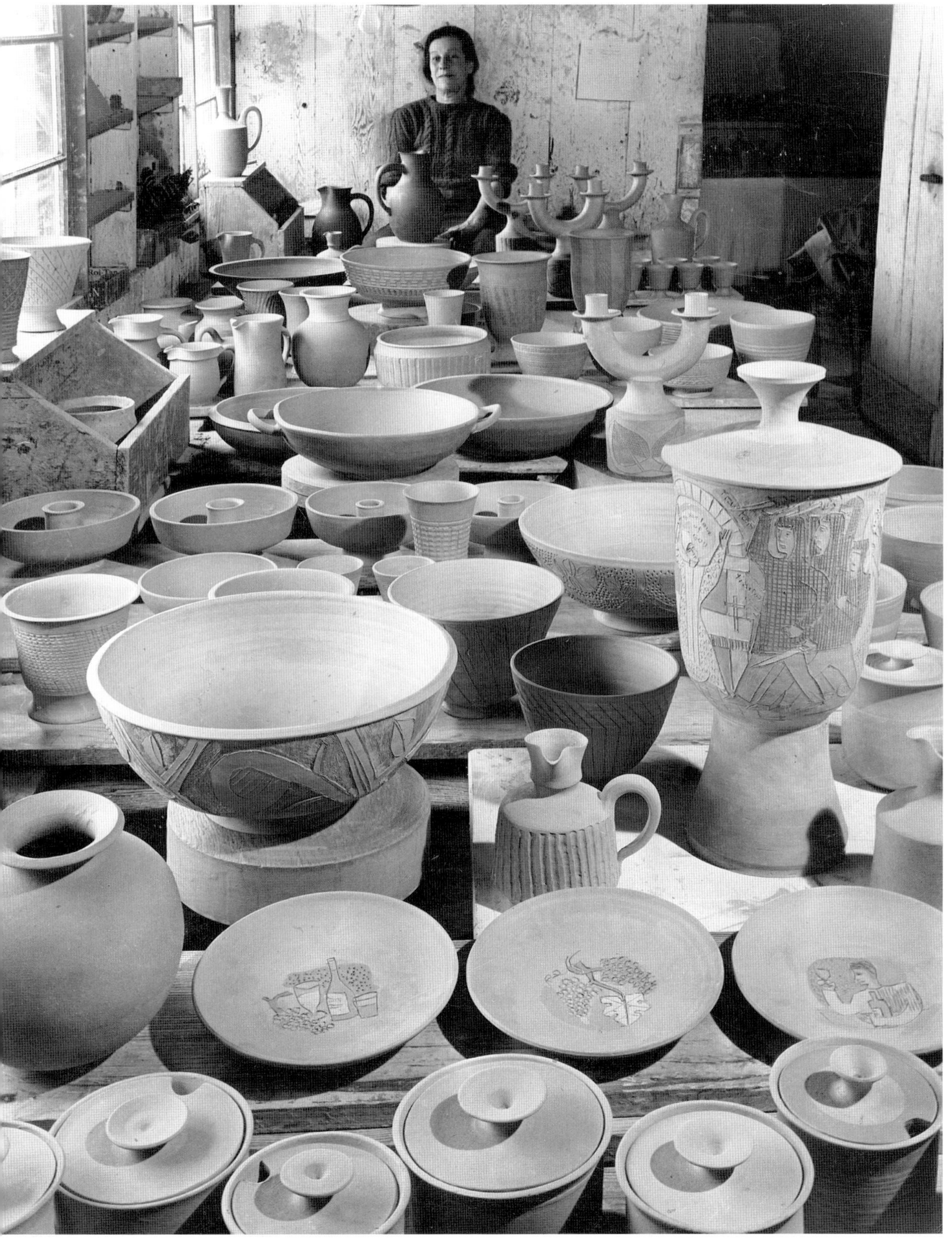

As an extreme form of living, Pond Farm exceeded any previous limits of labor and discomfort in relation to the production of studio pottery in mid-century America. Yet Pond Farm was also restorative, a way of beginning again, on all levels: for urban students unused to roughing it and for Wildenhain herself, in a new country and culture. In its initial years, Wildenhain provided basic barracks-type lodgings on her property, nearly replicating her experience at Dornburg; students were to learn to fend for themselves without a modern kitchen, pooling cookware and food to prepare common meals cooked in a fire pit.

For most of Pond Farm's existence, however, students rented rooms in Guerneville, an affluent summer resort town, or camped outside the town, which sat on the edge of a redwood forest. Roy Behrens, a former student who went on to teach graphic design at the University of Northern Iowa, authored a brief self-published memoir of his time at Pond Farm. Recalling the summer of 1964, he wrote:

> By way of Marguerite's fine map, we not only found the Ridenours [a local couple] that day, they also readily agreed that we could set up our two-person campsite (at no cost, I recall) in a deep ravine of redwood trees, adjacent to their hilltop house. What an exotic way to spend the summer, in a grove of these towering giants—one of which (thank goodness in terms of our bathroom needs) was wide and dead and hollowed out. It truly was primitive living: our only refrigerator was an ice-filled Styrofoam cooler, which was constantly coated with moisture. Among the most repugnant sights of the day (every morning) was to wake up to find the outside of the cooler crawling with dozens of huge green slimy slugs.[8]

The rudimentary nature of camp living would not have impressed Wildenhain, as she herself did not own a refrigerator until the early 1960s, one of the many American luxuries of which she disapproved—in her memoir, she disdained it as the "holy-holy of American life."[9] Compare this with the Alberses pride in suburban living. Nicholas Fox Weber reports that the couple enjoyed showing off their modern Sears Roebuck appliances to visitors and friends.[10] But this was in keeping with Wildenhain's distaste for convenience, which led to the laziness and shortcutting she found to be so pronounced in American students. The students who came to Pond Farm, however, were broken of these and other poor habits. The purposeful deprivations experienced at Pond Farm merged with the rigors of slowly, meticulously mastering the medium of pottery.

The distance between Guerneville and Pond Farm was a daily four-mile hike uphill on unpaved roads. Upon arrival, students gathered under the peach tree before 8 a.m. to observe a ritual silence. Afterward, students were expected to spend seven to eight hours per day at the pottery wheel, throwing forms dictated by Wildenhain. While she herself was a studio potter, making singular objects for the art or luxury consumer market, her teaching methodology, like the Bauhaus, was geared toward industry, or what is known as production pottery, making dozens of wares to be sold as complete, handmade sets:

> Students were thoroughly instructed on wheel techniques and on how to develop a critical eye. The workshop was as demanding as one could find. Beginning students started with the infamous "doggie dish"—a shape that is somewhat troublesome of beginners. After making ten or twenty of these, permission was given to move on, and the students worked through about fifteen basic forms such as flower pots, bowls, bellied coffee pots, spouted pitchers, footed bowls, cups, plates, and eventually teapots.[11]

Students brought their lunch and ate together at long wooden picnic tables behind the work barn. Monday through Friday the workday ended at 4 p.m., and each Wednesday afternoon was devoted to drawing from nature. Wildenhain would frequently hold outdoor seminars and discussions, lecturing on various topics, showing her own work or that of other artists, and frequently read literature aloud, often poetry or from the journals of Van Gogh, Rodin, and Delacroix (figure 2.3).[12] At the end of the day, students trudged back down the hill into town, unless invited to drink sherry with Wildenhain in her garden. Three or four times a session, evening parties would be held at the beach.

Working and living in the natural environment, with an aged Bauhäusler directing the experience, such meditative precision cultivated not only an intensity among her charges but also perhaps a kind of unspoken honor for the dead, for the fledging utopia that had been rudely crushed by Nazi culture, a moment of Wildenhain's own youth that could never have been fully explained but could indeed be imparted through process: the forms themselves could live on, made and remade by generations of rudimentary potters, eager students grasping and marveling at the complexity and variety of simple forms.

The disappearance of the potted object, and its subsequent passage into immateriality runs parallel to the greater avant-garde ethos of the 1960s. The anti-object stance of performance, video, and land art, and to a lesser degree minimalism, which rejected the hand-produced or individually wrought object, challenged the hegemony of the materialist tradition in artistic production. It also functioned as a critique of the market and its easy commodification of the art object. In doing so, artists were able to propose new relationships between objects and viewers, artists and audiences, and artists and institutions. Ceramics has been left out of this history. However, Pond Farm's anti-object stance troubles the dominant narrative, marking an important and overlooked moment in modern craft.

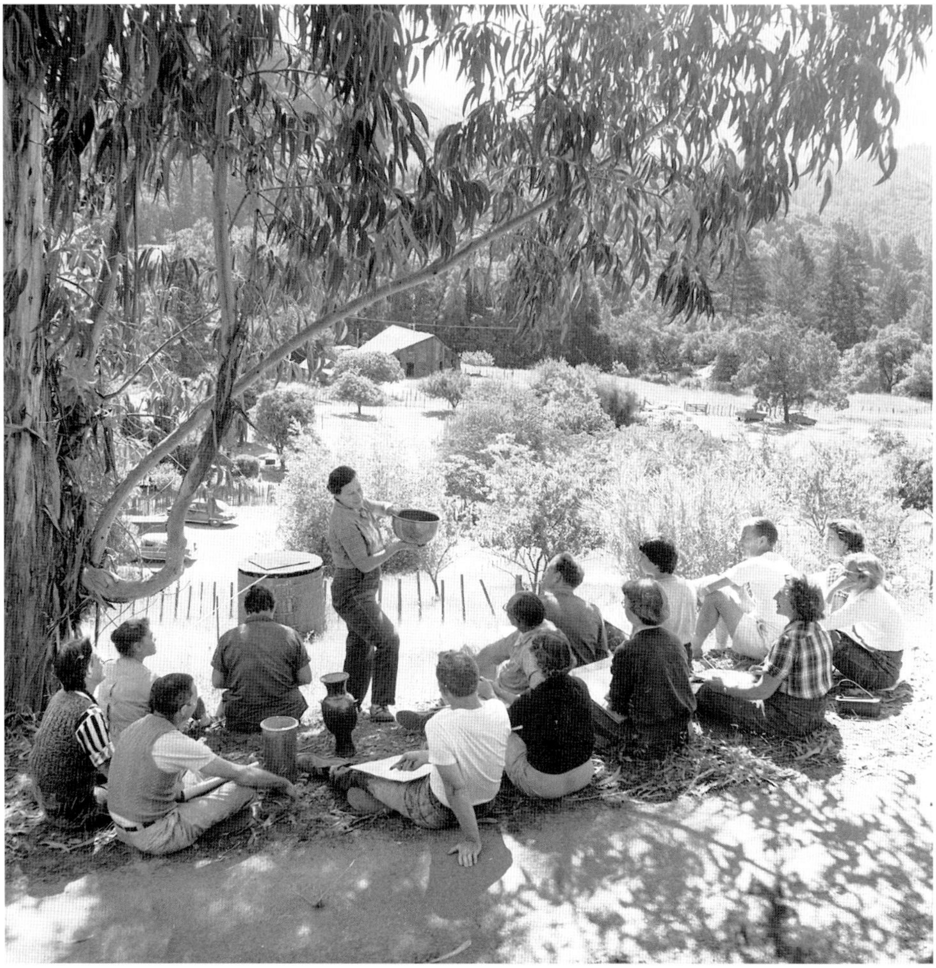

Figure 2.3 *Pond Farm, Marguerite Wildenhain,* 1952. Photo by Otto Hagel. Collection Center for Creative Photography. ©1998 Center for Creative Photography, The University of Arizona Foundation.

Nathan McMahon, a student who first came to Pond Farm in 1954, described the destruction of the pots at Pond Farm as an act of submission, in which the objects and their makers were broken down—only to be rebuilt more robustly:

> More often the pots were destroyed while the clay was yet plastic, cut through with a wire to gauge the thickness of the walls and to evaluate the quality of our throwing technique. It was a sound practice, offering lessons that would otherwise have been impossible to learn, and they helped separate us from the tendency toward regarding every standing pot as worthy or precious. It helped to speed up the learning process by undermining the persistent ego.[13]

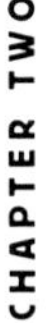

This ritualized destruction contained within it an exchange of power: students giving themselves over to the master, allowing themselves to be trained according to her will. Summer workshop participants came as seekers, looking for an encounter with craftsmanship through subordination. No doubt such a collective experience of discipline was also a means of undoing the excesses of individuality endemic to the abstract expressionist era: ego, personality, and free expression. The documentary photographer Otto Hagel, a close friend who made primary images of Wildenhain's work throughout the 1950s, captured the collectivity of the student enterprise at Pond Farm when he captioned this particular image with the following interpretation: "The problem for all students was to make a pitcher with a rim. Different personalities, different conceptions, different pots" (figure 2.4). But in a setting of noncompetition, students also had to learn to trust each other's skills and their communal ability to coax the most that they could out of the raw clay. In this way, the non-object pot became symbolic, the subject of a focused reading, or deconstruction, and then obliterated, smashed down and made anew, a *live form* engaged in an endless process of renewal, led by Wildenhain.

She herself modeled this resiliency through her own wartime narrative: making the best of what was and starting anew, loyal to the precepts of the Bauhaus, and continuing the modernist tradition without a crisis of faith. The intertwining of clay's elasticity and Wildenhain's own resilience functioned as a unique pedagogical model that remains unreplicated by other key pedagogues of her generation.

In his 1973 treatise on pedagogy, *Fellow Teachers*, the cultural sociologist Philip Rieff asks: "Would you like to know how to recreate authority? You would have to again begin outside yourself. A true interdictory authority must be taught to us; it cannot be thought up by us."[14] Rieff's quote is rhetorical, of course, establishing that authority itself is rooted in an established power dynamic, one that is implemented through the ideological apparatus that shapes a subject during his or her lifetime. For Wildenhain, Dornburg became the shape she mapped onto Pond Farm, an authoritative space inscribed by the intensities of exile and fear.

Weimar: The "Wrong" Bauhaus

Owing to the school's pervasive institutional sexism, Wildenhain, like other Bauhäusler women, was allowed to enroll in only one of three workshops: textiles, ceramics, or bookbinding.[15] The ceramics workshop was led by two artists, potter Max Krehan (1875–1925) and the figurative sculptor Gerhard Marcks (1889–1981). This was a conscious split in accordance with Bauhaus Rector Walter Gropius's ideas: Krehan was the master of craft, while Marcks was the

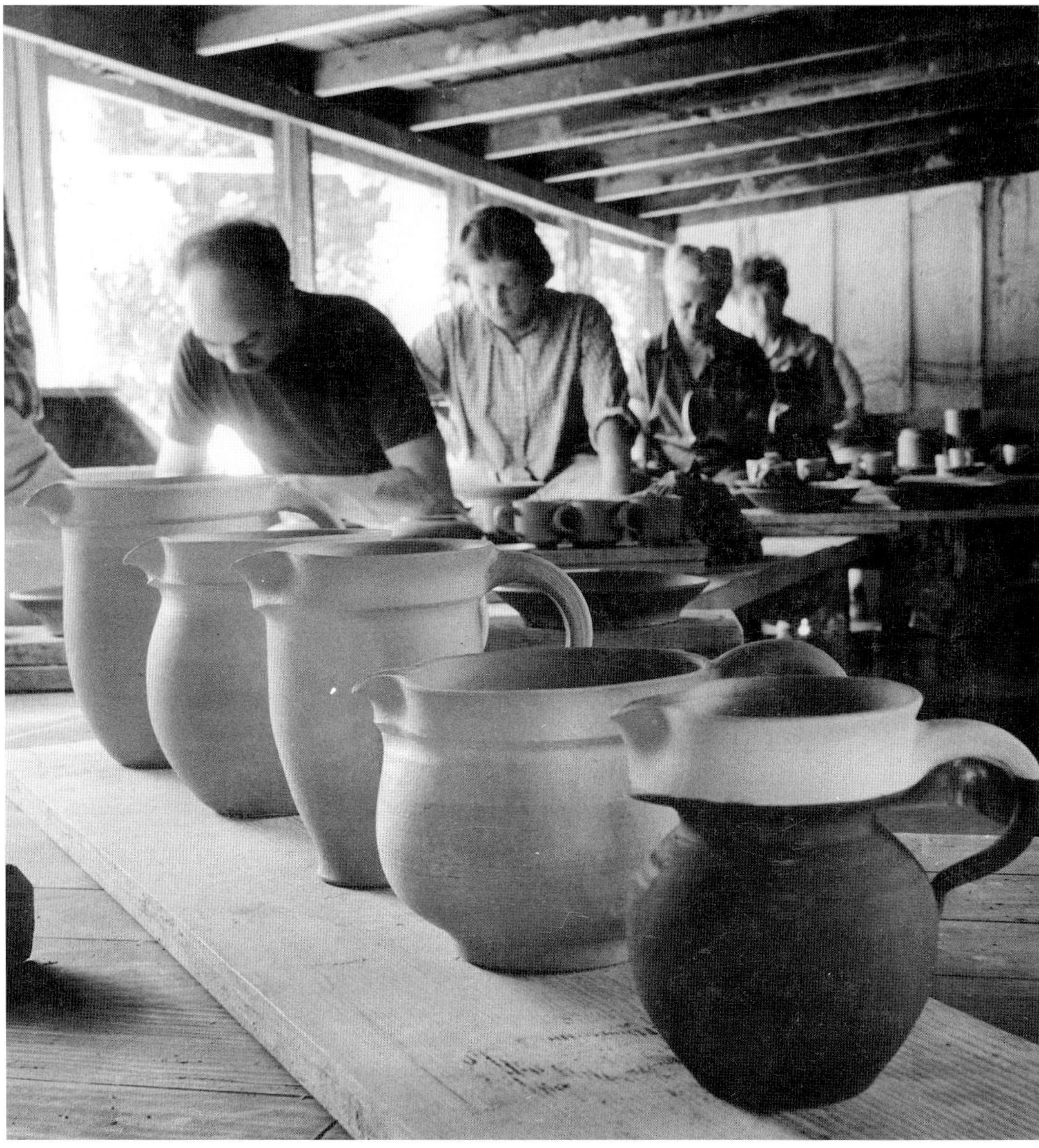

Figure 2.4 *The problem for all students was to make a pitcher with a rim. Different personalities, different conceptions, different pots,* ca. 1952. Photo by Otto Hagel. Collection Center for Creative Photography. ©1998 Center for Creative Photography, The University of Arizona Foundation.

master of form. Gropius saw the camaraderie of craft as a natural progression toward the collaborative practice required of architecture. In his 1923 treatise "The Theory and Organization of the Bauhaus " he wrote, "Training in a craft is a prerequisite for collective work in architecture. This training purposely combats the dilettantism of previous generations in the applied arts."[16] Gendered female, this so-called dilettantism betrays the fear of amateurism that plagued Gropius after many more women than men initially enrolled at the Bauhaus, resulting in the exclusion of women from all but the three Weimar-era craft workshops.

Rather than eradicating traditional media hierarchies—as the Bauhaus is often credited—Gropius, in actuality, reinforced both the low status and the feminine associations of craft, appropriating it as a service discipline laboring on behalf of architecture, its masculine superior. Thus, Gropius's views paralleled the therapeutic craft endeavors described in chapter 1, in which female-taught handicrafts were utilized as a strategy for social reintegration among returning war veterans. In both realms, craft was relegated to the sphere of the precondition, widely regarded as a stepping stone en route to a more respectable, albeit "masculine," profession.

The 1924/1925 school year was pivotal in the workshop's history: Marguerite's future husband Frans Rudolf Wildenhain was the final student to arrive in February 1924; the following April, after the conservative party swept elections, the provincial government of Thuringia concluded that the Bauhaus was unprofitable and, accordingly, dramatically slashed its budget with the intent of terminating the masters' and director's employment contracts, effective December 31, 1925. In actuality, the Bauhaus was permitted to continue, albeit in a reduced and diminished capacity, reopening in the city of Dessau in 1926. However, on October 16, 1925, Max Krehan died suddenly, and in December, Gropius closed the pottery workshop permanently.[17]

Following his passing, Wildenhain made a figurative relief of Krehan's head to memorialize her teacher, an homage to the European tradition of venerating the dead through commemorative sculpture (figure 2.5). She treats the work as a death mask, depicting him with his eyes closed, at peace, and with the self-possession of a dignified, intelligent face: a long, narrow and sunken jawline offset by sharp, broad cheekbones and an expressive mouth. As a tribute to her former teacher, Krehan's face, laid to rest, is presented as an unbearable fact, a final tenderness. But his passing had significant repercussions for her artistic career: it also signaled the death of the ceramics workshop, a workshop that was truncated by Gropius at the end of 1925. This also resulted in the loss of the sculptural component of the workshop. Though Gropius could have reappointed Gerhard Marcks's position in Dessau, he did not, which can be interpreted as intentional.[18]

Along with Lyonel Feininger and Johannes Itten, Marcks was one of the original three faculty members at the Bauhaus. Marcks and Gropius had previously worked together at the 1914 Deutsche Werkbund exhibition in Cologne. Despite this longstanding relationship, Marcks's dismissal can be construed as a break

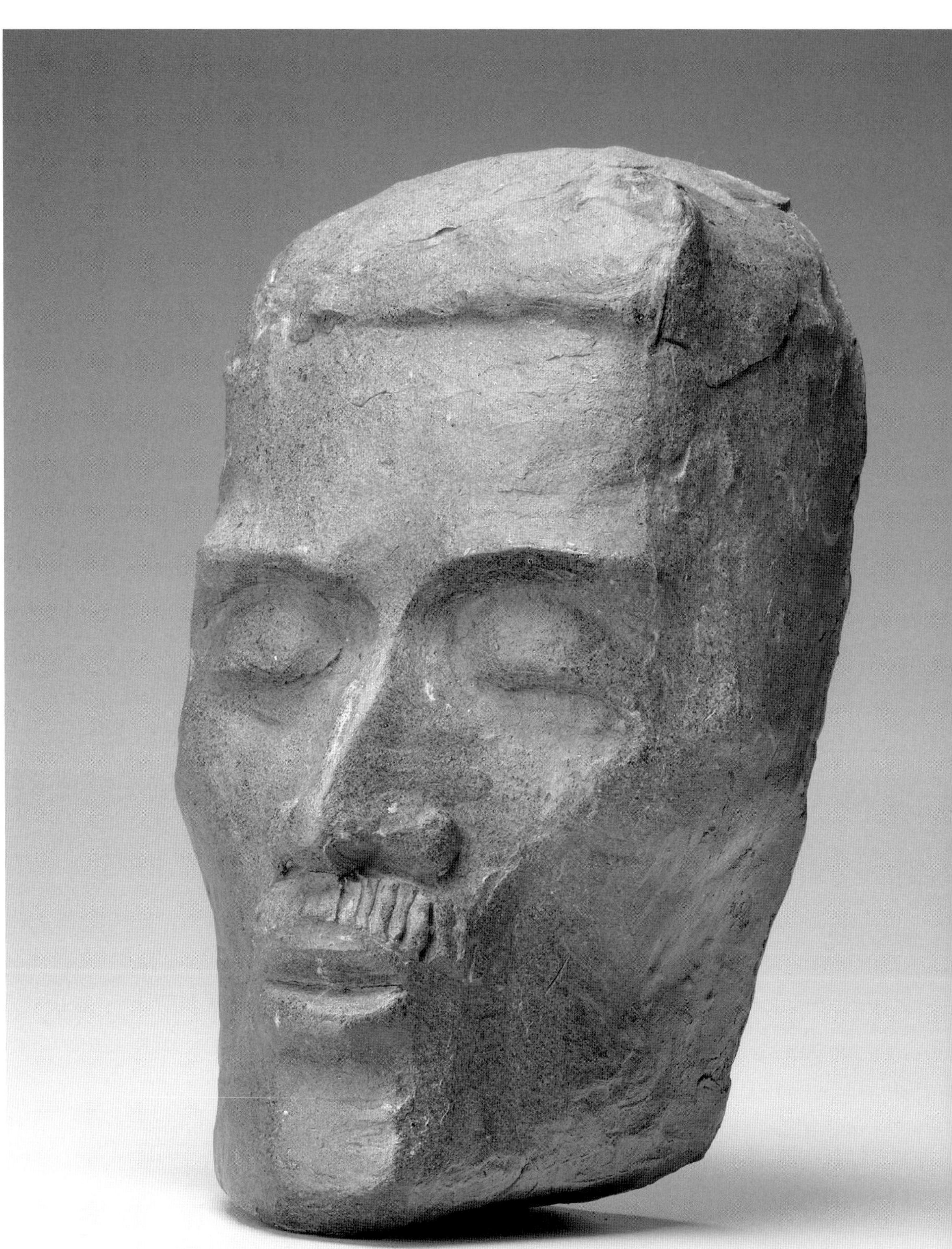

Figure 2.5 *Head of Max Krehan*, 1925. Earthenware, 8 3/4 x 5 x 4 inches. Collection of Forrest L. Merrill, Berkeley, CA. Photo by M. Lee Fatherree.

with the past, Gropius's pointed attempt to leave figurative sculpture behind, as though its idyllic expressionism belonged firmly to the romantic tradition of the nineteenth century. An imperial city, Weimar was at the center of German romanticism, a movement that strived for great emotion in its cultural production, most notably its music and literature. Weimar was home to Goethe, Schiller, Liszt, and, in the baroque period, Bach. Through the preservation of famous residences and much commemorative public sculpture—always figurative—Weimar's illustrious cultural history is woven into the fabric of the city.

Gropius seized upon the Bauhaus's forced move to Dessau as an unprecedented opportunity for radical change. As an affirmation of modernity, Gropius designed the school's brand-new building, which opened in December of 1926 and became the benchmark of the much-imitated "Bauhaus style," later known as the International style. He initiated the architecture workshop in 1927, which quickly became the dominant workshop. Thus, Dessau's new building branded Dessau as the internationally recognized Bauhaus—the "proper" Bauhaus. Weimar was largely forgotten, and its reputation suffered accordingly. This was only heightened after World War II, as the city itself became tainted by its close proximity to the concentration camp Buchenwald, only eight kilometers from Weimar's center.

In 1926, Marcks accepted a position at the Burg Giebichenstein School of Fine and Applied Art in Halle-Saale, Germany, and recommended that Marguerite be appointed head of ceramics, a testament to both Marcks's sense of equality and to Wildenhain's abilities.[19] Nearly ten years her junior, Frans apprenticed under Marguerite in Halle-Saale, a highly unusual arrangement for the era. The couple wed in 1930. Decades later, Frans reflected upon the extreme differences in their education:

> It is true it took me a long way to find myself and still I am stumbling further to be overcome myself. M's formative years where [sic] although under a more happier sky—She was 4 years at the Bauhaus with Marcks and Krehan.—I only a year. But the "Voukurs" [Foundation Course] was the most decisive passing through my lifetime . . . Coming from a family of craftsmen it was a naturalische [natural] affinity which we developed towards M. Krehan. He was the master anonymous. Under other stars he would have been a Hamada for Germany. Nobody escapes his Karma for German Mutterspresiche

> [mother tongue] vernacular. Later Dutch still in relation to it . . . In this way the short period with M.K.[Max Krehan] in D.[Dornburg] gave me the foundation—In a happier time (Germany, inflation, etc., lost war—political upheaval, ideas new and old hanging like a thick fog over people) I would have become perhaps only a good country potter—making pots people could use and in between something more.[20]

From Frans's observations, we can glean several things. First, happiness is a relative term: Marguerite's time at Dornburg was marked by poverty and hunger, yet, according to Frans, proved artistically rich. This is affirmed by his remark that had *he* experienced a similar "prosperity," he would have ended up just "a good country potter" rather than an artist. Certainly the implication is that suffering merits, while affluence produces complacency. Indeed, Frans did prosper: he won multiple prizes at the Ceramic National exhibitions and became a well-known ceramic sculptor who left California and taught for the majority of his career at Rochester Institute of Technology in Rochester, New York, from 1950 until 1970. Frans Wildenhain's sculptural pursuits were quite possibly a way of distinguishing himself professionally from his wife's own functionalism, yet being embedded in an academic atmosphere most certainly would have led to different and likely better opportunities, such as the large-scale ceramic mural commissions he completed at RIT during the late 1960s.[21]

Written in 1974, long after the couple was divorced and had become American citizens, Frans's ruminations offer a window to the collective psyche of "the Germans" at Pond Farm, Black Mountain College, and elsewhere and the potential for extreme misunderstanding between themselves and their American students. Even after decades of American citizenry, Old World values of hardship and uncertainty were not supplanted by a new world order. Rather, their extremity of experience bred a delicate combination of sensitivity and intolerance, which translated into a level of rigor previously unknown in American art schools and humanities-based colleges. As such, "the Germans" (more politely known as "the exile generation") were collectively revered as educators throughout the nation, at Bard College (Heinrich Blücher), Black Mountain (Josef and Anni Albers), Harvard (Gropius), and newly formed progressive institutions such as the New School in New York (Hannah Arendt, Erich Hula, Karl Mayer, Frieda Wunderlich, Kurt Riezler, Alfred Schutz) and the Institute of Design, Chicago (Moholy-Nagy, Mies van der Rohe).[22]

So, too, another Old World value: homage to the teacher, which Frans pays Krehan regarding his deeply imbued sense of tradition, which was imparted to both Wildenhains. As Frans suggests, the couple was perhaps responsible for spreading this influence among their Dutch peers. However, the recognition of a German vernacular heritage became untenable after the Third Reich's program of Aryan nationalism. Thus, Krehan's influence and reach had enormous, but ultimately unfulfilled, potential: unlike Japan, Germany was not permitted a Shoji Hamada, the Japanese potter who came to the Black Mountain Pottery Seminar, discussed in the next chapter. Hamada functioned as a sovereign figurehead with universal spiritual appeal.[23]

This was a repercussion of the vexing international issue known as the "German problem," a term that circulated widely among American intellectuals in the early postwar period, which assigned the tenor of the country's national disposition as aggressive and dominant, what Nietzsche famously called the "will to power," equating the essence of life with the need for mastery. In fact, Germany's persistent need to dominate was "proven" through the nation's Enlightenment-era heritage: the body of philosophical texts (Nietzsche included) that were appropriated by the Nazis themselves.

In 1945, Hannah Arendt described the international community's faulty logic in the *Partisan Review*:

> By identifying fascism with Germany's national character and history, people are deluded into believing that crushing Germany is synonymous with the eradication of fascism . . . Ideologically speaking, Nazism begins with no traditional basis at all, and it would be better to realize the danger of this radical negation of any tradition, which was the main feature of Nazism from the beginning (though not of fascism in its first Italian stages). It was, after all, the Nazis themselves who were the first to surround their utter emptiness with the smoke-screen of learned interpretations.[24]

Arendt points out the danger of equating Nazism's own false intellectualism with Germany at large. The fear of a Germanic association—either in form or content—explains why the International style, the quintessential architecture of modernity established at the Dessau Bauhaus, was so much more palatable as a cultural export than the truncated, and forgotten, ceramics production from the Weimar era. Excised of ornament, the symmetrical glass and concrete structures of Marcel Breuer, Walter Gropius, and Mies van der Rohe were without any trace of national association or heritage.

Moreover, "the International style" was an external designation, a term coined by the Museum of Modern Art's first director, Alfred Barr, and subsequent exhibition *The International Style: Architecture Since 1922*, organized by Philip Johnson and Henry-Russell Hitchcock in 1932. The popularity of the exhibition and its catalog codified the movement. In 1995, a late-in-life Philip Johnson, then the only living figure associated with the Modern's early years, and its unofficial historian, authored a new, anecdotal foreword to the exhibition catalog, in which he conveniently attributed the term "International style" to Barr, and conceded:

"The moral certainty of the three of us responsible for the direction of the book makes it sound preaching and schoolmarmish . . . . We knew what was right and we were very evangelical about it."[25]

Freed of its Germanic origins, the International style, then, could be attributed at will, according to rigorous standards wholly determined by a group of zealous Americans on the eve of the Berlin Bauhaus's closure in 1933. The show's impressive reception smoothed the transition for the Bauhaus's first wave of émigrés in 1933, who, with the help of Johnson, found immediate placement at Black Mountain (the Alberses) and Harvard (Gropius and, later, Breuer).

Thus, the cultural fallout from "the German problem" is partly to blame for Marguerite Wildenhain's largely unknown status in America. More German vernacular than International style, Wildenhain was invariably the wrong kind of "German": in both her chosen medium (folk-influenced pottery, associated with a doomed workshop) and her pedagogy (authoritarian in its top-down master-apprentice approach), she was to be forever associated with the "wrong" Bauhaus, Weimar's premodernist, decidedly humble beginnings.

Form Follows Nationhood

In keeping with core Bauhaus values, which emphasized that artistic and industrial production were conjoined, Wildenhain began designing professionally for industry in Halle-Saale as a means of supplementing her meager teaching income, working exclusively with porcelain. At Burg Giebichenstein, she established a productive and well-regarded workshop, retraining workers who had previously produced, as she described it, "mediocre vases and bowls and mass-produced tiles for stoves."[26] The models she developed were widely admired, and national periodicals of the era proclaimed her wares as "among the best modern craft work in the whole of Europe."[27]

In time, she developed a solid individual reputation as a designer, developing an extensive collection of tableware from 1928 to 1933 for porcelain manufacturer Royal Berlin. Sleek and modern, her Royal Berlin cup and saucer (ca. 1928–1933) (figure 2.6) is a prototype intended for in-flight dinner service, when air travel was the ultimate luxury, complemented by an elegant meal. Startlingly innovative, Wildenhain's saucer is hollowed out to allow the full, hot coffee cup to remain firmly in its base, preventing sloshing or scalding. The form is both efficient and crisp, offering a variety of possible visual delights to appear within the hollow, such as a shadow or tabletop texture.

Known by its official moniker, Konigliche Porzellan-Manufaktur (KPM), Royal Berlin was established in the mid-eighteenth century by Frederick the Great (1712–1786), when the king sought to further Berlin's independence from other

Figure 2.6 *Royal Berlin cup and saucer prototype*, ca. 1931, remade 1995. Marguerite Wildenhain (American, b. France, 1896–1985). Porcelain, 2 ¼ x 3 1/3 x 2 5/8 in. Collection of Forrest L. Merrill, Berkeley, CA. Photo by M. Lee Fatherree.

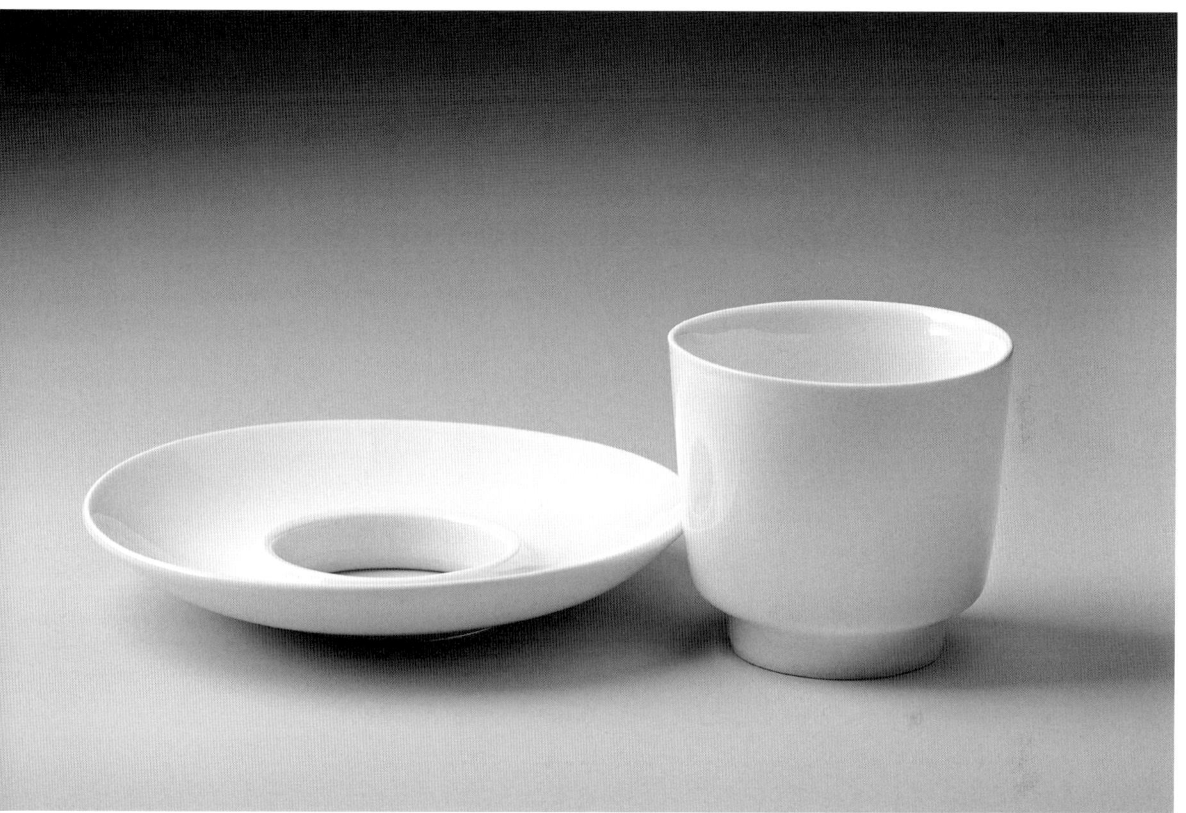

cities, both financially and aesthetically. The German historian Gerhard Masur writes:

> With porcelain becoming the vogue, Frederick [the Great] built his own factory, which took the royal septre as its trademark. The royal Berlin porcelain differed from Meissen or Sèvres; in keeping with the sober Prussian style, it specialized in white. To increase the sales of Royal Berlin porcelain, all Jewish brides were obliged to buy their dinnerware at the royal factory.[28]

Masur's specific point about Jewish brides is especially interesting in relation to Wildenhain's own personal history. Her promising career came to an abrupt halt when Adolf Hitler was elected chancellor of Germany in January of 1933. Historian Richard Bessel maps the brutality perpetrated against Jews during first six months of 1933, arguing that the attacks were secondary in terms of accruing or galvanizing political power and, instead, were sadistic in nature and "... appear[s] to have satisfied emotional urges, the desire to humiliate and harm people who were alleged to have enormous power and influence but in fact were largely defenceless in the face of violence."[29]

In March of 1933, the first concentration camp at Dachau was established to receive Nazi Party dissidents, communists, and socialists. In April, the Nazis passed the Professional Civil Service Restoration Act (*Berufsbeamtengesetz* laws), only the beginning of the restrictive prohibitions on Jews, which denied them civil service or government positions, jobs that included teachers, professors, or judges. Shortly after, Jews were also barred from practicing the middle-class professions of law, medicine, and accounting.

In response to the new state-enforced anti-Semitism, Wildenhain was asked to resign her position. She recounted the experience in her memoir, positioning herself as a martyred figure:

> Everywhere schools were being purged of so-called "non-Aryan" teachers, until one day the mayor of our city came to me, and with tears in his eyes, asked me if I would do the painful favor to resign. I was the only Jewish teacher at Burg Giebichenstein, and he hoped to be able to save the school if I would leave. I did not hesitate one minute, and left the next day after having said goodbye to my students and helpers. That afternoon, K., the handyman, brought me some fruit from his own garden and said "I shall never forget you," and burst into tears. Many have since said exactly the same words to me, but I shall never forget the first time I heard them![30]

It is likely that her tone of bravery was a way of masking the utter fear and despair she must have felt at being forced to leave; even as late as 1973, it was perhaps necessary to perpetuate the goodness she inspired, rather than recount her vulnerability as the lone Jew and only woman on the faculty, made powerless in the face of the Nazis' malicious social policies. Her work with Royal Berlin also ended tragically, but of this she writes unsentimentally, marking it a collective experience: "At that time none of us could know by mistake the Allies in World War II would hit the porcelain factory, instead of the railroad station near it, completely destroy it, and that would be how my experiment with industrial design would end."[31] That did not mark the end of her porcelain's production, but her association with it was terminated. As a Dutch manufacturer wrote of it in 1945, this work "... was so characteristic and representative that years later under (the) swastika flag, they still came out with it on [sic] the Utrecht trade fair, though her Jewish name was not mentioned of course."[32] To this day, she is still left off the KPM website as an artist who designed for the company, whereas

Marcks and her non-Jewish Bauhaus colleagues such as Theodor Bogler are included.[33]

As an act of loyalty and civil disobedience, Gerhard Marcks also resigned his post at Burg Giebichenstein. Her husband Frans, who was also on the faculty, retained his, presumably out of the couple's financial need. Marcks's bold gesture potentially earned his inclusion in the Nazi exhibition *Degenerate Art* (*Entartete Kunst*), held at the Munich Haus der Kunst in 1937.

Marcks had a longstanding national reputation as a sculptor and had served on the Front during World War I. In 1933, he fled Halle-Saale in favor of a low-profile country life, settling in the poor Northern city of Mecklenberg. Like other German artists of his generation, Marcks suffered immensely under Nazi rule, both personally and professionally. According to art historian Peter Geunther, Marcks's "fate under the Nazis is an example of the complexities and contradictions of a tyrannical art 'policy.'"[34] In 1937, the same year as *Degenerate Art*, an attempt was made by his peers to elect him to the prestigious Prussian Academy of Arts (Preussische Akademie der Künste). However, his financial prospects were diminished by the lack of opportunity to either teach or exhibit artwork.

His inclusion in *Degenerate Art* consisted of two sculptures, both figurative works of young men, one of the archangel Gabriel and the other titled *Standing Boy* (ca. 1924) (figure 2.7). The wall text below the latter read: "Especially singled out recently by the literati."[35] Marcks's work was further singled out for condemnation by Hitler himself during a tour of the exhibition.[36] Both sculptures had been raided from the Museum Folkswang in Essen.

The most immediate consequence was the cancellation of Marcks's impending appointment at the Dusseldorf Akademie, as well as a one-man exhibition that was being planned at Galerie Bucholz in Berlin. Professionally, he fared better than others, in that the Nazis never officially censured his artistic production. Personally, however, he sustained devastating losses. In 1943, his son died on the Russian front, and Allied bombs leveled his home and studio. After the war, he discovered that a stockpile of his bronzes, which he had buried for safekeeping, had already been purged, melted down to provide metal for munitions.[37]

Emigration and Craft

Marguerite fled Halle-Saale for the relative safety of Holland, where her youngest brother, Henri Friedlaender, a printer and typesetter, had already resettled. She spent several months in Holland scouting for a suitable location, while Frans continued to work at Burg Giebuchenstein. By strange coincidence, the exact day

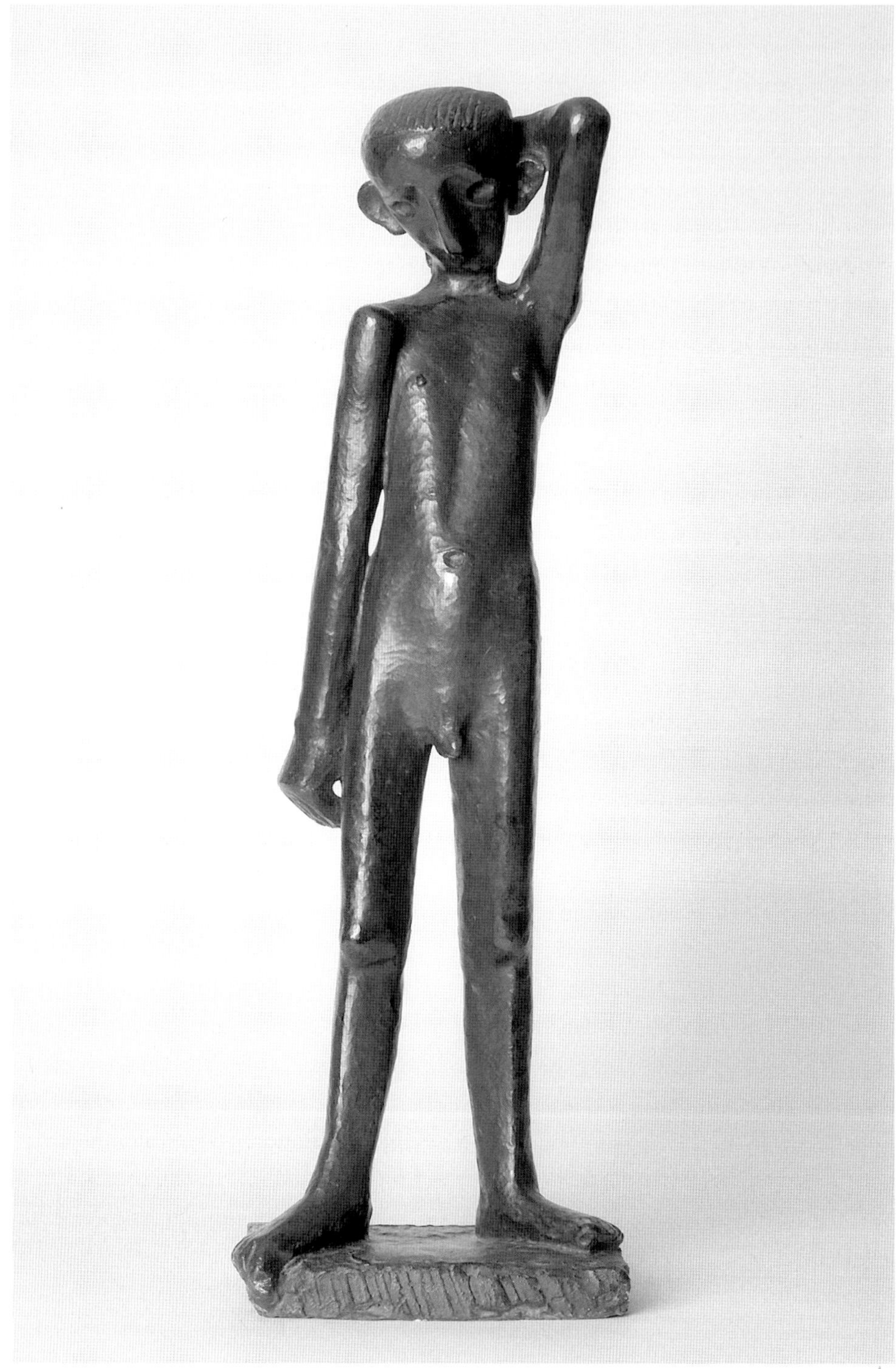

Figure 2.7 *Stehender Junge* [*Standing Boy*], 1924; bronze cast, 1948. Gerhard Marcks (German, 1889–1981). Courtesy of the Kulturhistorisches Museum Rostock, Rostock, Germany. Photo by Thomas Häntzschel (fotoagentur nordlicht). © 2015 Artists Rights Society (ARS), New York/VG Bild-Kunst, Bonn.

Marguerite sent for him via telegram, the remainder of the faculty was dismissed and the school closed.[38] The couple settled in the small town of Putten, Holland, sixty kilometers from Amsterdam. They rented a failed commercial pottery with a small adjoining home and renamed it "The Little Jug" (*Het Kruikje*). In a 1982 interview, Marguerite stoically recounted, ". . . I learned pretty early to have to do things on my own . . . . One doesn't find a place; one makes a place."[39]

Wildenhain was just one among thousands of Jews who flocked to the Netherlands. The most enduring example of this phenomenon is Anne Frank's family, whose patriarch, Otto Frank, moved his family from Frankfurt-au-Main to living quarters above a small pectin factory in Amsterdam. Like the Franks, the majority of Jews were of the merchant and educated classes. Most arrived in the Netherlands seeking economic opportunity, rather than political asylum, never believing that their lives would be at risk just a few short years later.

As a craftswoman, Marguerite Friedlaender Wildenhain stood outside this privileged population of émigrés: though multilingual (she had grown up speaking French, English, Hebrew, and some German), her Bauhaus training counted as a vocational, rather than elite, education. Historically, there were few Jewish craftsmen, as the medieval guild system had excluded Jews from the apprentice system, barring them from skilled labor. In the late eighteenth century, however, artisanry became central to the ideology of Jewish emancipation as a form of occupational restructuring, in which Jews became "useful" members of society in order to facilitate reciprocity with the German majority culture.[40] By the nineteenth century, however, swept up in the secular ideals of romanticism, German-Jewish intellectuals repudiated this strain of thought and instead emphasized social integration through assimilation.[41]

After the rash of anti-Semitic laws were passed in 1933, the urban-based German Jewish community responded to the crisis by initiating manual training programs with the intent to prepare Jews for emigration, to Palestine, South America, or the United States. A network of social agencies developed, including Jewish farm schools, focused on vocational objectives. This was compatible with the aims of the contemporaneous Zionist movement, which emphasized agriculture and handicrafts as a way of fostering self-sufficiency among the "pioneers" who would establish settlements in remote desert regions of Palestine. Originating in the nineteenth century, Zionism was a Jewish nationalist movement that sought the restoration of a Jewish state as a means of political and social emancipation, ultimately returning a diasporic people to their ancient homeland of Israel (then Palestine).[42] By the end of 1937, 535 people (eighty-one families) had been successfully resettled in Argentina.[43] Writing in 1939, Rudolf Stahl, a German Jew in charge of the training program in Frankfurt, recounted: "In the early excitement it was thought that every former lawyer could without much ado become a locksmith or cabinet maker and that every youngster leaving school could become a farmer or artisan."[44]

In 1924, the Ministry of Culture in Weimar concluded that out of a total population of 200 students at the Bauhaus, there were only seventeen Jews.[45] Far and away, the most well known was Anni (née Fleischmann) Albers (1899–1994), who, like Marguerite, always emphasized her secular origins. Among the faculty, the Hungarians Marcel Breuer and László Moholy-Nagy were also secular Jews. Such a small number attests to the stark class divisions held dear by the German Jewish middle class, among whom a rigid social hierarchy prevailed, emphasizing upward mobility. Indeed, both Wildenhain and Albers had wealthy merchant fathers peripherally involved in design, the former a silk merchant and the latter a furniture manufacturer. In her memoir, Wildenhain details her early life, replete with tutors and private schools. But she had willingly forsaken such advantages to study at the Bauhaus. Thus, the privations she had experienced at Dornburg were atypical for a young woman of her background. Likewise was her rejection of the traditional constraints of her gender. Rather than marrying well and raising children, Marguerite pursued a career and enjoyed her sexual freedom: she modeled nude on multiple occasions for her former teacher and colleague (and possibly lover) Gerhard Marcks. This was not done in shame or in secrecy. In this particular sketch (figure 2.8), she also modeled unclothed alongside her dear friend and, later, colleague Trude Guermonprez. Finally, in a reversal of sexual politics of the era, she eventually married her student, Frans.[46]

Marguerite's rather conspicuous sexuality underscores her secularism, since observant Judaism puts a high premium on female virginity and gender separation. Additionally, her given name was not just secular but intensely, heroically German: as Faust's beloved, Marguerite was Goethe's personification of romantic love. However, Marguerite grew up speaking Hebrew. This confirms the likelihood that her parents, Oscar and Rose Friedlaender, were Zionists, a then-minority movement in Germany.

In fact, it is difficult to imagine a family speaking Hebrew in the early decades of the twentieth century who were *not* Zionists—Hebrew as a living, spoken language was reintroduced by German Zionists only in the last two decades of the nineteenth century.[47] After the Holocaust, resettlement efforts on behalf of European Jews intensified, gaining international political momentum that culminated in the recognition of Israel's independent status as a nation in 1948. Among practicing Jews, this passage of return, known as making *aliyah*, is viewed as a moral imperative, one of the highest forms of ethical, and ultimately religious, expression. Marguerite's own brother, Henri, eventually made *aliyah*, leaving Holland to resettle permanently in Israel in 1951.[48]

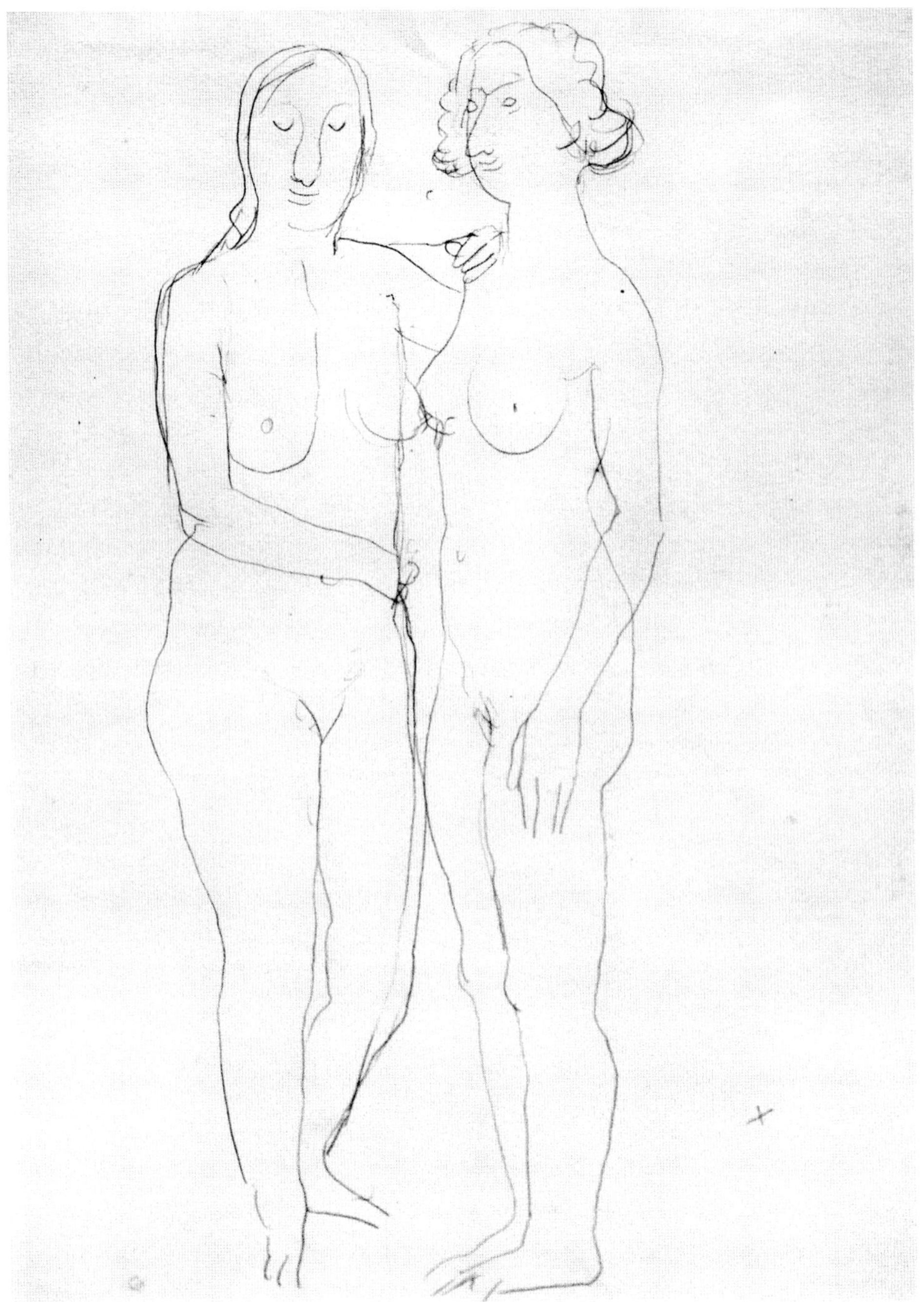

Figure 2.8 *Marguerite Wildenhain and Trude Jalowetz*, 1930. Gerhard Marcks (German, 1889–1981). Graphite on paper, 14.062 x 10.062 inches. Marguerite Wildenhain Collection, Luther College. Courtesy of the Luther College Fine Arts Collection, Decorah, Iowa. © 2015 Artists Rights Society (ARS), New York/VG Bild-Kunst, Bonn.

Zionism's tendency toward gender equality would surely have influenced Marguerite's predisposition to view herself as a leader or member of an elite group who must persevere in the face of adversity (Jews and craftspeople alike). This may account for her characteristic self-sufficiency and her quest to outperform men (such as her husband), casting herself in the role of the able-bodied provider and moreover, the consummate craftswoman.

So, with an unwavering commitment to her craft, Marguerite Wildenhain was undoubtedly more prepared than most Jews for emigration. Unafraid of hard physical work, Marguerite and Frans painstakingly rebuilt the Dutch pottery. One of the only surviving works of this era is a colander-like bowl, slatted white and cinnibar tinged, outlined in a delicate blue line, a nod to the famed blue-and-white delftware of the region (plate 4).

Consciously or not, the workroom they created is strikingly reminiscent of the pottery at Dornburg, with its solid plank workstations configured so the couple worked in a row, rather than side-by-side. Always matter-of-fact, she detailed their struggle:

> So Putten was a new start. Again it meant new materials, the new kilns to cope with, new glazes to work out, etc. At the Bauhaus we had fired with wood a salt glaze kiln; in Halle the two muffles and the porcelain kiln were fired with coal; and here we had to fire with peat . . . . Also now it became essential to find out how to allow the widest range of colors and textures inside a workshop that had to be self-supporting. It meant all this in a new country with a language we did not know and in a time of terrifying political unrest. But here, too, we made it.[49]

Only this next-to-last sentence betrays the reality of their lived experience, the fear and anxiety of uncertainty, and then just as quickly Marguerite returns to her façade of strength. During their seven years in Putten, she and Frans eventually learned Dutch. Frans tried, and failed, to obtain Dutch citizenship. The couple established a good reputation, sold work, and reconnected with old friends, namely Trude Jalowetz Guermonprez (1910–1976), a Jewish, German-born weaver of émigré parents who had been a student at Burg Giebichenstein and also a model (and lover) of Marcks. A generation younger than Marguerite, Trude was the daughter of Heinrich and Johanna Jalowetz, Black Mountain College's resident "Germans," the musicians who were most vocal in their disapproval of John Cage, and, in actuality, Austrian nationals.

Additionally, friends came through Holland: Marcks visited in 1938, but it was virtual strangers who were the catalysts in a series of life-changing events. In 1939, a wealthy bohemian couple from San Francisco by the name of Gordon and

Jane Herr arrived in search of interested artists who would form an art colony in Northern California. A sometime jeweler and architect, Gordon Herr was influenced by the seductive combination of the Arts and Crafts movement and D. H. Lawrence's writings, which championed passion and instinct. Fortunately for him, Jane Brandenstein Herr was an heiress from a prominent Jewish family in San Francisco, willing to support his vision. Coincidentally or not, Herr described his ideals to his wife in Zionist terms:

> We need a return to the land—in many ways. Just as the Jews have done so in Palestine—so we must do. They are forewarnings we must heed . . . When I look at a piece of old beaten copper, done in a half superstitious half religious ecstasy of creative impulse it strikes me with terrible force and says more of significant living truth than a thousand contemporary works. I think back to parts of the work I saw at the Wildenhains' at Putten. Some of their designs were in the direction of recreating basic truths—and these stirred up some response in me.[50]

Herr's plans were tenuous, a series of half-formed ideas, rather than a concrete strategy. Without actual land on which to build, or any sort of secure income, the venture seemed untenable. Then there was also the Wildenhains' professed skepticism about the United States, which Marguerite described:

> At that unquiet period, many Europeans were interested in leaving the old continent, just to escape the Nazis, but we, at that time, had never actually considered going to America. England, yes, but not the United States. It was too far! They [the Herrs] left, and we did not think of it much more.[51]

But, as the European situation worsened, they did think of it. So much so that in August of 1939, Marguerite shipped a trunk of ceramics to the Herrs and asked for help in obtaining teaching positions for herself and Frans in America. As a French citizen, Marguerite was able to secure a visa to emigrate. As a German citizen, Frans was rejected. Exceptions were made only for clergy and teachers at accredited colleges.

On March 3, 1940, Marguerite left alone on a Dutch ship that set sail for New York, the last, incidentally, that would leave Holland before the German invasion. This would mark the second time that Marguerite was forced to flee without her husband. Intent on finding Frans a teaching position, Marguerite wrote Walter Gropius within days of her arrival in New York. She asked for his help in establishing their artistic reputations.[52] Gropius was very generous in his reply:

Dear Mrs. Wildenhain,

> Naturally I remember you very well and I am glad that you were fortunate enough even without your husband, to come to this hospitable country. I have given it some thought how to help you. Your idea to gain some publicity through some displays is the right one. It just so happens that one of my former students, Mr. Eliot Noyes, Director of the Department for Industrial Design in the Museum of Modern Art has just landed this job. I recommend that you go and see him with the enclosed letter in hand, or better yet send him the letter ahead of your visit. If he would be able to schedule a display of your work in due time in the Museum of Modern Art, such a turn of events would be most favorable to you. I do recall some of your pieces that you had made for the Royal Porcelain Manufacture [KPM] and can only hope that you have been able to maintain that line of work. Such pieces would definitely find approval here if their display could be staged in New York. Please let me know, so that I could give you additional addresses of people who could have sincere interest. Here I am thinking of directors of industrial design schools. But please be prepared to face difficulties in finding a job for your husband as long as he is not here. We have found our experiences over here are not very positive in the last year since immigration has gone up.[53]

The Modern show never materialized, nor did a teaching position at Columbia University, for which Gropius recommended her. He also wrote favorably on her behalf to the Refugee Committee.[54] But there seemed to be few leads, even within the American Bauhaus network. The United States was at war, and, as Gropius points out, xenophobia was rampant. There was also the problem of legitimacy, since the Bauhaus had conferred a diploma only, its stature based upon a master-apprentice model. As Marguerite recalled:

> The time was wrong, the schools had all their teachers already, and we did not have the required "degrees." The Bauhaus had never given degrees. Gropius, I suppose, had thought, rightly, that the students would prove with their work and their ideas the value of Bauhaus training. Also, I believe that I really did not know how to present our case. In the States one seems always to have to "sell oneself," a thing that I have always detested and at which I was never adept.[55]

In the end, there was little Gropius could offer her, other than a bit of hospitality in her new country; she thanked him gratefully for the opportunity to lunch with him and his wife Isa at their home in Cambridge.[56]

On May 10, 1940, a mere two months after her arrival in the United States, Germany invaded Holland, cutting off news from her husband and her brother. She would not hear from Frans for months. From Henri, it would be years. Frans continued to live in Putten and make pottery until 1941, after which he moved to Amsterdam and taught briefly at the School for Applied Arts. He managed to keep a low profile until January of 1943, when he was conscripted into the Nazi Army. All of his postwar English language biographical materials gloss over this, politely referring to his time in the "German Army."[57]

This euphemism carries the historical burden of what can only be described as Frans Wildenhain's shame: an unwilling perpetrator whose own wife was a potential victim. For both parties, their situation must have been horrifying and bitterly ironic; Frans had suffered alongside Marguerite, enduring the financial and emotional strain of persecution and fear, leading to her eventual flight from Europe. Perhaps Frans had even felt abandoned and residually angry; though had Marguerite remained in Holland and been found, he would have been executed as a Jewish sympathizer. Indeed, the Wildenhains' bore witness to such an ordeal: Trude's husband, Paul Guermonprez, who had been a photography student at the Bauhaus, became an active member of the Dutch Resistance. He was arrested and executed not long after the German invasion.[58]

Like the Frank family, Henri Friedlander also went into hiding to escape the deportation and certain death that awaited Holland's Jews. For three years, he remained sealed in the attic of his Dutch fiancé's childhood home, never seeing daylight.[59] After the war, he and Maria Bruhn were married and emigrated to Israel. In 1997, two years after her death, Maria Bruhn-Friedlander was honored by Yad Vashem, the international body established in 1953 to commemorate, document, and preserve the legacies of the Holocaust. As a "righteous gentile," a non-Jew who risked her own life, Bruhn-Friedlander's name was added to the "Righteous Among Nations" Memorial, both a physical site in Jerusalem and a virtual Wall of Honor.[60]

No such honor was accorded to Frans. In contrast, his Nazi past continued to haunt the couple into the postwar period: the US State Department would not allow him to enter the country. However, it was recommended that he overcome his "inadmissibility," by

> establishing by positive evidence from reliable sources that he believes in democracy in the American sense, that he did not voluntarily serve against the Allies and that he was not voluntarily a member of the Nazi Party. Such evidence should be presented by the Applicant to the Consular Officer concerned.[61]

Still an émigré herself, without citizenship, Marguerite responded by gathering declarations on his behalf. In the form of a brief notarized statement, Henri came forward with a rather moving testament to his and Frans's intertwined fates:

> I herewith take pleasure in declaring that during the time that I as a Jew had to live hidden, F.H. Wildenhain, who at that time was on military service regularly provided me with foodpackets taken from German army stock. During this time I regularly held contact in writing with him and I am fully convinced of his anti-fascist feeling and his political security.[62]

Trude Guermonprez also responded, confirming Frans's secret, and illicit, help:

> Mr. F.R. Wildenhain was known to me as a [sic] Anti-Fascist already before the war. During the years of occupation, when I performed underground for my husband (Leadership Committee of Resistance) Wildenhain often assisted me and I considered him of complete political security.[63]

Publicly, though, Frans was neither exonerated of his Nazi ties nor celebrated for the personal risks he took for his Jewish friends and relations. Perhaps he did not crave recognition. But to have been forced to serve an enemy army was still a defeating experience. First in Holland, and then again in America, Marguerite had foraged ahead, a pioneer in a strange new land, seeking out opportunity and location and then resettling her younger husband, Frans, who once again had little say in determining his own fate. Due to the nature of their crisis, there was precious little room for macho posturing; instead, it was Marguerite who operated in a state of gender fluidity, assuming the typical responsibilities of male protector/provider *and* the decidedly female role of the domineering mother.

An extreme version of this dependency was transposed to America, where Frans arrived in New York on September 22, 1947, with nothing more than the simplest of goods: a mattress, a stool, a teapot, family photographs, and some personal effects.[64] Penniless and unable to speak English, he arrived wholly reliant upon Marguerite. Frans's redemption, then, came at the price of his sovereignty. Marguerite met him in New York, and the couple drove cross-country back to Pond Farm, the art colony that had been established in his absence. They had not seen each other in seven years.

Pond Farm's Early Years

Unable to find work or artistic success in New York, Marguerite had taken the Herrs up on their offer. She arrived in San Francisco just a few weeks after Holland's invasion, on May 27, 1940, and eventually found a teaching position at the California College of Arts and Crafts (CCAC) in Oakland. She intended the job for Frans, pressing the school repeatedly for his hiring papers, while offering her own courses in exchange for little money and a tiny studio. Despite her personal hardships, she continued to produce beautiful works during this time, including a delicate vessel made of thin-walled stoneware with vibrant white and blue abstract designs set against a dark brown glaze (figure 2.9). The Old World versus New World culture clash was, early on, apparent; Marguerite complained bitterly of the lack of rigor:

> What the students wanted, and what the College supplied, was not to actually learn to make pots decently, but just to dabble in clay long enough to bring "something" home and of course later get a degree so that they could teach . . . The obvious mediocrity of most of the results was not only the fault of the students, but also of an insane curriculum, totally unrelated to the requirements of a craft that made it impossible even for the gifted ones to do anything in any way related to art. I was deeply miserable, and worked myself to shreds trying to get the impossible from a group of students who never really wanted more than the least.[65]

The generation gap asserted itself in Marguerite's pedagogy. This puts her squarely on the same page as that other, more famous German, Josef Albers, who similarly pushed his American students to disconcerting levels of expectation, albeit to widespread acclaim. Albers is best known for his rigorous formulations of color and cognition, visualizing the physicality of color through his geometric compositions. He imparted this knowledge to his students first at Black Mountain College and, later, Yale University and finally through his writings, including his classic text on color theory, *Interaction of Color* (1963).[66] But, as Eva Díaz concedes, "he was not anti-institutional, though he lambasted the inattentive habits reproduced in institutions and in culture."[67]

Marguerite Wildenhain, however, was anti-institutional. Distinctly so, condemning the deep-seated American structure that encouraged young craftspeople to teach full-time and depend on institutional facilities for space and equipment, permanently weakening the social and entrepreneurial structures necessary to ensure a healthy network of independent craftspeople. After just two years at CCAC, Wildenhain quit and turned her attention to implementing a Bauhaus-style workshop in Sonoma County, California, where the Herrs had purchased 160 acres of farmed land, which they renamed Pond Farm.[68]

Known as the Pond Farm Workshops, the program was short lived, running from 1949 to 1952. Gordon Herr (architecture), Marguerite Wildenhain (pottery), and Frans Wildenhain (ceramic sculpture) were joined by two other German-Jewish refugees: Trude (Jalowetz) Guermonprez (weaving) and Victor Ries (metals). Additionally, the Greek collage artist Jean Varda (who had previously taught at Black Mountain) came from nearby Sausalito to teach once a week, as did the San Francisco-based sculptor Claire Falkenstein. Like Varda, Guermonprez had

Figure 2.9 *Cup*, 1940–1942. Marguerite Wildenhain (American, b. France, 1896–1985). Stoneware, 5 5/8 x 4 1/8 inches diameter. Collection of Forrest L. Merrill, Berkeley, CA. Photo by M. Lee Fatherree.

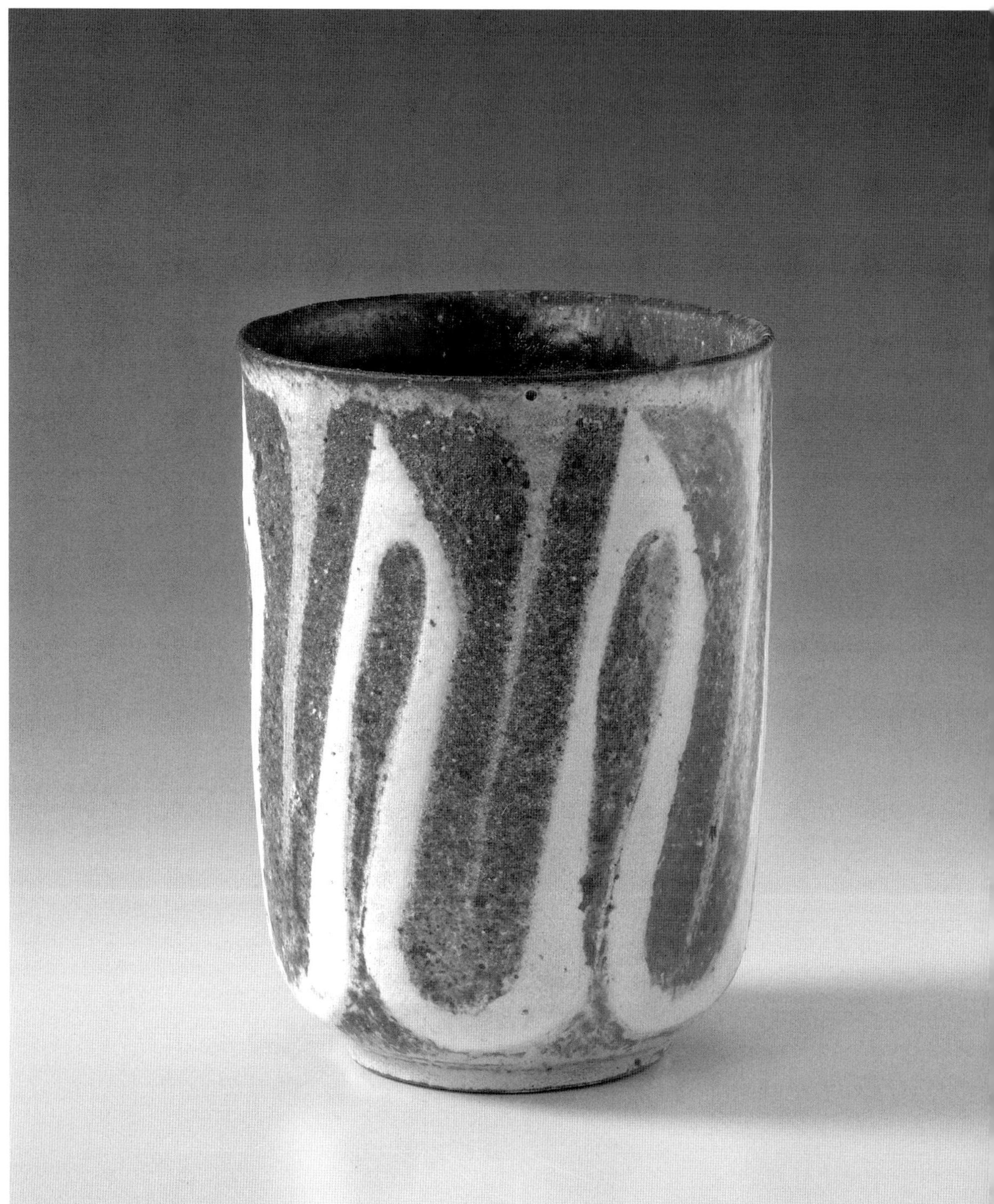

also already taught at Black Mountain between 1947 and 1948, first filling in for Anni Albers, who was on sabbatical until the winter of 1948, and then teaching alongside her. During this time, Trude and Marguerite visited each other in their respective locations. So Marguerite had experienced Black Mountain firsthand, in 1947, and Guermonprez must have seen something promising at Pond Farm, since she opted to leave Black Mountain for the fledgling artist's colony.

As Guermonprez recounted:

> It [Black Mountain] was such an intense place that it . . . wasn't good to stay there very long for faculty. My parents stayed there because I suppose it was a matter of need, and they were the older ones . . . My father [Heinrich Jalowetz] apparently from what I hear and read suffered greatly from the personal, you know, sort of currents and anti-currents that constantly are present in such a situation.[69]

Concurrent to Black Mountain College (which is discussed at length throughout the next two chapters), Pond Farm is best interpreted as a site somewhere in between: neither a college with a traditional humanities curriculum nor a purely communal utopian venture. There were odd coincidences: like the Dreiers at Black Mountain, the Herrs also lost a young child early on in their new community, their five-year-old son Jan, who died of accidental mushroom poisoning in 1944. Pond Farm was also routinely misconstrued as a haven for "communists, nudists, and anarchists," as the local mantra went, confirming that the xenophobia Marguerite had first encountered in New York was bicoastal and longstanding.[70] Finally, both Black Mountain and Pond Farm had a German widow lurking in the background as a kind and motherly Old World presence: at Black Mountain, this was Jalena Jalowetz, whose daughter, Trude, was at Pond Farm in the presence of the *other* German widow, Gesina Peterhans, widow of Bauhaüsler Walter Peterhans, master of the photography workshops at Dessau. Peterhans had initially come to study with Marguerite, but severe arthritis kept her from throwing pots.

As at Black Mountain, Pond Farm had struggled to build during the war years, using scrap and salvaged materials. Gordon raised chickens to avoid the draft, and the poultry buildings eventually were turned into artists' studios. Here, too, the planned vision exceeded the reality of the situation, where very little actually came to fruition. Black Mountain's odd corrugated metal structure, known as the Studies Building, became a central gathering point, with mixed usage over the years, accommodating both academic and living spaces. Similarly, Gordon Herr designed and built the workshop building as well as Hexagon House in 1949, an eclectic hexagon-shaped wooden structure with what was to be a gallery space, a three-story atrium, and a peaked roof held aloft with gigantic Douglas fir poles, and flanked by cabins that would house students. In actuality, these primitive structures, wooden cabins built on concrete slabs, without running water, housed Guermonprez, and Victor Ries and his wife, Esther.

The school failed for various reasons: first and foremost, Gordon Herr's poor leadership, which asserted itself as top-down dominance, rather than the cultivation of a collective vision. The artists' insistence upon communal ownership of the school, and with it some sort of long-term deed or lease, went unheeded. Additionally, Marguerite Wildenhain had financed the construction of her own house on the property but was not offered a mortgage in return. However, Victor Ries attributed many of Pond Farm's issues to Marguerite's gender "problem":

> He [Frans] was teaching painting, pottery, sculpture together with his wife [Marguerite]. But he couldn't make it with his wife. He was a real man, but she was a "man" too. And this didn't work. And she gave the whole tone on Pond Farm, since Gordon was not able to handle the situation. In meetings, which were mostly evenings, there was a terrible situation between Gordon Herr and Marguerite Wildenhain. They hated each other! They fought each other. He sometimes took his hunting knife, which he always had here, and threw it in front of her into the floor. And things were going on that way all the time. Unbelievable. Anyway, Marguerite actually dictated what has to be done and what has not to be done on Pond Farm, and he had no word whatsoever. But we had nothing to say, too. Marguerite was the one who dictated about everything.[71]

Unwilling to acquiesce to either her husband or Gordon Herr, Marguerite's "manhood" became a source of disgust to the men in the community, who equated leadership with masculinity. Frans was pitied, cast in the role of the cuckolded husband.

However, it is questionable if female authority would have ever been possible for this particular generation of men who could not envision true parity for women artists. Even her old teacher Gerhard Marcks had praised her for her masculinity, concluding, ". . . she really had the mental and physical strength of three men."[72] Several students, in their written testimonies gathered by Dean Schwarz, a former student himself, recounted that Marguerite, even into her sixties, would occasionally wrestle her students, pinning them before an excited group of their peers.[73] This thirst for total domination, intellectual *and* physical, makes her seem intent on enforcing the supremacy of her position. Thus, Marguerite's ambition and strength were considered male virtues, since it was inconceivable for a woman to embody such character traits. Similarly, assigning Marguerite a male persona made her easier to dislike, or discredit, as Ries's statements express, portraying her as a severe and unwelcome authoritarian.

But Ed Rossbach, the Bay Area fiber artist who visited Pond Farm as a student on a class trip during the late 1940s, described being differently affected by her pedagogy, awestruck by her absolute power, which he identified as distinctly feminine:

> [Her] approach, which is the antithesis of what I believe in now. The absolute antithesis. Very rigid and authoritarian. I hate authoritarian approaches to things. This woman knew, she knew. It's marvelous that people know anything. And she had these ideas that you threw, and I don't know how many years or months or whatever, you threw, and you threw, learning to throw these certain [forms] well, they're really traditional forms, although they've become identified with the Bauhaus as Bauhaus forms. She was there [at Pond Farm], a very vigorous sort of Earthmother type of woman. You probably think I'm pushing this too much, but I just felt like an innocent moving through this scene of sophistication and what? It's more than sophistication. It's different moral standards . . . . I was just agog, you know, at everything I was seeing.[74]

In the end, Pond Farm proved vulnerable to the same in-fighting that took its toll at Black Mountain. Yet Pond Farm's demise as a utopian community is markedly different, in that it offered a snapshot of the collective aftermath of the Holocaust. Marguerite Wildenhain, Trude Guermonprez, and Victor Ries were all Jewish refugees with disparate wartime experiences: Wildenhain had fled Holland in 1940 on the last boat; Guermonprez had spent the war years in Holland, hidden by righteous gentiles; Ries had left Germany in 1933 and spent twelve years abroad in Palestine before arriving in California. As exiles sequestered in near isolation, struggling with the burden of their own experiences, their situation can only be described as charged, with the potential for either collective estrangement or catharsis. Tensions were likely compounded by the presence of Frans, the gentle Nazi, and Jane Brandenstein Herr, the American Jew at some remove, likely conscious of her role in financing a kind of refuge for her European kin. Indeed, by 1950, Frans had moved apart from the others, to a remote cabin on the edge of the property. Yet the war was never discussed. As Ries affirms: "We did not talk about it [the war]. What for? We were not Jews at Pond Farm, we were artists."[75] His point is key: marked as Jews, and persecuted as Jews, in Europe, they had been one thing only. In being forced to be Jews, they had ceased to be artists. At stake was the ability to reclaim selfhood and individual identity. However, as German modernists, trained to simplify centuries of decorative excess, *any* narrative strain of artistic production *would have been* excess. To make work steeped in form, then, was to self-actualize. Perhaps this is why Pond Farm workshops failed to cohere, the fierce quest for individuality exceeding the collective enterprise.

Things did not go as planned: just three years after their reunion, Frans left Marguerite for a local woman and resettled permanently in upstate New York, teaching at the School for American Craftsmen, a division of the Rochester Institute of Technology, for the remainder of his career. Finally, Jane Herr's death from cancer in 1952, at age 40, was a tragic blow to the already-fragmented community. Both Guermonprez and Ries departed soon after, and Gordon Herr, grief stricken, quickly lost interest, preoccupied with the immediate needs of his two small children, who were being cared for by Gesina Peterhans. Only Marguerite remained. Released from the burdens of both male authority and collectivity, Pond Farm, instead, took a drastic turn, toward the camp.

Camp Life

The specter haunting Wildenhain's camp is, of course, the concentration camp. The personal truth of Wildenhain's life was the narrowness of her escape—twice, from the jaws of Hitler's Final Solution. Yet this survival was overdetermined, as she was not subject to the most distinctive and disturbing aspects of the Holocaust: camp life. She was not marched, she was not tattooed, she did not contract body lice or typhoid, she was not beaten or broken through grueling physical work, she did not starve, perishing slowly, in a death camp. Marguerite did not even suffer as her younger brother had, starving in hiding, waiting for Frans to deliver food surreptitiously. As a Jewish refugee whose Zionist background and hardscrabble endurance had empowered her to flee, Marguerite Wildenhain arrived in the United States relatively unscathed, one of the lucky ones. Like Hannah Arendt, who arrived a year later in 1941 and continued to promote and translate Martin Heidegger, her mentor and former lover, Marguerite, too, maintained deep ties to Germany.

Those who survived—and flourished—confronted a different sort of burden, often known as survivor's guilt. As Holocaust scholar Lawrence Langer has written, "the postmodern replacement of the death sentence that dominated Western thought from Freud to Camus is the life sentence."[76] Though she wrote three books in her lifetime, like many survivors of her generation, Wildenhain never wrote of the Holocaust. She never spoke of it. She never mentioned it by name. Nor did she abandon *her* German mentor: poor but proud, throughout the 1950s, she managed to scrape together enough money to send care packages of food and goods weekly to Gerhard Marcks and his family. Like so many others, Marguerite did not consider herself to be Jewish. After her divorce from Frans, she did not reinstate her Jewish maiden name; years later, when Marcks refers to her as a "Jewess," she retorts:

> Whatever you think of me as a "Jew" is completely wrong, I believe. Mother was 100% an Englishwoman, and Father had lived in Lyon since he was eighteen years

old. He was almost fifty when I was born there and was 100% French then. We never celebrated any "Jewish" festivals or anything else. Father wanted up to belong to the "whole world." He was well traveled, so were we children for that time. I have been in a Catholic church much more often than in a synagogue (in my whole life hardly six or eight times, if even that often).[77]

In his book *Homo Sacer: Sovereign Power and Bare Life* (1999), Giorgio Agamben theorizes the concentration camp as the paradigm of modern nonjuridical, or sovereign, power in the twentieth century. In its convergence of the body and the law, it is a site of exceptionalism and crisis, imposing the threshold of existence, or what he terms "bare life": the abandoned body stripped of all its identities, without recourse to life, death, or autonomy. Pond Farm, then, reinstated the logic of the camp, where the intensity of "bare life" was experienced symbolically, where personal freedom was suspended willingly, in service to a larger moral purpose, that is, collective skill building. "Bare living" was a term Wildenhain herself employed, seen in chapter 1, to refer to a strained or impoverished existence. Thus, as a nonschool and a noninstitution, Pond Farm reflected the camp paradigm of a structured existence through the adversity of exceptionalism and crisis.

Its site was simultaneously a *campus*, a topography of learning, the grounds that comprised the buildings of an educational facility, and an *encampment*, a series of provisional, makeshift structures, utilized seasonally, for a temporary amount of time, where most of life is lived outdoors. It also approximated the traditional notion of summer camp, which would constitute the *enchantment* of encampment, where leisure is achieved through fellowship and collective skill building.

Camp life is intimate and always public; at Pond Farm, Wildenhain's allegiance to Dornburg's workshop-based training established a situation of dominance predicated on "discipline, restraint, orderliness, and control."[78] These values were reinforced publicly as well: each day in the studio, Marguerite would begin by "tearing apart" the work of the previous day. For each student, labor was a crucial act of self-formation, performed and reperformed as an intricate exercise of power and compliance. Foucault wrote that ". . . discipline produces subjected and practised bodies, 'docile bodies.'"[79] Rather than seeking freedom from Wildenhain's self-styled cultural imperialism, American potters were eager to partake in the perceived supremacy of the Bauhaus's workshop tradition, becoming docile subjects obedient to Wildenhain's sovereign power, molded, in effect, by her forming.

This self-formation is consistent with the "clay body," a ceramics term which refers to the material composition of the vessel, the mixture of clays combined for specific properties, such as translucence or durability. If the "clay body" is a body both predicated on, and commensurate with, form, then the craftsman's form-of-life is an indivisibility between the format of production and the production of form. That is, the craftsman is an intensely productive subject of his or her own labor, reinscribing the body within the replication of hundreds and hundreds, and over a lifetime, thousands and thousands, of clay bodies. An army of clay bodies. This energy of near-constant production manifests what Deleuze and Guattari call "the process of self-cure that brings. . . . the production of a new humanity."[80] In teaching craft as an ethical imperative, Wildenhain endowed her students with a new humanity, pressing them to accept the "different moral standards" that Rossbach had glimpsed during his brief visit.

The experience of the camp, in any of its forms, from Auschwitz to government internment, to basic military training, to eight weeks of overnight camp at the age of twelve, is marked by two conditions: transformation and trauma. Despite time and distance, the residue of the camp persists as a powerful mark of lived experience, and as a divider of the self into the person before and the person after. Such splitting is what constitutes the traumatic experience, the unseen evidence of exposure and endurance. In the following analysis, I've embedded, in italics, the testimonies and memories of Wildenhain's students, as collected by Billie Sessions.[81] *Pond Farm is so much a part of me that it is hard to see where I stop and it begins.*[82] Deeply altered, the self is a dislocated entity, set apart from others, but intimately joined to those who lived through the same experience. *For me, Marguerite and Pond Farm were everything. I learned my craft there, proposed to my wife there, and was married there. I met most of my dearest friends there and learned the meaning of life there. In my heart I will always be there.*[83] "There" is the site of Pond Farm itself, enveloped and embodied by the student, the wound seared upon the heart, a happy wound, but a wound nonetheless, pleasurably burdened by its difference. This wounding represents the process of mental and emotional scarification, the violence inflicted upon the body, and the permanent reformulation of the soul. *I feel very strongly about the absence of Marguerite from my art life. She left a void in all our lives that is next to impossible to fill and that is not readily understandable by those who were not there.*[84] The suppression of individual identity in favor of a collective identity ("those who were there") is to pay tribute to insularity, reenacting the Pond Farm community as a site of exception, so as to achieve an afterlife that functions as an afterglow, its purpose and its achievement conferred upon a special group of people. This is similar to many survivor and veteran communities, who gather to commemorate and honor their wartime experiences.

In her work regarding the teaching of Holocaust survivor narratives, literary theorist Shoshana Felman asks the question, "Is there a relation between trauma and pedagogy?"[85] This takes us full circle, where modern American craft begins as a social service, a therapeutic endeavor for returning veterans, and is transformed into a boot camp, breaking the student-potter and remaking him or her through intense training and discipline. What Wildenhain offered her students was a format for living, or what Agamben calls a "form-of-life," writing, "If we give the name 'form-of-life' to this being that is only its own bare existence and to this life that, being its own form, remains inseparable from it."[86]

If we take Wildenhain at her word, where form is taught as a life-skill rather than as a lesson, as a survival mechanism rather than a novel experience—then Pond Farm, as a site of unproductive labor, was indeed a site of exceptionalism, making and remaking exile as a form-of-life. Felman notes:

> . . . if teaching does not hit upon some sort of crisis, if it does not encounter either the vulnerability or the explosiveness of an (explicit or implicit) critical and unpredictable dimension, it has perhaps not truly taught.[87]

For Marguerite Wildenhain, this crisis was the crisis of freedom. To be without discipline was to be without a form-of-life, and therefore without a framework on which to structure artistic practice. But her militant and transformative pedagogy was more than just a cathartic response to the trauma of exile or a solution to the inadequacies of the American situation. It was also a reenactment of Dornburg, the Bauhaus's original site of exception and crisis. In the relationship between exceptionalism and exclusion, Pond Farm invariably became Black Mountain's camp: forgotten, overlooked, and even occasionally mocked.

Marguerite Wildenhain and Voulkos

Josef Albers has been noted for the productive antieffect he had on a generation of now-canonical 1960s artists, most notably Robert Rauschenberg (who proclaimed himself Albers's "poorest" student) and Eva Hesse (who "did not like" Albers).[88] In comparison, Wildenhain's students remain relatively unknown, mainly teachers like Frances Senska, the Montana potter who unwittingly trained Marguerite's most important tormentor, Peter Voulkos.

Rose Slivka praised Voulkos for his embodiment of ". . . the action of men—ruggedly individual and vernacular men (the pioneer, the cowboy) with a genius for improvisation."[89] Voulkos's hypermasculine persona was exactly what Wildenhain railed against when she wrote that "with an art education that is neither sufficiently basic or competent in its techniques, we have fostered a generation of 'Rock 'n Roll' craftsmen who float in a sea of violent and misunderstood 'self-expressionism,' disregarding all essential laws of human and artistic integrity."[90] Yet Voulkos was the inheritor of Marguerite's rigor. Andrew Perchuk has argued that despite the critical support for Voulkos as a second-wave abstract expressionist-style sculptor, his work retained the specificity of ceramics in its "unspontaneous rigor" and "its insistence on process."[91] This makes Voulkos into more of a disciple than he would have cared to admit. Unbeknownst to his students, the two had even worked closely together at Archie Bray in the fall of 1953, when Voulkos hosted Wildenhain for a five-week workshop. While no documentation exists from their time together, it is reasonable to assume that a strong gender dynamic would have been in play, reminiscent, even, of Marguerite's relations with Frans. While she never remarried, Wildenhain clearly enjoyed the company of younger men, always selecting male studio assistants, one of whom worked alongside her for sixteen consecutive summers.[92] By turn, Voulkos was a skillful lothario. In playing the role of the eager disciple, his platonic flattery may have fully charmed Wildenhain, converting aesthetic tension into sexual tension, a situation that is only possible when there is some kind of tacit equality at stake. Such an art becomes thrilling, a game of reversals. Wildenhain's final studio assistant, Dean Schwarz, recalled a Voulkos demonstration in which he poked fun at Marguerite, commenting on her "big juicy rims," ostensibly about her pots, which no doubt doubled as a sexual reference.[93]

Indeed, Senska attributed her own throwing technique to Wildenhain's discipline and structure.[94] But, as she described it, such skills were always passed from teacher to student:

> SENSKA: Another guy came up to me and said, "You know, you hold your hands just the way I hold my hands, and I learned from Peter Voulkos." And I said, "Yeah, and Peter Voulkos learned from me." So this is what education in the ceramic arts is all about. You learn from each other.
>
> FORBES: Great lineage, yes. That's right.[95]

Voulkos's lineage, then, was *women* potters. For a clay artist working on the precipice of sculpture, in a craft medium with relentlessly feminine associations, such a legacy was entirely emasculating. Perhaps it was even humiliating. Sometimes referred to as the "mother of Montana ceramics," Senska, along with her "friend" Jesse Wilber, a printmaker with whom she lived and taught alongside at Montana State College, were beloved figureheads in Montana, childless (and lesbian) caretakers, known for their enormous generosity toward students, always opening their home or dinner table to those in need.[96]

It would have been vicious, and unwarranted, to attack Senska. More than likely, Voulkos felt entirely protective toward a woman whose homosexuality made her an easy target and whose own parents, Greek émigrés in Montana, might have been subjected to a similar kind of "outsider" bias. Such deep-rooted loyalties do not sufficiently explain why Voulkos and his students were so cruel when it came to Marguerite and her methodology.

But proximity does. In 1959, Voulkos was fired from Otis and began a teaching appointment at the University of California, Berkeley.[97] Marguerite was already an established Bay Area potter. Among California artists, with fierce allegiances to the strong cultural differences between north and south, such a move was unprecedented and perhaps even initially unwelcome. Such close proximity between the two potters—the old modernist and the young upstart—amounted to an intergenerational "turf war," predicated on the already-contentious northern versus southern regionalism that permeated California's postwar period.[98] Los Angeles art was overwhelmingly "Pop"—bright, garish, facile, with a Hollywood-induced veneer (David Hockney, Ed Ruscha, the "Finish Fetish" of Billy Al Bengston and Craig Kauffmann); while San Francisco was "Beat"—dark, earthy, hermetic, whose avant-garde was more literary (Lawrence Ferlinghetti, Diane di Prima) than visual, but for a smattering of assemblagé artists (Wallace Berman, Bruce Conner, Jay deFeo) whose work was psychologically complex.

Such regional and stylistic antagonisms could have only intensified when Voulkos arrived, though, as part of the craft world, both Voulkos and Wildenhain circulated outside of these dominant strains of production. However, Voulkos's severing of both traditional form and conventional ceramics instruction can be seen as an undoing of female pedagogy, and Wildenhain's specifically, as her authority loomed large in American ceramics. As one critic wrote in the 1990s, "At one time a summer course with Marguerite Wildenhain was an essential rite of passage for studio potters."[99] So was visiting Pond Farm. As Voulkos himself recounted, in a noted British documentary:

> I think when I was going to graduate school down here, I went up to see her one time and she read her little treatise, and little paper, that sounded all well and good, real Bauhaus, but when you have a heavy German accent, anything sounds great and if you read it on a piece of paper . . . it was just ZZZZ, ZZZZ, zeit zeit [mimicking her accent].[100]

Voulkos's public disavowal of Wildenhain's pedagogy continued on throughout the 1960s. But acknowledged or not, her influence on him was apparent; as Senska herself once noted, "Sometimes, when you watch Pete throw, you can see Marguerite Wildenhain."[101] This lineage continued: the famed Funk ceramist Robert Arneson, trained by Voulkos, arguably remade Wildenhain's portrait at Pond Farm as an entire sculptural parody. Titled *Smorgi-Bob, The Cook* (1977) (figure 2.10), Arneson portrays himself as a semiserious chef, shirtless but for a tall chef's hat. His own likeness defiantly meets the viewer gaze, trapped

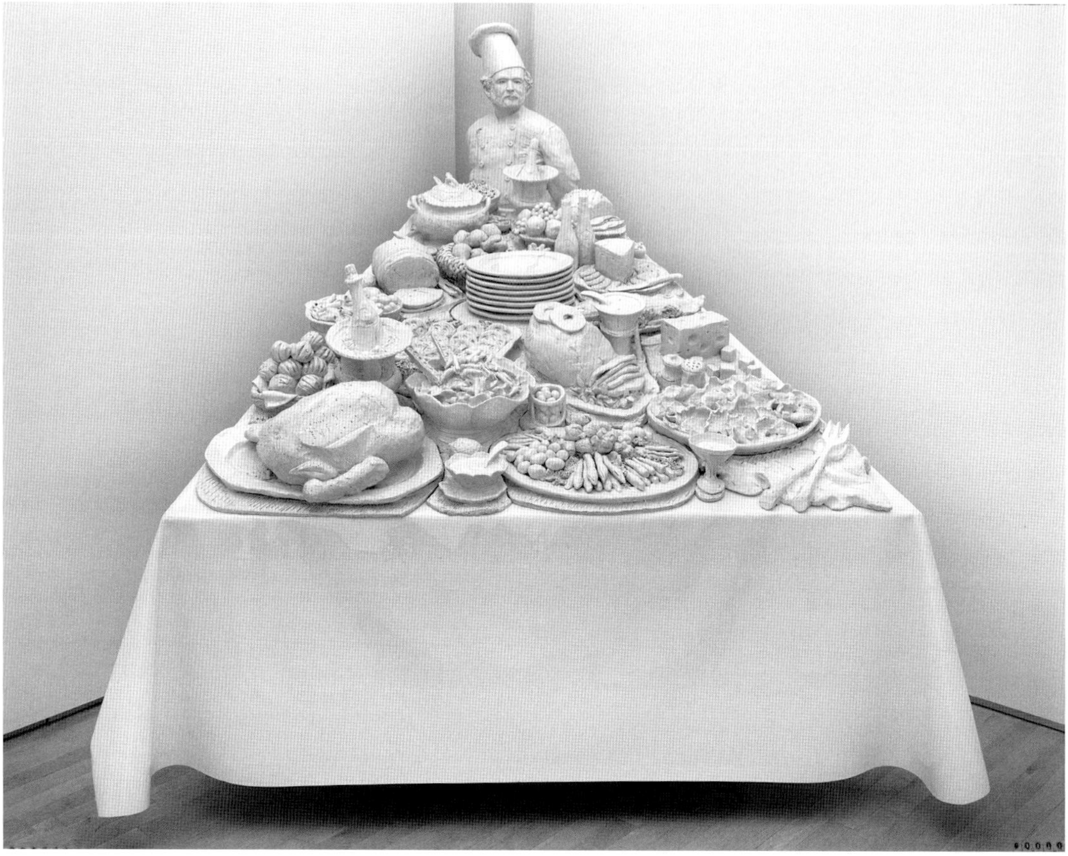

Figure 2.10 *Smorgi-Bob, The Cook,* 1971. Robert Arneson (American, 1930–1992). White earthenware with glaze, vinyl tablecloth, and wood table, 73 x 66 x 53 inches. Collection of the San Francisco Museum of Modern Art. Art © Estate of Robert Arneson/Licensed by VAGA, New York, NY.

behind a long, narrow table reminiscent of Wildenhain's. This one, too, is laden with ceramic wares, but Arneson's are all white porcelain and filled with an embarrassment of foodstuffs, as if lifted from a Dutch *vanitas* painting: roasted turkey, a ham with pineapple garnish, plates of vegetables and potatoes, and a saucer heaped with nuts. The gleaming, all-white installation, complete with the formality of a tablecloth, is a visual pun on the formalism and monochromatic artworks found in minimal sculpture. Arneson literally feasts on the gesture of parody on many layers: poking fun at not only Wildenhain but also the hyper-serious minimal artwork that the sculptors of his own generation were making. Arneson maximizes his visual vocabulary, including an obscene amount of functional ceramic wares and food sculptures, embedding his own artistic persona into the smorgasbord through the title.

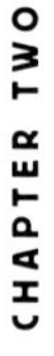

The German Problem

Even among her most devoted students, Wildenhain had a "German problem." That is, her Bauhaus training and her uncompromising fierceness became one singular heritage. But here lay the contradiction of Marguerite: marked by Dornburg, faithful to her master, post-Holocaust, it seems she divested herself of one of the most fundamental aspects of her background, her Germanness. A former student recalled her sharp reaction to his first, rather stereotypical, impressions:

> I had rather expected to see someone much larger and fiercer, no less a Valkyrie than Brunhilde . . . . Her accent was decidedly German and with the name Wildenhain I naturally assumed that she was German. I suppose the gods were with me that day when I suggested as much. That she hadn't cast her Valkyrian spear must have been an auspicious omen with which to begin the summer, for had she not like me a little, she most certainly would have proclaimed, fiercely, on the spot, that her roots were French. Her mother was British, and her father was German, but she grew up in Lyon, France. She was, therefore, indisputably French! The first protocol of Marguerite.[102]

Unquestionably, to be German during the early postwar period was to risk being ridiculed or, worse, treated as an enemy within the United States at large. Moreover, Wildenhain's emphasis on her French origins accounts for a sublimated recognition of her Jewishness. But as Ron Nagle, postmodern potter and former Voulkos student, noted:

> She was the enemy. Absolutely. She was the carrier of the torch for the community of clay. But there was a very specific kind of look and form and so forth that she thought you had to make to be legit as a potter, you know? It was just dogmatic! And the real cornball romance with her and the toe, kicking that wheel, and I just thought, you know man, that says it all, the artisan and her wonderful *tool* ready to kick that piece of stoneware into immortality.[103]

Nagle's mockery is nearly a throwback to Victor Ries's assertion that Wildenhain was a "man." Wildenhain's "wonderful tool"—her wheel—becomes a phallus, despite its unquestionable roundness. Based on her uncompromising authority, Nagle could only conceive of her in masculine terms, a grudging affirmation of her importance. His implicit critique, though, also reflects the intergenerational clash: a bunch of Otis hipsters who used clay as a vehicle of experimentation and rebellion versus Wildenhain's belief in an art form that was imperious and eternal. In hindsight, some of her own students believe she had misunderstood their intent. As Hunt Prothro, a former student, charged, "She couldn't see the work of, say, Peter Voulkos or Ken Price, as a corroboration of her own impulse toward a renewal of form and structure. She couldn't see the humor or irony in their work either."[104]

A key component of the Otis group's legacy is the mythology surrounding their outrageous behavior toward Wildenhain. Curator Mary Davis McNaughton recounted the cruel parody that the Otis group performed before a meeting of American Ceramics Society; a well-circulated Wildenhain adage was her analogy about basic skills, comparing potters to doctors who were not below learning how to bandage. John Mason threw a pot and asked for assistance in decorating it, Ken Price volunteered, and the two of them shredded the pot to pieces while another assistant, dressed in a nurse's outfit, stood by handing them implements as they called out for "sponge," "needle," and "scalpel."[105]

Everything about the spoof might have provoked and embarrassed Wildenhain: the male student in nurse's drag, the shredded pot, which mimicked Wildenhain's own technique of cutting through student pots, and finally, the satire itself, the use of a surgical theater to literally dissect her core beliefs, performed by students who had never experienced Pond Farm firsthand. Compare this to the nearly slavish devotion expressed by Wildenhain's own students, who sang the song, excerpted below, at the 1971 summer session final party:

> But we'll keep throwing till Judgment Day
> Let no man accuse us of playing in the clay
> 'Tis sad but true that time will tell
> Just which of the potters will go to hell
> There is a place down there reserved for those
> Who took their talents lightly and never arose
> Above the level of the dillitant [sic]
> They'll pay for ever in the devil's haunt
> For those who never rose above the dillatant [sic].]
> All kidding aside we do agree
> You've shown us how a life of dignity
> Of belief in work and the love to truth
> Can make a sort of heaven right here on earth
> And on departing we may raise a fuss
> But Marguerite we'll take a part of you with us
> And on our departing we may cry and cuss
> But we'll always have the spirit of Pond Farm with us
> But we'll always have the spirit of Pond Farm with us.[106]

Like a battle hymn, the song expresses the fidelity of these particular students to the bitter end, all the way to Judgment Day, with Marguerite as the oblique central figure, or judge, who had, like a religious leader, shown them the true way. But such lyrics also pronounce judgment on that *other* group of potters, the Otis group, those who play in the clay and "wasted" their talents on such play, permanently condemned to hell as amateurs, or "dilettantes." In this way, hell becomes a religious allegory for history, where only time will tell the "truth."

But the truth did not bear out quite as Wildenhain would have liked. In time, Voulkos's reputation as a teacher surpassed her own, crystallizing into a kind of permanent and legendary origin myth, in which Voulkos and his Otis students invented non-functional ceramics, forcing both the art and craft worlds to acknowledge their hybrid practice. Marked by its substantial size and formless expressiveness, mixed media ceramic sculpture, as it came to be called, superseded the rigid functional modernism promoted by Wildenhain. It also didn't hurt that a number of Voulkos's students, including John Mason and Jerry Rothman, became nationally known, while Ken Price and Paul Soldner achieved international reputations. During the era in which it took shape, post-abstract expressionism, Otis-era ceramics was untouchable in terms of its glamour.

So Wildenhain's students labored, as in an image of a student making a multitude of lids (figure 2.11), while Voulkos's students played. But it was precisely this kind of controlled play that was the ethos of Black Mountain College and that Josef Albers, in particular, had encouraged. Using cut and colored paper and objects like found leaves, Albers had advocated a relentless experimentation with materials that resulted in collage studies, rather than finished work. Wildenhain also reinforced process, albeit in a much more rigid way. As Bauhäuslers, Albers and Wildenhain modeled their respective pedagogies on two entirely distinct eras of the school's history. Albers's, of course, won out: his Bauhaus legacy was transposed to the mythos surrounding Black Mountain College, which only continued to grow after the college closed, echoing the Bauhaus's own fleeting, but influential, existence. Albers's fame, even posthumously, must have been as bitter for Wildenhain as it was her mentor, Marcks. Three years after Albers's death, and the same year as his own, he wrote her from his hospital bed in Cologne, in 1981:

> And you will laugh to learn that the damned Bauhaus has even followed me here. The main room is encircled with nothing by colorful squares by Josef Albers! True Germans, the nice nurses don't even look in there. They are interested only in music. They are not at all at home with the fine arts, so no lectures will help here either; e.g., that these colorful pictures are art theory and application, but not art.[107]

Figure 2.11 *A student tries to make fitting lids, ca. 1956*. Photo by Otto Hagel. Collection Center for Creative Photography. ©1998 Center for Creative Photography, the University of Arizona Foundation.

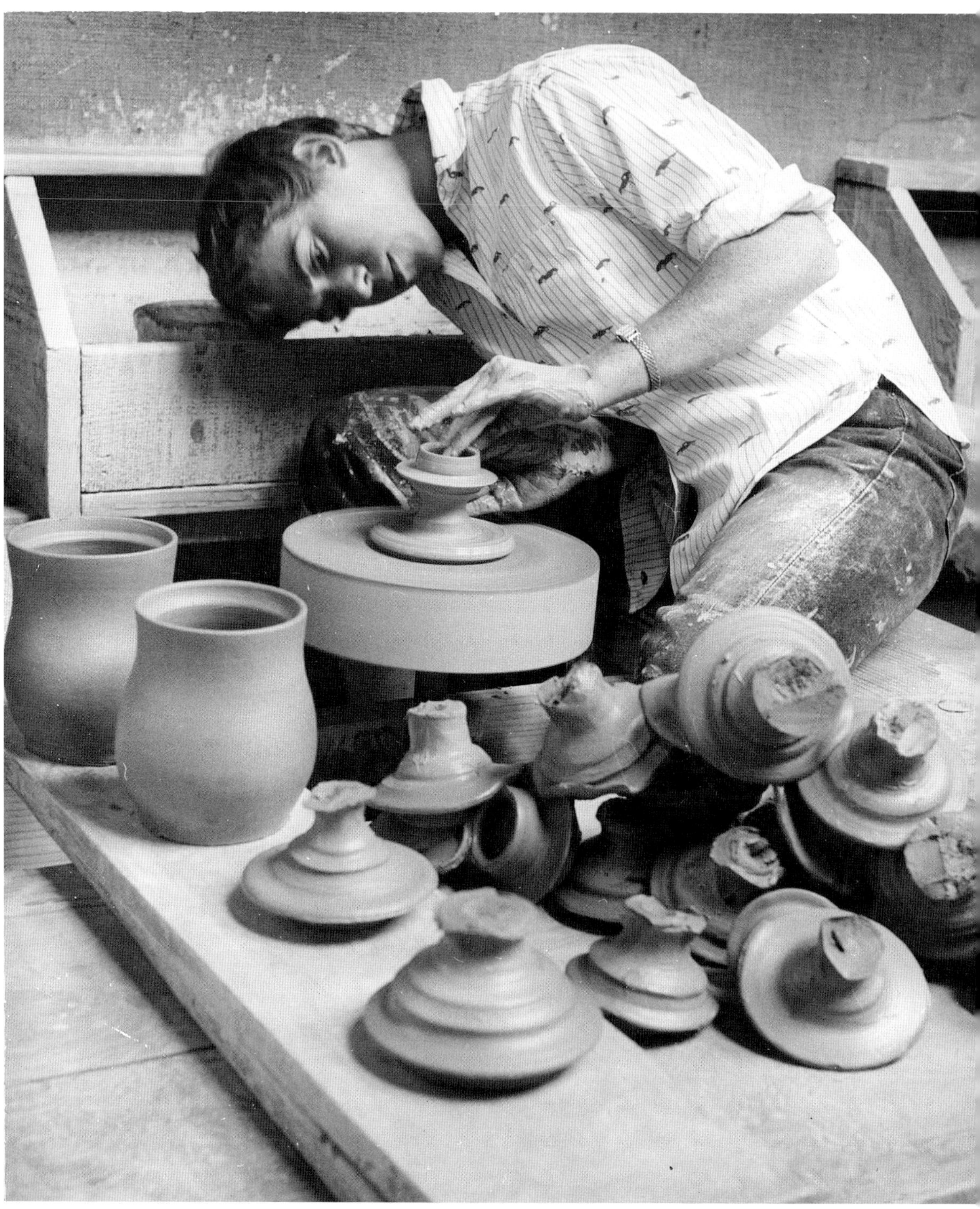

Marcks's critique of Albers, that his artwork was merely theory, and not art, is supported by a 1950 essay published in the *College Art Journal.* Titled "Kindergarten and the Bauhaus" (1950), its author, Frederick M. Logan, a professor of art education at the University of Wisconsin, Madison, traces the creative expression encouraged at the Bauhaus, as well as its media-based workshops, to specific lessons drawn out from the little-known history of kindergarten, established in mid-nineteenth-century Germany. The kindergarten lessons he points to are specific activities in sewing, drawing, perforating, intertwining, folding, and cutting.[108] Such creative activities involved working with specific, basic geometries and spatial perception and mirror many of the lessons taught within the Bauhaus's foundations course, taught by Johannes Itten.

While Logan's theory was more observational than research driven, his larger point is pertinent: the same cognitive play that stimulated young children could be implemented to even greater effect on young adults, sparking freshness and clarity just by encouraging structured playtime with color and shape. It was exactly such unrestricted freedom to play *with* and among forms that Wildenhain disallowed at Pond Farm. Such resistance to play was, inherently, a refutation of collectivity. Surely, Wildenhain would have agreed with Hannah Arendt when she wrote, "There can hardly be anything more alien or even more destructive to workmanship than teamwork."[109]

Instead, Wildenhain dictated, and her dictation made a powerful, lasting impression. Even Frans voiced his disapproval, writing, "I don't tell my students—I listen to them. Perhaps there are no answers only questions but I know that's not your cup of tea. Discussion closed."[110] But as a woman, Marguerite Wildenhain did not have the luxury of fostering play in the manner of either Albers or Voulkos. In postwar America, for a woman to achieve Wildenhain's level of authority was perhaps permissible *only* in the role of a teacher, and it was only through teaching that Wildenhain was permitted the role of the master artist.

Training her students in her own image, Wildenhain herself represented a maverick persona: financially independent, emotionally self-sufficient, unconstrained by either traditional domesticity or institutional responsibilities. Indeed, she had a special aversion for women students who had chosen marriage. As Kate Huffman wrote:

> My fondest memory of Marguerite is that it took me seven summers of begging her to let me come because I was a wife. She had already had my husband Jerry for four summers and she continually wrote me these lovely letters saying "No, no, no. Wife doesn't compete. You just support your husband." And I finally got that letter that said okay, come. And unfortunately that was pretty close to her last year so I was just known as "that girl," she couldn't quite put a name to me.[111]

It was only through her severity and perseverance—and self-internalized sexism, as witnessed above—that she was able to accrue the cult of personality that Voulkos had by his bad-boy image alone. And unlike Albers, lauded for his seriousness of purpose, as a woman, Wildenhain was without access to the same system of rewards: she was neither a master at the Bauhaus nor the rector of an American college. Rather, she was an independent potter whose successes were modest, whose triumphs were incremental, and whose freedoms were hard won.

Exile and World Building

In 1963, an unexpected crisis arose for Wildenhain: the California Park system intended to accession 4,000 acres of virgin redwood forest and coastal lands to create a protected wildlife and nature refuge that came to be known as the Armstrong Redwoods State Preserve. As the California Property Acquisition wrote, "Included in the Wilderness Park project are certain lands owned by you."[112] While Wildenhain would be compensated for her 7.83 acres, she would have to leave behind the life and livelihood she had worked so hard to build: her home, with her beloved gardens and grape arbor, her studio, her kilns, and her showroom. Though Wildenhain had left all these things before, many times, what she had not yet left behind was a school of her own making. And it was this that she could not bear. Pond Farm was a place unto itself; what Wildenhain had planted would not grow back elsewhere. Her response is revealing in its rare acknowledgement of vulnerability, tinged with defeat:

> . . . As you may possibly know (and will hear from Mr. Kishbaugh) my personal situation is very different from all the other people involved in this transaction. Not only have I lived here for 23 years, but I have founded and developed at Pond Farm a workshop and a school of pottery that have attracted students and visitors not only from all over our nation, but as far away as India, China, Lebanon, and Nigeria. My own work and my school have, without being conceited, a world reputation. Anybody in the field of the arts and crafts will acknowledge that.
>
> Excuse me for mentioning all this, but the situation that is facing me at my age (nearly 67) is definitely tragic and possibly outright destructive for the whole rest of my life. To leave what I have built up painfully during 23 years nearly single-handed, is to condemn me to end my life at a time where I could still have about 10 years of richly creative work, both teaching and producing. One thing I do not feel I have any more is the physical strength, nor the mental stamina to start all over again somewhere else. I have had to move 6 times in my life, twice due to Nazi persecutions, and I just cannot do it another time any more.[113]

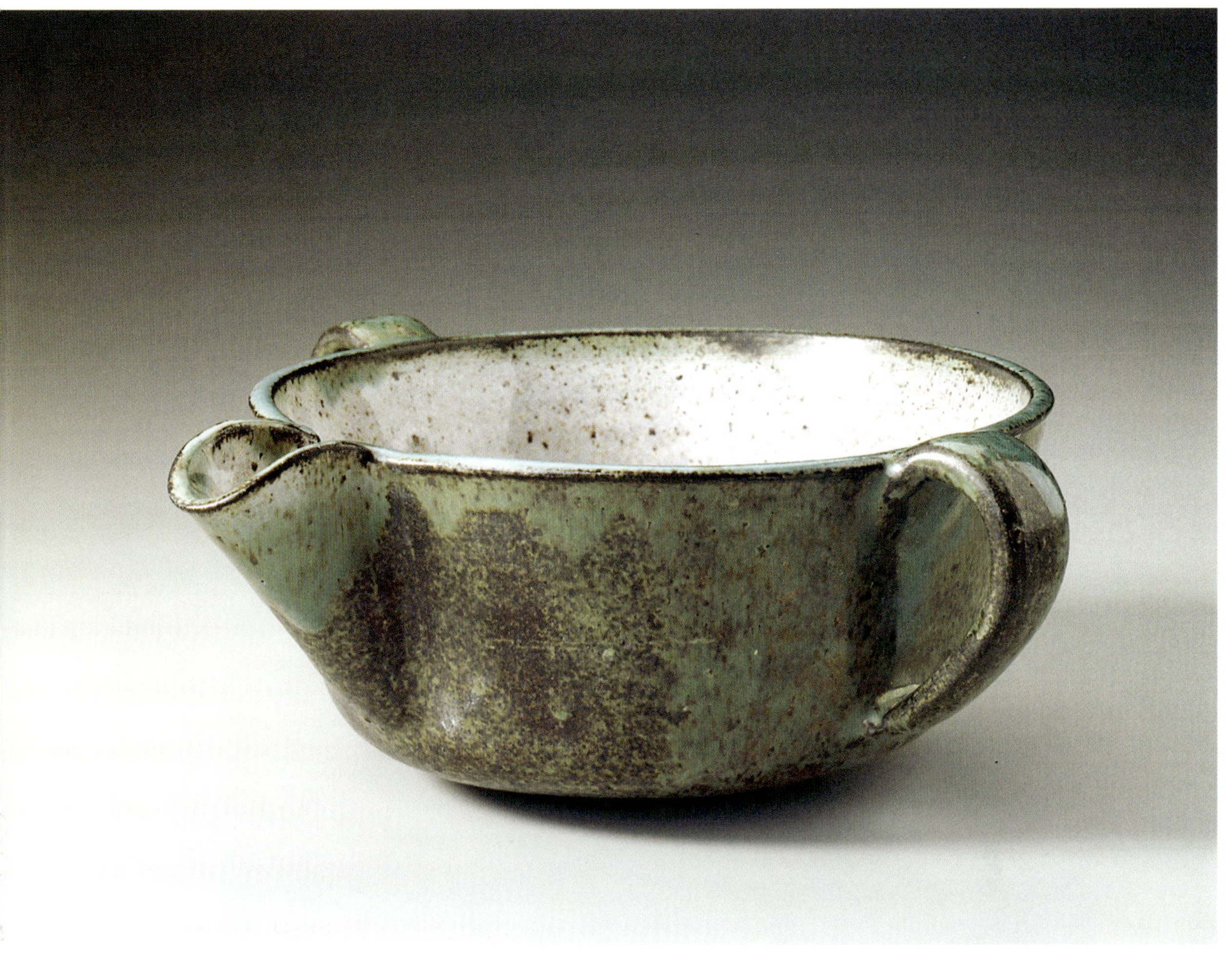

Plate 1 *Batterbowl*, ca. 1960. Marguerite Wildenhain (American, b. France, 1896–1985). Stoneware, 3 1/2 x 9 1/2 x 8 ¼ inches. Collection of Forrest L. Merrill, Berkeley, CA. Photo by M. Lee Fatherree.

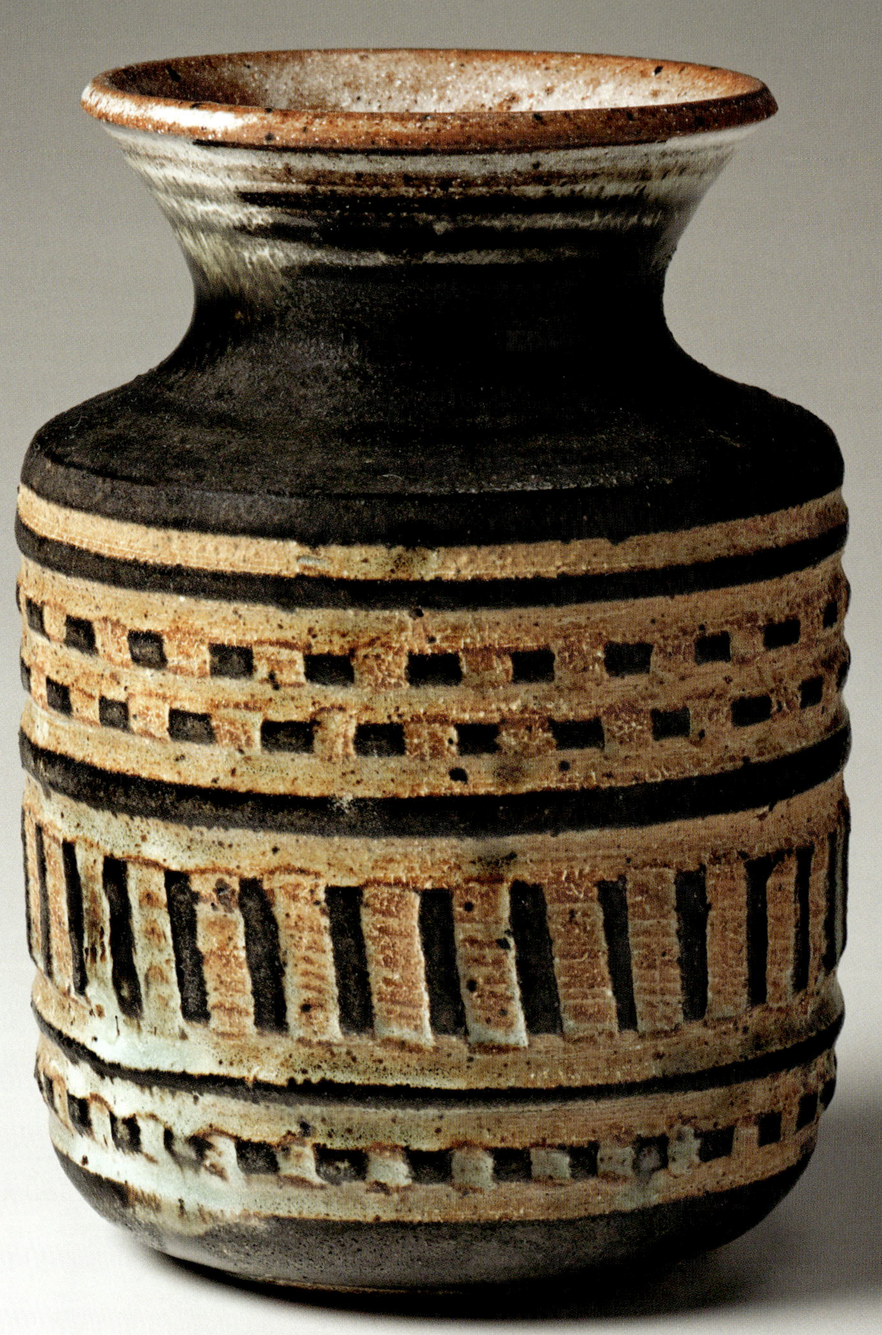

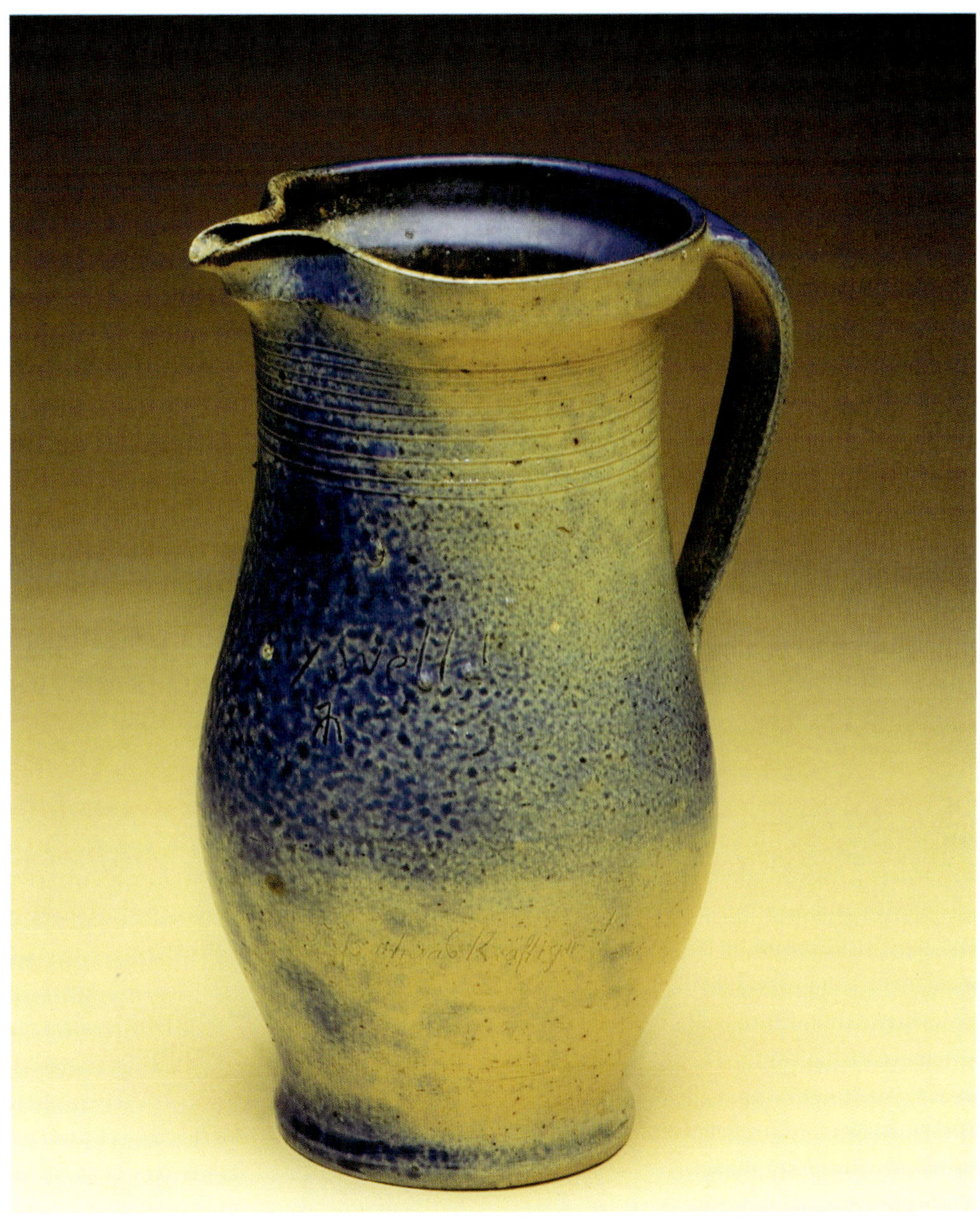

Plate 3 *Milk Pitcher*, ca. 1923. Marguerite Wildenhain (American, b. France, 1896–1985). Stoneware, 9 x 6 x 4 1/2 inches. Photograph by Robert Whitehead, courtesy of the Robert and Frances Fullerton Museum of Art, California State University, San Bernardino. Collection of Forrest L. Merrill, Berkeley, CA.

Plate 2 *Vase*, 1960s. Marguerite Wildenhain (American, b. France, 1896–1985). Stoneware, 5 3/4 x 4 inches. Collection of Forrest L. Merrill, Berkeley, CA. Photo by M. Lee Fatherree.

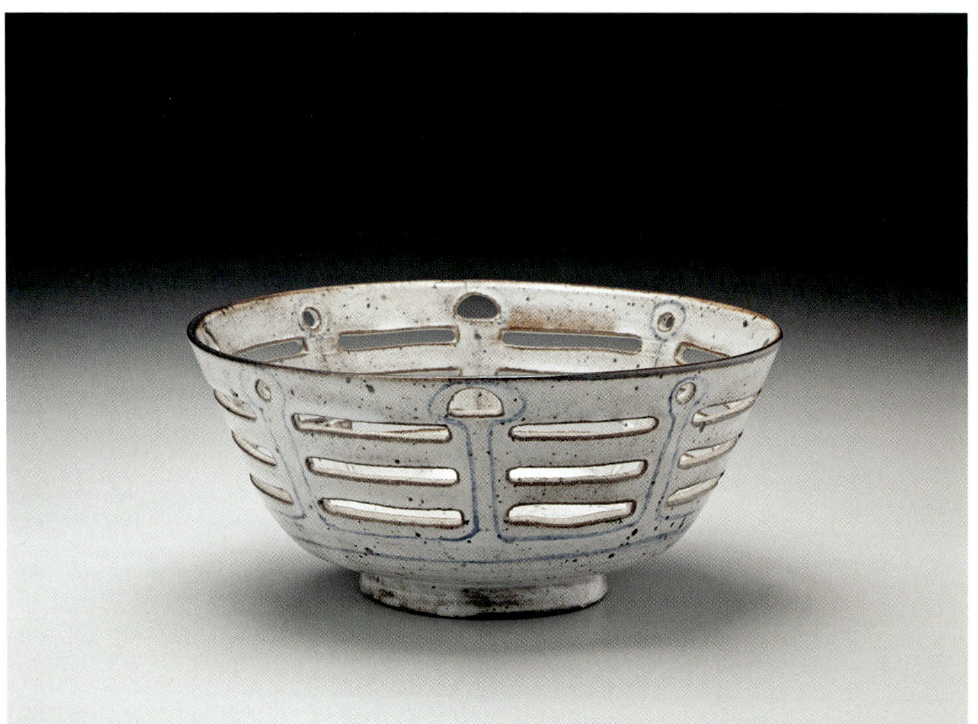

Plate 4 Open form bowl, ca. 1933–1940. Marguerite Wildenhain (American, b. France, 1896–1985) and Frans Wildenhain (American, b. Germany, 1905–1980). Porcelain, 3x 6 1/2 inches diameter. Collection of Forrest L. Merrill, Berkeley, CA. Photo by M. Lee Fatherree.

Plate 5 *Untitled*, ca. 1960s. M. C. Richards (American, 1916–1999). Stoneware, 19 1/2 x 8 x 5 inches. Collection of Forrest L. Merrill, Berkeley, CA. Photo by M. Lee Fatherree. Courtesy of the Estate of M.C. Richards.

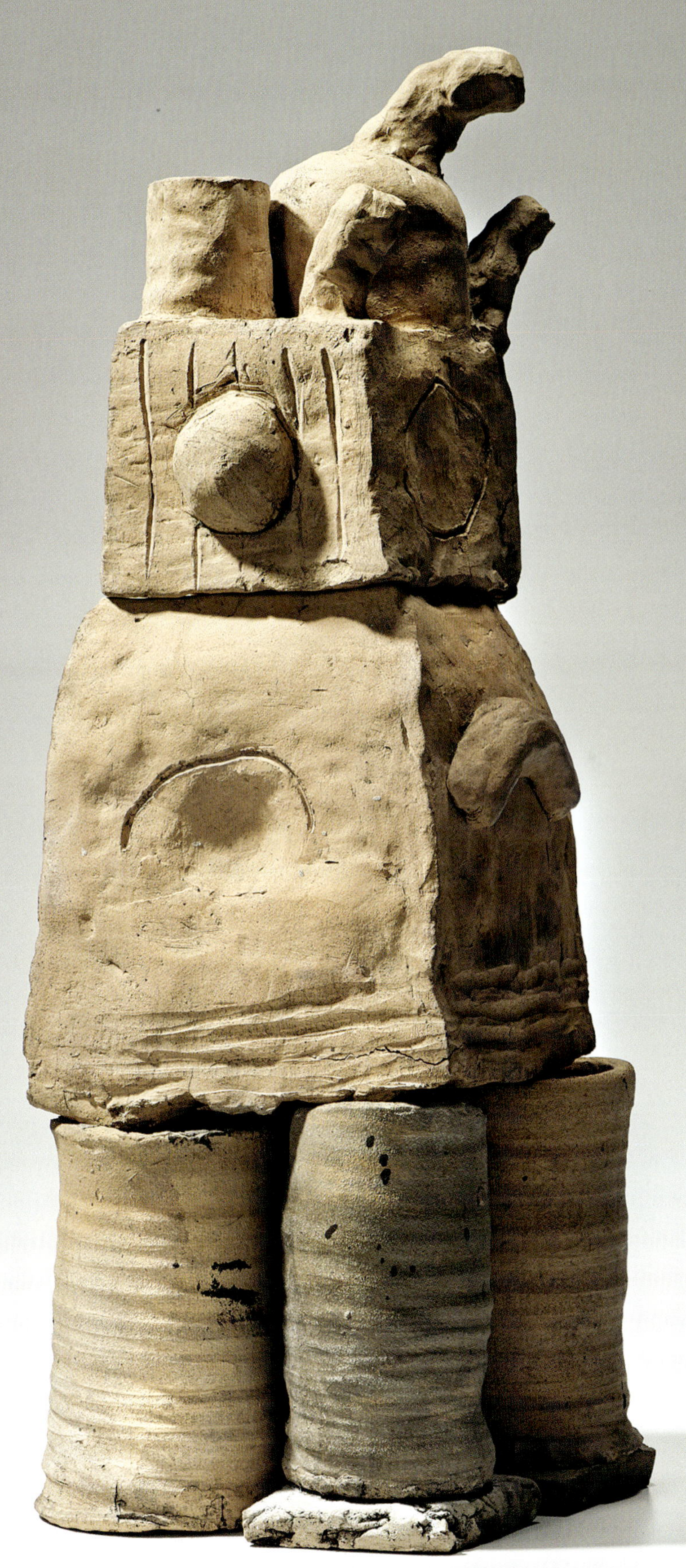

Plate 6 *Dragonfly Tiles*, ca. 1960s. Stoneware, four tiles, 16 x 12 3/4 inches in total. Collection of Camphill Village, Kimberton, PA. Courtesy of the Estate of M. C. Richards and the Jeffrey Spahn Gallery, San Francisco.

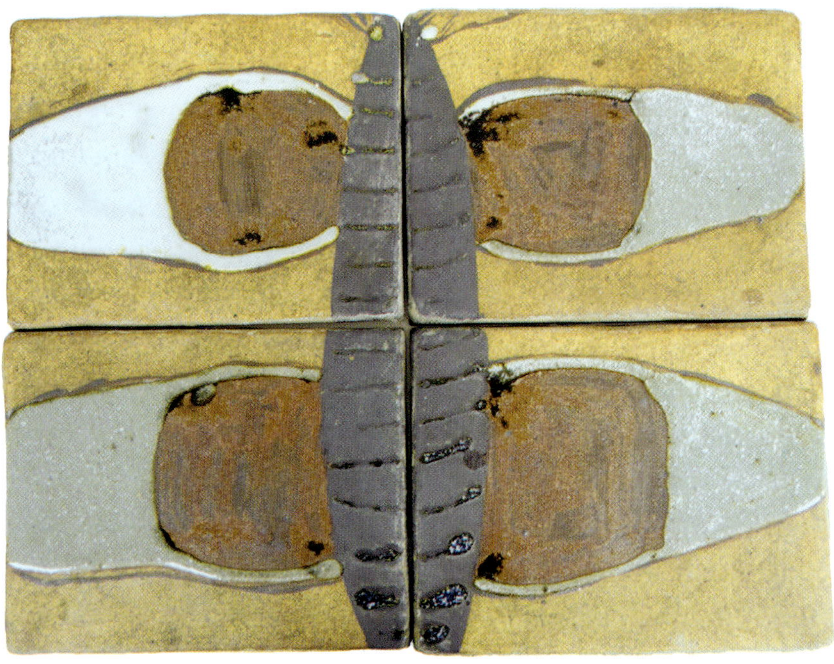

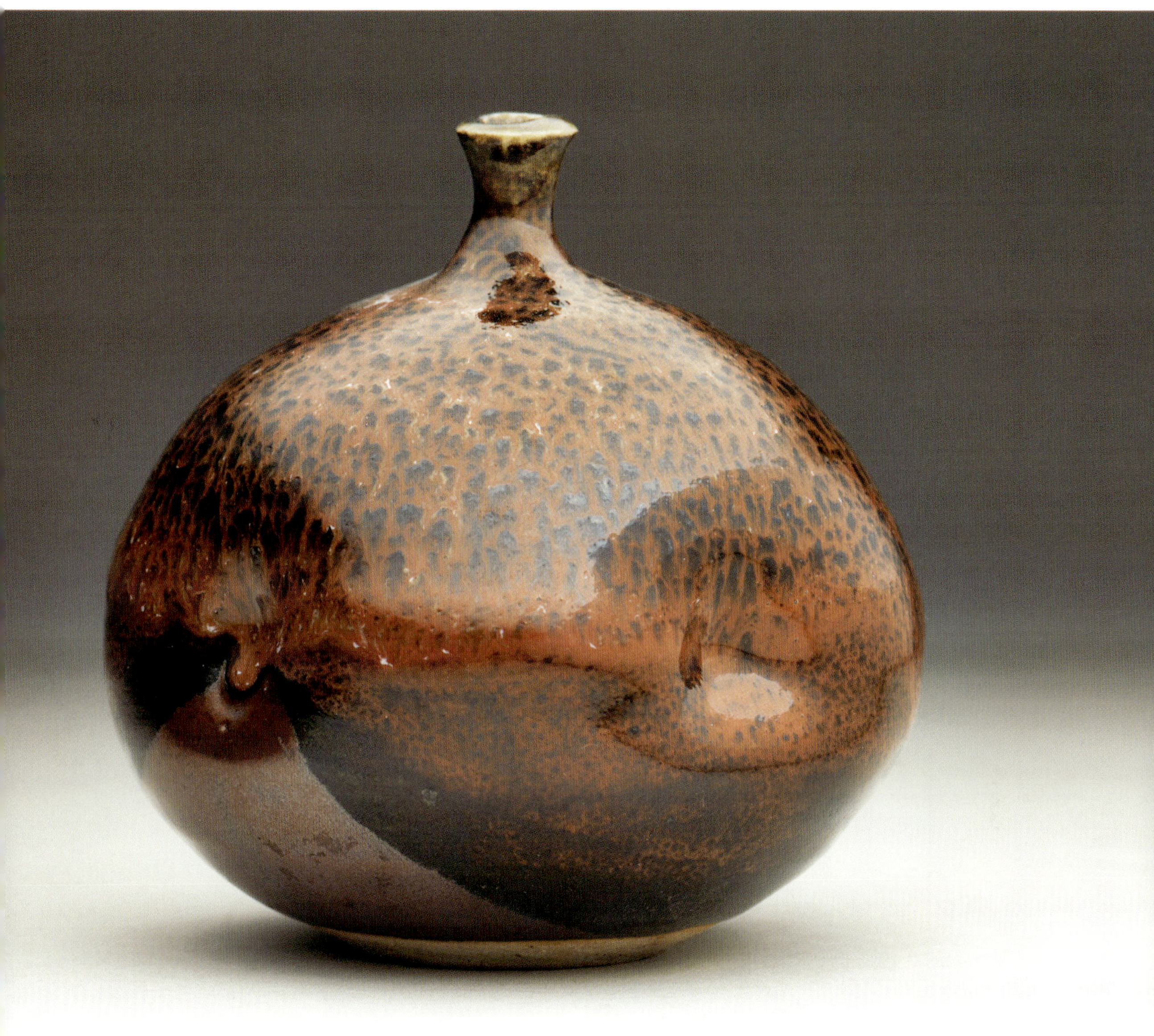

Plate 7 *Bottle*, undated. Susan Peterson (American, 1925–2009). Stoneware, 4.625 x 13.625 inches. Nora Eccles Harrison Museum, Utah State University, Logan, Utah, gift of the artist. Photo by Andrew McAllister. Courtesy Jill Peterson Hoddick, Jan Peterson, and Taag Peterson/the Estate of Susan Peterson.

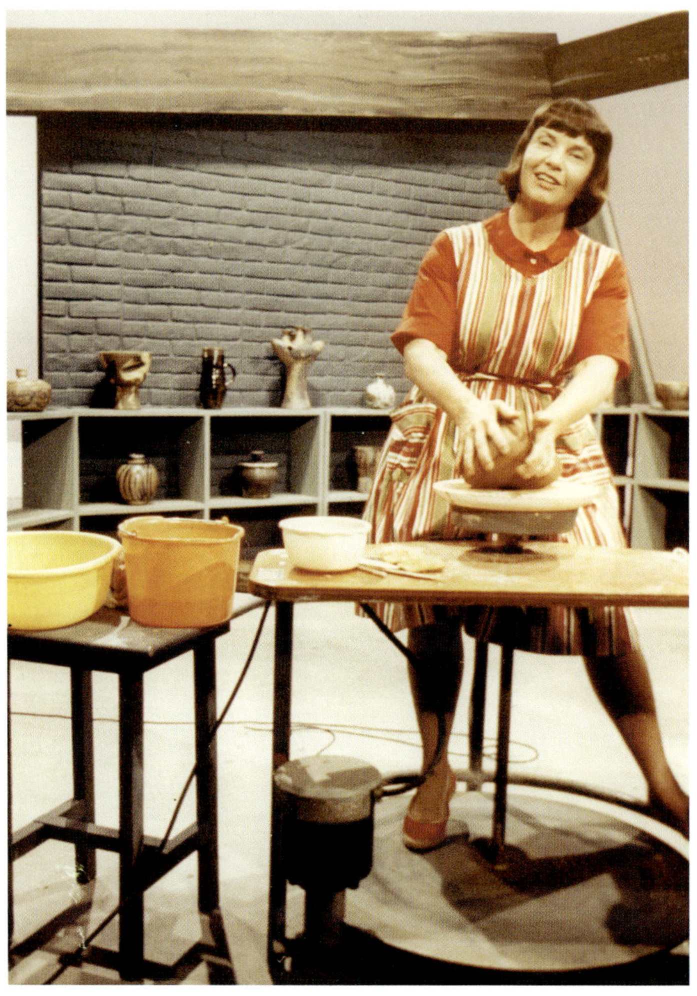

Plate 8 Publicity still, *Wheels, Kilns, Clay*, 1964. Susan Harnly Peterson Archive, Ceramics Research Center, ASU Art Museum, Arizona State University, Tempe. Courtesy of Jill Peterson Hoddick, Jan Peterson, and Taag Peterson/the Estate of Susan Peterson.

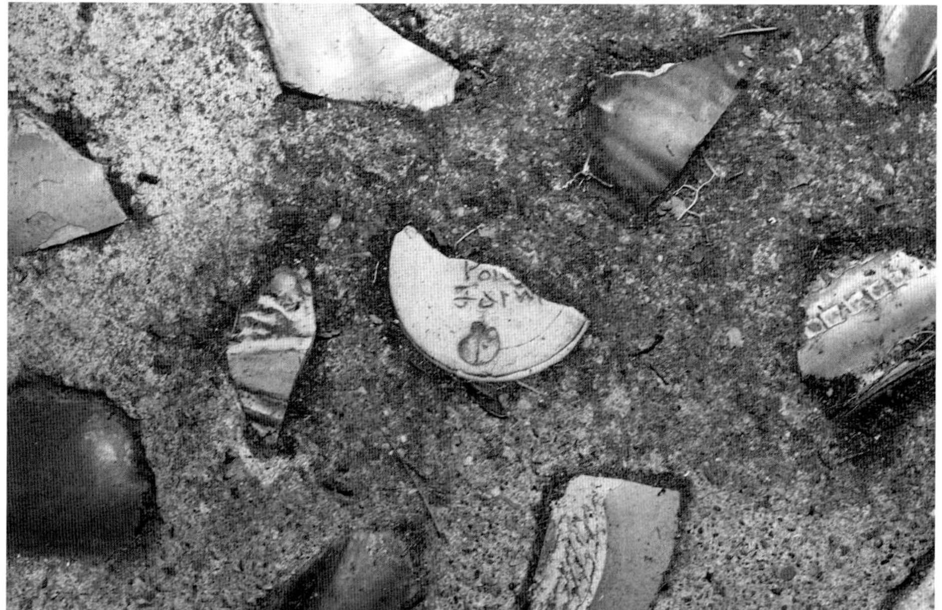

Figure 2.12 Pond Farm walkway, 2015. Photo by Cheri Owen-Sorkin. Courtesy of the author.

Marguerite had finally been broken, forced to admit her human frailties. But it was not by any of her enemies, real or imagined. Rather, it was the State of California, the homeland she had enthusiastically adopted, forcing her hand. To American-born college students, the tremendous loss and destruction that Wildenhain had experienced would have been unfathomable—indeed, untranslatable—without Pond Farm as a conduit. Now that Wildenhain's way of life—and, by extension, theirs—was under imminent threat, she reached out to her students and colleagues.

Throughout the spring and summer months of 1963, nearly 200 letters of support were sent to then-Governor Pat Brown, advocating on her behalf. Fiercely supportive, their actions must have been exceptionally moving to Wildenhain, reaffirming her own belief in the power of the master-student dichotomy for life. Richard Petterson, assistant professor of ceramics at Scripps College, wrote one of the most lucid and heartfelt letters:

> Dear Governor Brown:
>
> I speak for hundreds of artists, teachers and potters who feel, as I do, that Marguerite Wildenhain, of the Pond Farm Pottery School, and cultural center in Guerneville, is an outstandingly valuable and unique asset to California and the nation. . . . Marguerite

> Wildenhain, as a person, is so much a part of Pond Farm, and it of her, as a result of twenty-three years of unremitting work there; her studios, workshops, kilns, home, and associations, that one cannot conceive of separating her from all that she has created there, and destroying one of the most important cultural and artistic centers in the country. . . . The reason for writing you, specifically, is that she faces eviction as a result of the proposed expansion of park lands in Guerneville . . . It would be a constructive act, I feel (and one which the world of art-interested people would praise), if Marguerite Wildenhain and her uniquely valuable lifework, the Pond Farm Pottery and School, could be allowed to remain and continue, during her lifetime.[114]

In August, the letter finally came: Wildenhain was granted what the state called "life tenancy," allowed to live out the remainder of her natural life on the property.[115] Wildenhain even offered the state her rock and mineral collection, volunteering to erect a small building at her own expense, which would house the collection along with "museum-quality pieces," of her own work. The state declined her offer. The land, after all, was meant for recreation, not culture. After her death, the land would be allowed to return to its natural state, with the Parks Service reserving the right to dismantle the buildings. Pond Farm would cease to exist (figure 2.12).

In 1985, Marguerite Wildenhain passed away. Pond Farm, as a site, ceased to exist. But having created Pond Farm, Wildenhain had engaged in what Hannah Arendt termed "world-building," in which the structure itself is more permanent than the goods and things produced *within* the structure. In *The Human Condition* (1958) Arendt writes, "The human condition of work is worldliness . . . Labor insures not only individual survival, but the life of the species."[116] Through Pond Farm, Wildenhain bred a particular species of potter based on the belief that one's humanity is realized only through one's artistic labor. Bound by their collective experience of hardship, students arrived as artists, as individuals, and left as anonymous artisans, productive subjects bent to the will of form itself. Pond Farm stands as a testament to the power of craft not as object but as a form-of-life, a template for living as an indefatigable *process of making*. For Marguerite Wildenhain, Pond Farm was a site of exceptionalism and exceptional despair.

ZEN VETERANS AND THE VERNACULAR

The Black Mountain Pottery Seminar

CHAPTER THREE

Black Mountain College (1933–1957) was a short-lived experimental liberal arts college in rural North Carolina, structured around community and cooperative living, with the arts at the center of its curriculum. Through a combination of misfortune and mismanagement, the college closed for lack of funds and a waning enrollment.

Held for two weeks, from October 15 through October 29, 1952, Black Mountain College's Pottery Seminar has been an overlooked moment in the college's history, bringing together three of the most legendary performers and pedagogues of postwar ceramics. Each represented a distinctive tradition of pottery production, identified with definitive aesthetic "styles": Shoji Hamada, a Japanese potter renowned in the West as a Zen practitioner; Bernard Leach, his translator, the self-described "courier between East and West," a Hong Kong-born Englishman who idealized the East; and Marguerite Wildenhain, whose work exemplified the symmetry and aesthetic perfection of Bauhaus modernism (figures 3.1 and 3.2).[1] In addition to the three potters, Sōetsu Yanagi (1889–1961), the founding director of the Imperial Folk Museum in Tokyo and the leader of the Japanese *mingei* movement, also participated as a lecturer on Zen aesthetics. Cultivated in prewar Japan, *mingei*, literally "people's art," or what came to be known as the folk-craft movement, was an artistic movement spearheaded by Yanagi, who advocated for an aesthetic theory rooted in the appreciation of preindustrial folk crafts made communally and by anonymous artisans. As an event that brought Japanese figureheads to the rural South, the importation of Eastern aesthetics is particularly significant, in that less than a decade had passed since Japan had been an official enemy of the United States during World War II. At the time

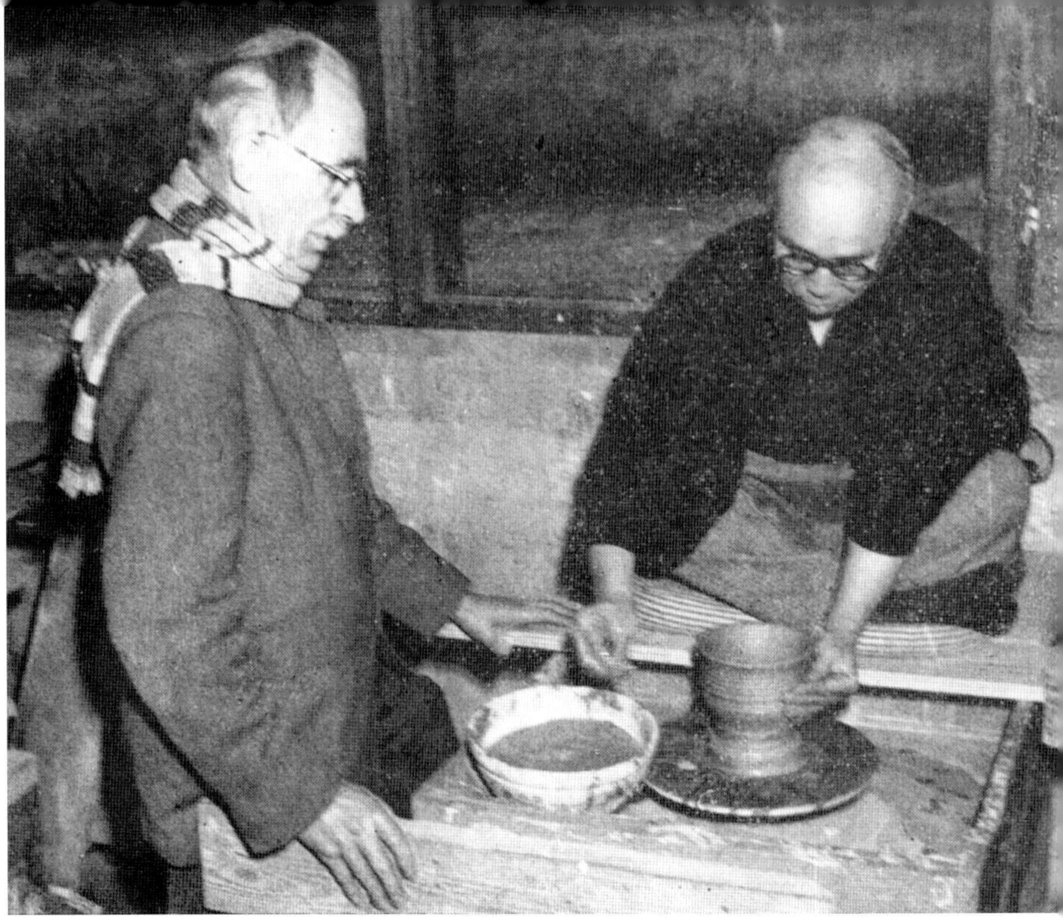

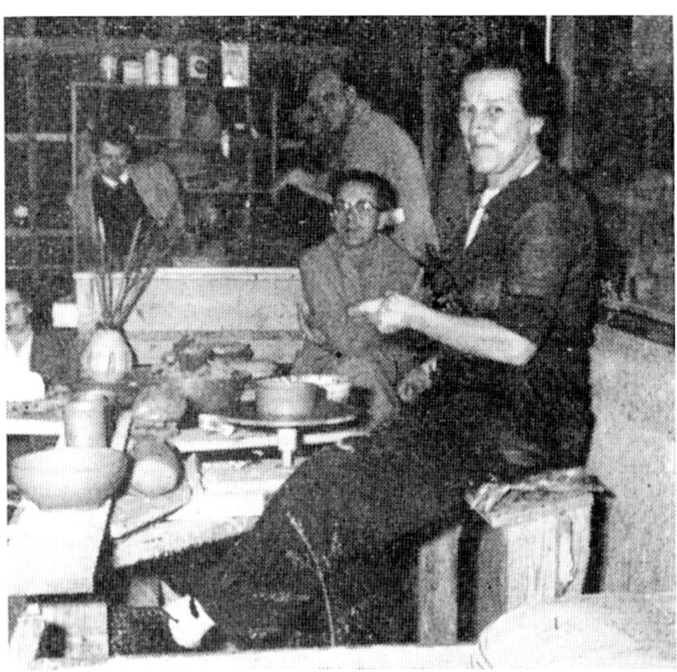

Figure 3.1 Bernard Leach and Shoji Hamada at BMC Pottery Seminar, October 1952. Photo by Robert Diffendal. Printed in *Ceramic Industry* (December 1952), page 103. Courtesy of *Ceramic Industry*, Columbus, Ohio.

Figure 3.2 Marguerite Wildenhain at the Pottery Seminar, reproduced with caption, 1952. Photo by Robert Diffendal. Printed in *Ceramic Industry* (December 1952), page 104. Courtesy of *Ceramic Industry*, Columbus, Ohio.

of the Pottery Seminar, Japan was only newly released from the occupation of American troops, which had lasted until April 1952.[2]

The seminar therefore stands as a pioneering moment, when the therapeutic properties of ceramics were integrated into an avant-garde context utilizing Zen philosophy, rather than a discourse of welfare. This is particularly significant against the backdrop of rural North Carolina, Black Mountain's immediate vicinity, an Appalachian state with a history of vernacular craft initiatives driven by strong social mandates. Largely established by women, it is previously unknown that these local craft groups were in attendance: through archival evidence, the Pottery Seminar's legacy can be further recovered through the pedagogical contributions of women, which converged with Eastern ideas to alter the medium's reception in the United States.

Ceramics at Black Mountain: 1949–1953

While Black Mountain's influence on midcentury avant-garde art production was vast, less well known is the institution's troubled relationship to craft and, in particular, ceramics. Black Mountain's history can be broken into three periods, shaped by three distinctive personalities, with wholly oppositional agendas: its foundational years, 1933 to 1938, established by founder John Andrew Rice (1888–1968), a classics professor influenced by John Dewey's educational philosophies; Josef Albers (1888–1976), rector from 1938 to 1949, who spearheaded Black Mountain's renowned visual arts curriculum; and its final years, 1949 to 1956, in which the poet Charles Olson (1910–1970), as the primary intellectual force, steered the college from the visual toward the literary arts in all of its manifestations, from drama to small press endeavors.

With Albers's Bauhaus antecedents, one of the enduring myths about Black Mountain College is its presumed lack of disciplinary hierarchies. Best known as a painter and printmaker, Josef Albers had experimented with found and stained glass and headed the glass workshop at the Bauhaus, where his future wife Anni had studied under Gunta Stözl in its weaving workshop. Upon his appointment to Black Mountain in 1933, Albers had fully intended to model the Bauhaus curriculum, with production workshops in weaving, photography, wood, metalwork, printing, sculpture, and clothing design. Over the years, nearly all these were implemented at least on a temporary basis but, given budgetary and equipment restraints, proved unsustainable. Though North Carolina was home to an important local vernacular ceramics tradition, and clay was plentiful and cheap, ceramics was an orphaned medium, despite the fact that it was among the original workshops at the Weimar Bauhaus. At Black Mountain, Josef taught color theory, design, and drawing, and Anni became the head of the weaving workshop. Marilyn Bauer, Anni's most advanced weaving student during the late 1940s, characterized Anni as

> . . . very much against what she called "arty-crafty" stuff. Any sort of pottery was "arty-crafty." And the stuff that I later found does sell, handwoven stuff, — was, to her, "arty-crafty," and indeed it is — table mats and idiocies like that.[3]

This vehemence was not Anni's alone. Ceramics came very late in the college's pedagogical trajectory, due in no small part to Josef Albers's hostility to clay, which he condemned as the most "easily abused" material by beginning art students.[4] In 1949—not by mere coincidence, the same year the Alberses left Black Mountain—the potter Robert Turner lent the college the money to build the Pot Shop and left Alfred to serve for two years as faculty.[5]

Black Mountain's late investment in ceramics was part of the larger do-it-yourself ethos of living off the land, sweat, and poverty in the service of art and self-enlightenment. Undeniably, this was the medium's primary attraction, a spiritual and intellectual alternative to not only the excessive consumer culture of the 1950s—the era of the first McDonald's hamburger and the drive-in movie—but also the celebrity of the New York School and their brooding, alcoholic ways. They were flabby boozers painting shirtless in their gritty cold-water flats, city men who lacked the sheer physical strength to perform the laborious tasks necessary to rural pottery (laying brick for a kiln, chopping wood, wedging clay) or, moreover, the psychic concentration to perform, before a student audience, the perfect amalgamation of bravado and humility most typically associated with the role of the potter.

The Pottery Seminar was termed the "Eastern Center for Interchange of Work & Ideas: East to West." As noted in the event's promotional flier, among the themes to be considered were "development of form ideas," "design for mass production," and "relation of craft to art" (figure 3.3). This flier—also a mailer—is itself an important visual marker. By taking a letter-sized piece of paper, and Orientalizing it—the mailer is meant to be folded and handled as quasi-origami, rather than viewed tacked up as a poster, replicating Western-style painting—its design performs the anxiety that these two dichotomies, East vs. West, Art vs. Craft, produced. Additionally, the mailer's stark black and white design is a deliberate pastiche of a Franz Kline painting, an inversion of his signature black abstract slashes against a highly textured white ground. Such a usage is pointed: a visual assertion of the perceived distance between 1950s ceramics and abstract expressionist painting, the dominant strain of artistic production in the United States, embodied by Black Mountain's very own Franz Kline. Kline's painting *Nijinsky* (1950) elides the movements of the famed Ballets Russes dancer with

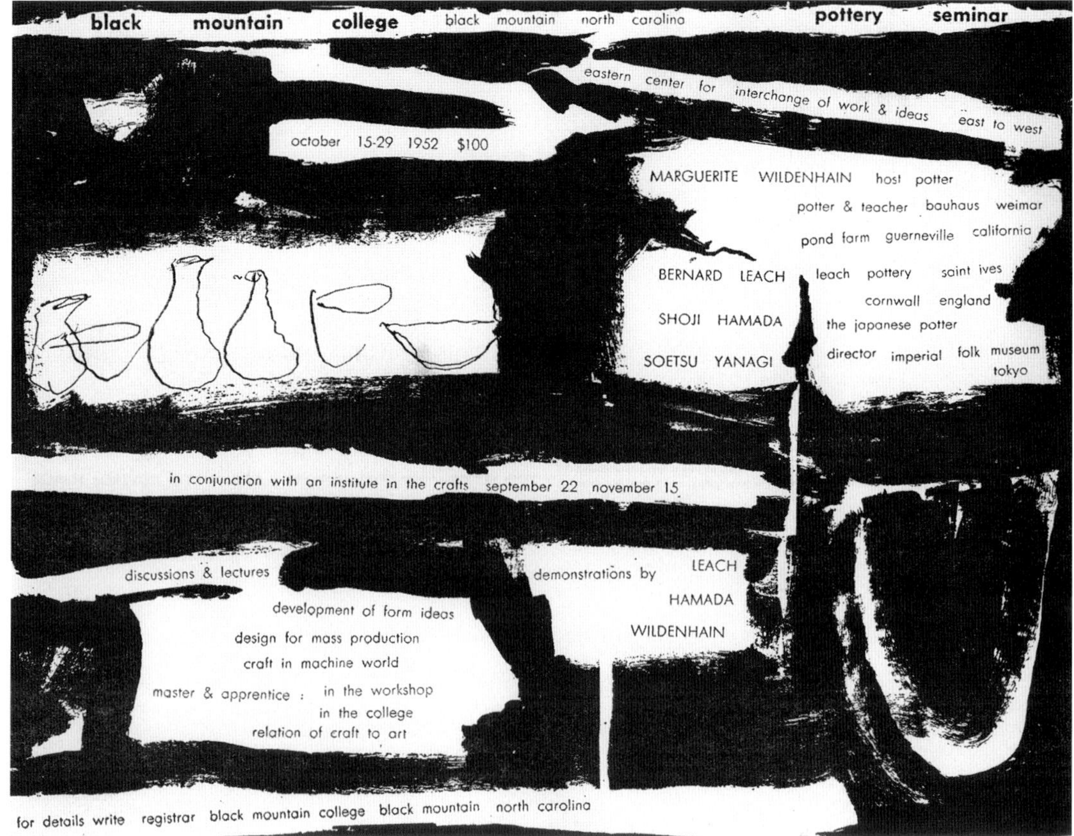

Figure 3.3 Pottery Seminar, 1952 Flyer. Black Mountain College General Files, box 28. Courtesy of the Western Regional Archives, State Archives of North Carolina.

the gesture of the brushstroke itself (figure 3.4). The mailer, then, pokes fun not just at the generic abstraction of abstract expressionism but also at the very real and revered pedagogy of Kline himself.

Alongside Charles Olson, Kline was the other significant alpha male at Black Mountain, teaching in the summer session of 1952, just a few months prior to the Pottery Seminar. In his memoir, the novelist and Black Mountain student Fielding Dawson referred to the pair as "his fathers," but he also devoted a separate book to Kline, titled *An Emotional Memoir of Franz Kline* (1967). Dawson makes an adjective—*Klineness*—of its main subject, using the *Klineness* of Kline to describe the distinctive spareness and overpowering quality of both Kline and his monumental paintings.[6]

The term *Klineness* is a stand-in for the man himself, a way of denoting the physical execution and psychic gesture of the canvas as one complete unit. In Japanese calligraphy (*shuji*), the formation of character (*kanji*) is predicated

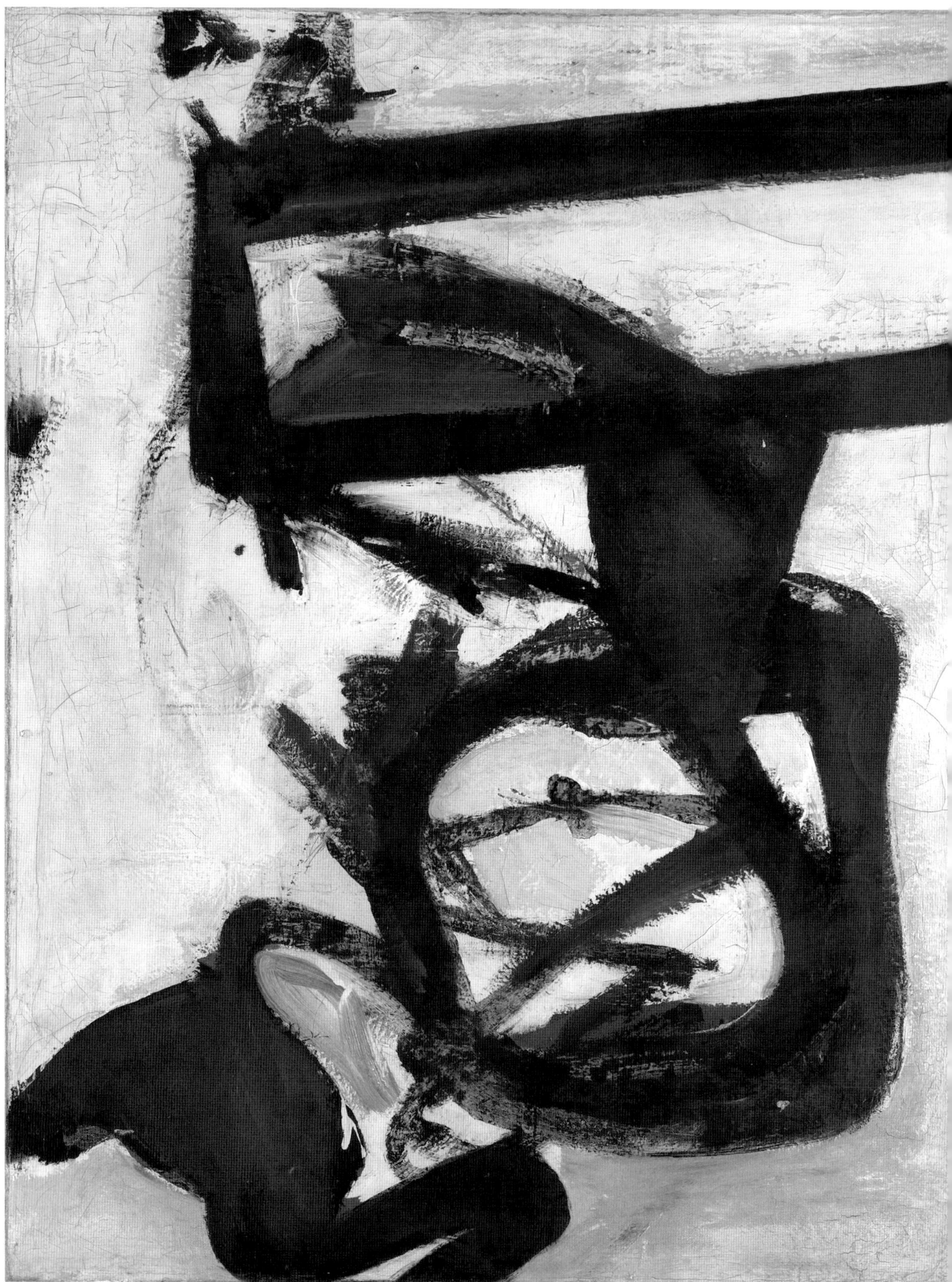

Figure 3.4 *Nijinsky*, 1950. Franz Kline (American, 1910–1962). Enamel on canvas, accession number 2006.32.28. 45 1/2 x 34 7/8 inches (115.6 x 88.6 centimeters). The Muriel Kallis Steinberg Newman Collection, gift of Muriel Kallis Newman, 2006. Collection of the Metropolitan Museum of Art, New York. Image copyright © The Metropolitan Museum of Art. Image source: Art Resource, NY. © 2015 The Franz Kline Estate/Artists Rights Society (ARS), New York

on spiritually driven precision, hence *Klineness* could also be thought of as the *Japaneseness* of Kline's work, in which the perfectly formed character begets the formation of personal character. Kline's *Japaneseness* was noted but refuted throughout his brief career, most famously by Clement Greenberg in his essay *American-Type Painting* (1958):

> Kline's apparent allusions to Chinese or Japanese calligraphy encouraged the cant, already started by [Mark] Tobey's case, about a general Oriental influence on "abstract expressionism." This country's possession of a Pacific coast offered a handy received idea with which to receive the otherwise puzzling fact that Americans were at last producing a kind of art important enough to be influencing the French, not to mention the Italians, the British, and the Germans. Actually, not one of the original "abstract expressionists"—least of all Kline—has felt more than a cursory interest in Oriental Art. The sources of their art lie entirely in the West; what resemblances to Oriental modes may be found in it are an effect of convergence at most, accident at least.[7]

Art historian Bert Winther-Tamaki has noted that the earlier version of Greenberg's essay, from 1955, reads as an affirmation of Kline's Japanese sensibility—the phrase *least of all Kline* was originally written as *except Kline*, meaning Greenberg had, in effect, eradicated the characteristic that he had previously emphasized and singled out as particular to Kline.[8] Thus, Kline's abstraction was unmistakably Oriental in its sensibility and became a liability sometime between 1955 and 1958, when postwar politics intervened: the freedom of individual expression represented by abstract expressionist painting was utilized as a tactic of showcasing cultural repression elsewhere.[9]

Recognizing Kline's *Japaneseness*, the (most likely) student designer of the Pottery Seminar flier cribbed it for the purposes of a Japanese-themed seminar, using the artist's calligraphic style as the primary design motif of the Pottery Seminar advertisement. A small thing, perhaps, but in 1952, at the height of abstract expressionist mania, such an inversion of art and craft is striking: painting becomes a *background* that *foregrounds* craft, privileging the "development of form ideas" seemingly unrelated to the discourse of Greenbergian formalism, demoting painting to the realm of decoration, reproducing, in graphic form, in effect, a cartoon of the New York School's own anxieties regarding the public reception of their work. Kline's *Klineness* becomes a graphic, a branding apparatus, used to readily identify and dismiss the discourse surrounding formalist painting, favoring instead an aesthetic dialogue that accounted for the circulation and reception of functional objects (a row of pots, as seen here, rendered as three-dimensional objects) rather than the disappointing flatness of the singular art object (i.e., the painting, or in this case, the expressionistic pot) made of one sweeping brushy gesture, in the lower register of the mailer.

Robert Turner's departure precipitated the need for a replacement in time for the 1952 summer session. Marguerite Wildenhain was Charles Olson's first choice. She declined the position. In a continuation of their correspondence, Olson wrote to Wildenhain:

> A note to thank you for your certain advice on the Alfred potters. . . . Harder's strongest recommendation, in which he was joined by [Robert] Turner . . . was actually Karen Karnes, whose husband, David Weinrib, is the one who gets the official post—this was Harder's suggestion—and we get both. It for the summer only, those 8 weeks . . . I shall have the Weinribs, in any case, see that clay, etc. according to your request, is taken care of, at the end of August.[10]

But however forward-thinking Black Mountain College was, as a Southern outpost for the New York avant-garde—once wryly described by Franz Kline as "downtown Manhattan"[11]—it was very much of its time with its clearly enforced gender roles, as was the case with the appointment of potters Karen Karnes and David Weinrib, freshly minted MFA graduates of the ceramics program at Alfred University.

Though Karnes was the recommended instructor, she was also a woman. In 1952, the prospect of a salaried wife and her unemployed husband might unduly damage the male ego. So Olson hired both but paid only Weinrib. Yet there was little need for even one instructor, as the pot shop was remote, situated in a field far from the central cluster of buildings on Lake Eden. Few students ventured there, and Karnes described it as being "perfect for the studio."[12] At Olson's behest, Wildenhain agreed to serve as the "host-potter" for the duration of the seminar, playing the gracious American hostess, catering to a group of men who were, on the whole, rather patronizing to women.[13] But Wildenhain was unfit for such a submissive task, leading to Karnes's unflattering assessment: "Marguerite sat in the back and grumbled. She didn't do much."[14]

Arriving on a campus with a compulsory work program and a working farm, Karnes and Weinrib were committed to demonstrating the value of the production potter to the community. In their first weeks at Black Mountain, they made serving bowls and platters for the dining hall, which went into immediate circulation, and just as immediately were defaced, piled high with food scraps and cigarette butts by students hostile to the encroachment of craft in their avant-garde enclave, students who, as amateurs themselves, felt compelled to mock the amateur quality of ceramics, through a literal rendering of the Josef Albers's epithet "ashtray art."[15]

The amateur associations of ceramics were not lost on Olson, who, as we will see, sought a new revenue stream for the overburdened college. Even today, amateurs are the bread-and-butter of summer craft programs like Haystack Mountain School of Crafts in Maine or Penland School of Crafts in North Carolina, where the experience of making is sold to urban professionals and hobbyists as a temporary reprieve from the demands of contemporary life, offering wholeness and community in a beautiful pastoral setting, with no shortage of small luxuries such as organic food and wireless access.

But Olson had little previous experience with ceramics. Without M. C. Richards's guidance, the Pottery Seminar could never have come to fruition. From 1945 to 1951, Richards was an important pedagogical force at Black Mountain, as its English professor and chairman of the faculty from 1949 to 1951. But she also became a committed potter while in residence, studying first with Robert Turner and then with Peter Voulkos, and eventually went on to teach ceramics and write the pottery classic *Centering: On Pottery, Poetry, and the Person* (1964). On Richards's interest in clay, Turner commented, "She was definitely involved [in clay] and in a way much more deeply involved than perhaps I realized at the time."[16] First and foremost an intellectual, Richards had begun learning the medium as early as 1948, making a point of seeing a show of Picasso's ceramics in Antibes while on sabbatical.[17]

Richards and Pedagogy

At Richards's invitation, Olson settled permanently at Black Mountain in 1951. In her annual letter as chairman, she praised him as "a source of useable energy for the college."[18] And indeed he was. A large and forceful personality, Olson was a midcentury intellectual, poet, and pedagogue. In contrast to his predecessor, Josef Albers, and other German colleagues, Olson was a committed Americanist, a relentless champion of American arts and letters at a time when American culture was widely viewed as secondary to European modernism.

Richards wrote fervently, even religiously, of Olson as her ideal replacement:

Providence's hand has been in everything, that's for sure, because, as I see it, if things hadn't been so all-fired bad, you would never have come to be in the position you now are with the feeling you now have for the goings-on there and the readiness and the plumb weight. God has seen to your becoming rector and holding the cove.[19]

In its "plumb weight," as Richards describes, the trust and belief that the Black Mountain community chose to invest in Olson was staggering. In two separate memoirs, Fielding Dawson (*The Black Mountain College Book*) and Michael Rumaker (*Black Mountain Days*) have written of the desperation and vitality that characterized the 1950s era at Black Mountain.[20] A campus filled with German refugees and women in dungarees provoked the suspicion and paranoia of the locals, who believed the college to be a haven for communists. Dawson recalls how often FBI men came to interview Olson; Rumaker, on the other hand, reveals that the college granted logging rights in order to ensure its financial survival during the 1953–1954 academic year.[21]

Betsy Baker, a local young woman hired by Olson in 1954 to be in charge of publicity, characterized these last years as

terribly depressing . . . a tragic period . . . the people who were there then were talented, but there was parasitism. You know, they would come and camp and live off of the farm and really not produce much . . . It was a terrible tragedy to me. . . . they wouldn't give of themselves to the college. They wouldn't give the best of themselves in these latter years, they would take from it, they would just vegetate.[22]

So Olson had his work was cut out for him. Baker noted that by the time she arrived, students were no longer paying tuition.[23] A steady stream of newcomers, rather than students already enrolled, seemed the most promising way to insure tuition revenue. Olson enriched Black Mountain's curriculum with a spate of new offerings known as "institutes," brief sessions led by visiting artists, meant to capture the spirit and intensity of the earlier popular summer sessions, which had been a highlight of the Albers era. Ostensibly, the institutes, with their well-known faculty, would lure people far and wide to Black Mountain. This formalized a proposition Richards herself had dreamed up, writing in 1951:

One new thing that we mean to do this year is to have, say, four visiting artists, on 2-month contracts, feeding into our study and experience through the year. This scheme seems on the way to being fit for an abundant pedagogy. It seems to be a kind of extension of what has been the summer program here, of inviting guest faculty during July and August to add to what we do during the year for an 8-week period. It keeps

> turning any subject round for new examination and evaluation and practice and perfecting. And it tests the growing suspicion that perhaps the "statesmen" of our civilization are to be coming more and more out of the arts. The eye for "multiple observations," which is the poet's eye, may be the vision that will take us where we want to go. . . .[24]

Taught by high-profile artists, writers, and intellectuals, who were unaffordable, unwilling, or unable to commit to longer stays, the first official summer session in the post-Albers administration commenced the summer of 1951, during which photography, painting, and dance were taught by Aaron Siskind, Harry Callahan, Herbert Matter, Hazel-Frieda Larsen (photography), Ben Shahn (painting), and Katherine Litz (dance), respectively.[25] The photography institute was initiated by Larsen, a student turned faculty member, who invited whomever she was "curious about," as Callahan put it.[26] Like Richards's letter, Larsen's programming evinces the informal structures in which Black Mountain's women assumed responsibility and leadership without any expectation of power. However, through their continuous, enthusiastic outreach directed at male, rather than female, artists, it seems likely that they were unaware of their own participation in reinforcing traditional gender hierarchies, similar to Jane Hartsook's programming of Greenwich House Pottery, seen in chapter 1.

Richards resigned her post at Black Mountain before the seminar's occurrence, the reasons of which are detailed in chapter 4. The incomplete correspondence suggests that Charles Olson, her successor, engaged in negotiations with all the involved parties.[27] At Richards's suggestion, Olson turned to Daniel Rhodes (see chapter 1), as an influential voice in the field, to insure that the event would transpire. In the end, none of the Pottery Seminar's off-site coordinators (Charles Olson, M. C. Richards, nor Daniel Rhodes himself) attended the event, though Olson took credit for it.

To sell Leach and his colleagues on the Black Mountain experience, Olson urged them to value community over salary:

> We shall, of course, house & feed you while here, in the most comfortable way we can . . . none of us gets pay enough to cover cigarettes, gas & the Gerber's baby food I now need for a daughter born October 23rd—$35 a month, I get, having that daughter! It's crazy, but, in a fat time, I am of the conviction that such poverty-stricken ambience is the one fit for living, for working: & it is this factor of severity which makes me so much believe in this community of so few persons—14 of us as a faculty, at present a total people of about 75, I'd guess, all of us doing all of the work—administration, repairs, dishes, etc, etc, (of course we do have this wonderful farm of our own, on which we also all work).[28]

But Leach didn't buy what Olson was selling, insisting that he, Hamada, and Yanagi be guaranteed a minimum of one thousand dollars each for their visit. This was far more than any of Black Mountain's best-known faculty had ever been paid. In the end, it was agreed that Leach and company were free to accept gigs in other places while in America. The trio asked to be allowed to leave the seminar three days before it was actually over, a request unequivocally vetoed by Black Mountain's Board of Fellows.[29] Thus, they sailed from England on the Mauritania, and their ten-month American tour commenced at New York University, followed by Boston and Connecticut, before proceeding to Black Mountain College, and ending, finally, at the Chouinard Art Institute in Los Angeles, which was hosted by Susan Peterson.[30]

All this effort, for only six students—six!—a false rumor Robert Creeley memorialized in his introduction to the anthologized edition of the *Black Mountain Review*, as a means of illustrating Black Mountain's brazen effort to stave off its own decline:

> In hindsight it is almost too simple to note the reasons for the publication of *The Black Mountain Review*. Toward the end of 1953 Black Mountain College . . . was trying to solve a persistent and most awkward problem. In order to survive it needed a much larger student enrollment, and the usual bulletins and announcements of summer programs seemed to have little effect. Either they failed to reach people who might well prove interested, or else the nature of the college itself was so little known that no one quite trusted its proposals. In consequence a summer workshop in pottery, which had among its faculty Hamada, Bernard Leach, and Peter Voulkos, found itself with some six rather dazzled persons for students.[31]

Creeley's version of events (conflating 1952 and 1953) perpetuates the myth that ceramics held little interest for the community at large. What else to make of this disparity between the luminary faculty (mistaken as Leach, Hamada, and Voulkos) and the six "rather dazzled" students? To Creeley and many others, the college in the 1950s held little allure but for the presence of Olson.

Fortunately, the historical record does not bear out Creeley's easy dismissal of the Pottery Seminar.[32] It turns out that twenty-five students attended, described by Wildenhain as a "heterogeneous group including potters, students, and elderly ladies" (figure 3.5).[33] At the age of fifty-six, Wildenhain herself was the "eldest lady" of the group, as the participants ranged in age from thirty-two to fifty-four, with the vast majority in their thirties. Wildenhain's rather patronizing remark attests to her view of American craft as insufficiently rigorous. It was why she had declined to join the faculty and instead reinforced her own decision to turn

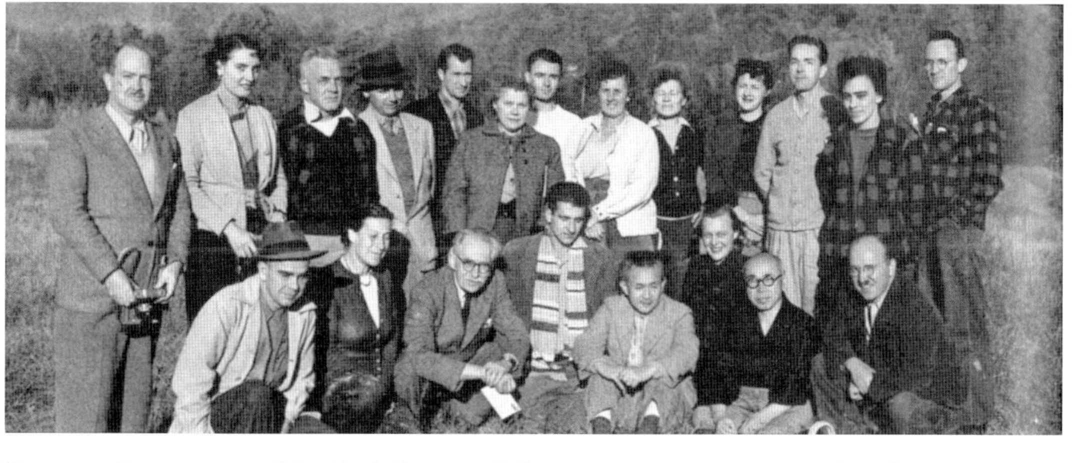

Figure 3.5 Group portrait of the Black Mountain College Seminar. Front Row (L to R): William Watson, Marguerite Wildenhain, Bernard Leach, David Weinrib, Soetsu Yanagi, Betty Edgar, Shoji Hamada, Karl Martz. Back Row (L to R): David Edgar, Helen Kruger, Allen C. Hoffman, E.L Holtzclaw, Douglas Ferguson, Frances Reps, Richard Hieb, Marjorie Dutton, Betty Crawford, Lillian Boschen, A. J. Spencer, Janet Darnell, Gordon Goehring. Photo by Robert Diffendal. Printed in *Ceramic Industry* (December 1952), page 106.

away beginning students at Pond Farm as she matriculated her first summer school class the following year, in 1953.

On a dilapidated campus that lacked regular maintenance and repair, eleven men and fourteen women came together at Black Mountain for two weeks of lectures and demonstrations. Such a small group was the unintended consequence of bad planning and poor administration, such as answering the mail. Had the college's staff been in less disarray, the Pottery Seminar would have been exponentially larger, as the interest level was high. There were numerous, excited inquiries from as far away as Maine, Idaho, and Chicago but, alas, no one to answer them. The registrar, Constance Olson, Charles's wife, had become ill, and in her absence, Hazel-Frieda Larsen was forced to take over the pressing administrative tasks. The seminar was scheduled to begin October 15, 1952, but eager letters from June, July, and August requesting travel information and a schedule of events went unanswered until October. Most of the letters Larsen sent were in a single batch, dated October 6, 1952, just two weeks prior to the workshop. Her replies were apologetic and hasty, encouraging even last-minute applications.

By this time, the college was in a terrible bind, cash poor and so desperate to fill the workshop that anyone, even those without prior experience, was virtually guaranteed a spot in the seminar. This was unfortunate, as many of the prospective students who wrote early in the summer were highly trained, among them the glass artist Harvey Littleton and Henry Okamoto, a Connecticut-based potter who had trained with Carlton Ball and went on to run the Clay Art Center in Port Chester, New York, a prominent community pottery. Other applicants were highly knowledgeable about modern art, such as Mildred Sinsabaugh, wife of the well-known Chicago photographer Art Sinsabaugh, who would have known of Black Mountain through Harry Callahan, her husband's former teacher at Chicago's Institute of Design. Neither Littleton, Okamoto, nor Sinsabaugh enrolled in the seminar, though, because nobody answered their letters.

The ability to attract participants across disparate media suggests the Pottery Seminar, in turn, had the potential to be a widely understood landmark event, expanding the ceramic medium's networks of influence and repairing the still-tenuous link between ceramics and other disciplines, damaged by the Albers ban. Further, local craftspeople are not documented on the final list of participants, yet the archive suggests they came as day students to the seminar. In the Pottery Seminar's only press coverage, it was confirmed that local groups were in attendance.[34] This is a crucial recovery that reconstitutes the Pottery Seminar as an important site of intersection for avant-garde and traditional craft practices, with the American craft revival as the backdrop to the historical record of Black Mountain's simultaneous gain and estrangement with regard to ceramics.

The Craft Revival

Disparate, simultaneous craft revivals infused with pedagogical models occurred in disparate international contexts with regularity throughout the 1930s—an era of great economic uncertainty worldwide. This is the moment that precedes both Yanagi's *mingei* and Albers's Bauhaus-infused art curriculum at Black Mountain. Craft and the rhetoric of autonomy, then, can be understood as having an increasingly strident nationalist agenda commensurate with specific political gains, including the Zionist interest in craft in Germany, discussed in chapter 2, the Japanese Peasant Art Institute in Tokyo, and the Southern Highland Craft Guild and John C. Campbell Folk School in North Carolina. All were self-liberation projects, fostering cultural unity and financial independence as part of an overarching nationalist (former) or regionalist (latter) imperative.

In 1923, the Japan Peasant Art Institute was established in Tokyo by painter Kanae Yamamoto (1882–1946). As an educational undertaking, its mission was to improve the finances of agrarian peasants by teaching them craft skills, as well as improving the general well-being of urban people everywhere by encouraging them to practice craft as a hobby. This was known as "self-sufficient crafts," open to students between the ages of fifteen and forty, offered at no charge above the cost of the materials.[35] The school became subsidized in 1925 by the government's Ministry of Agriculture and Trade, which led to courses beyond the city and the formations of craft guilds throughout the late 1920s and early 1930s. However, due to large debts, the institute closed in 1935.

Like British Arts and Crafts, *mingei* also developed into an urban (Tokyo-based) revivalist movement, popularized both nationally and abroad by Yanagi's various endeavors—writing, lecturing, organizing—which closely parallel those of William Morris (1834–1896), the nineteenth-century artist and ardent socialist, albeit a half century later. In the first quarter of the twentieth century, Japan was undergoing rapid modernization, absorbing the joint influences of industrialization and Westernization. Japanese peasant art and *mingei* addressed different populations: Yamamoto endeavored to provide poor and middle-class people with manual craft skills, while Yanagi's discourse was more cerebral, aimed at a circle of peers, the artist-intellectuals of the Shirakaba Group already engaged with issues of aesthetics, design, and production.

As a recovery project, Yanagi's *mingei* existed in contradistinction to the welfare-driven ventures already embedded in the American South. The period from 1890 to 1940 has been historicized as the Southern Craft Revival, a coordinated regional effort between civic-minded women; private philanthropists such as Olive Dame Campbell, who started the John C. Campbell Folk School in Brasstown, North Carolina; and government agencies such as the Women's Bureau of the Department of Labor, to initiate craft schools in order to develop highly skilled home workers, many of whom were women, thereby improving their economic prospects (figure 3.6).[36] Like Black Mountain's visual arts program, the Campbell School also had a European-influenced curriculum, as it was predicated on a mid-nineteenth-century Danish *folkehojskoler* (folk high school) model of secondary education, which offered formal but vocationally oriented schooling to peasants. The merging of vernacular culture with educational reform produced a craft revival that romanticized Appalachia as a keeper of Caucasian national values.

Black Mountain College was located a mere twenty miles from Asheville, North Carolina, the revival's epicenter and headquarters of the Southern Highland Handicraft Guild, which had been founded by Frances Goodrich (1856–1944) in 1930. Goodrich also fits the profile of well-educated, middle class women establishing philanthropic, community-based missions to uplift the poor. Of the same generation as Jane Addams (1860–1935), Goodrich attended Yale, then was dispatched to the South as a missionary and teacher for the Presbyterian Home Mission Board.[37] Through her first organization, Allanstand Cottage Industries, and later the Southern Highland Craft Guild, she applied the settlement house paradigm to a rural, rather than urban, setting, with craft as its driving force. The

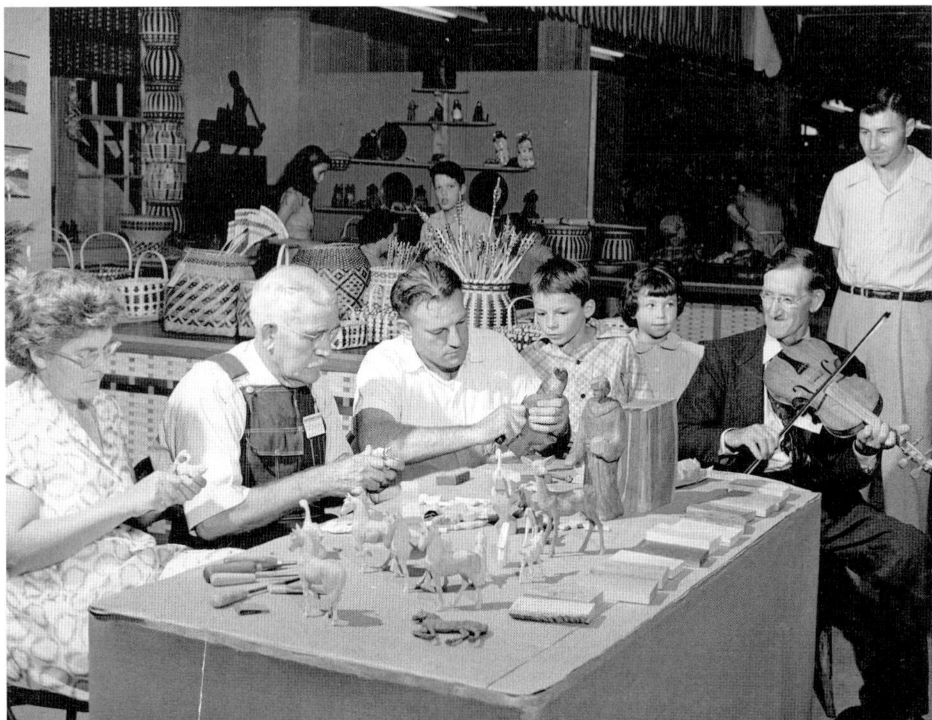

Figure 3.6 *Brasstown Carvers, Craftsman's Fair,* 1953. Image used with permission of the Hunter Library Digital Collections at Western Carolina University and the Southern Highland Craft Guild.

Southern Highland Craft Guild had a strong therapeutic dimension: the lofty goal of healing an economically depressed region by teaching handicraft skills to the rural poor to foster self-sufficiency.

The "Southern Highlands" refers to the southern sweep of the Appalachian Basin, a large forested terrain that encompasses the Appalachian Plateau, Valley, and Ridge, the Blue Ridge Mountains, and the Piedmont and covers portions of nine states: Alabama, Georgia, Kentucky, Maryland, North Carolina, South Carolina, Tennessee, Virginia, and West Virginia. Beyond mere location, it was also a promotional term used to brand the region known as Appalachia as a means of disseminating its homespun, folksy country charm through the production of saleable crafted goods—such as quilts, canned goods, honey, corn husk dolls, rocking chairs, and whittled objects—that were produced locally but destined for a national marketplace. Between World Wars I and II, craft became an important interwar industry that helped to revitalize the Southern Highlands through the production of goods that brought tourism and services to a region whose citizenry was economically and socially disadvantaged but culturally rich, full of the "old

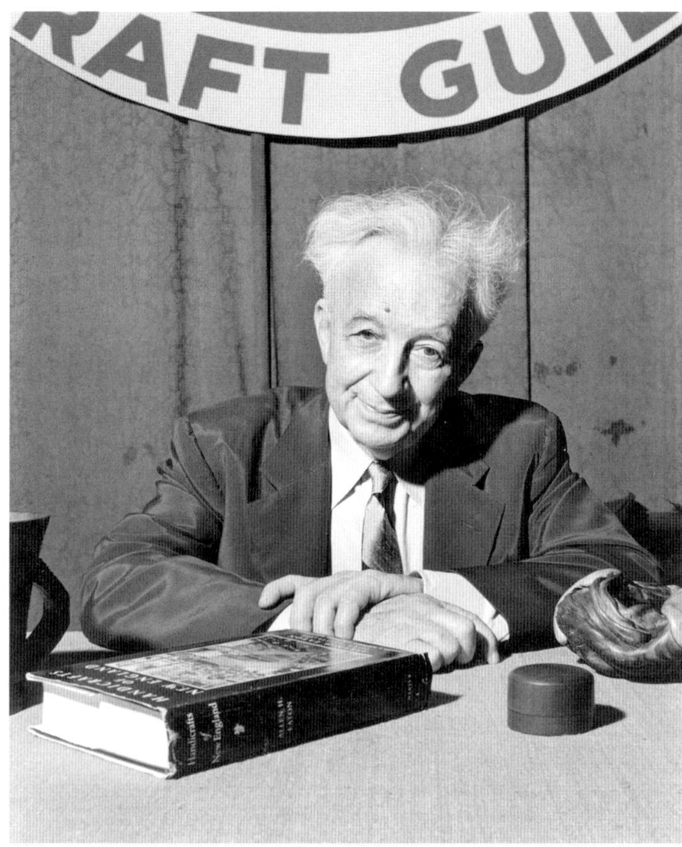

Figure 3.7 Allen Eaton at Southern Highland Handicraft Guild, 1955. Image used with permission of the Hunter Library Digital Collections at Western Carolina University and the Southern Highland Craft Guild.

ways," handicraft traditions threatened by early-twentieth-century industrial production and mechanized labor.

The Southern Craft Revival's greatest chronicler was Allen Eaton, who wrote a series of books that endorsed and advanced the handicraft cause (figure 3.7). In 1937, he published his first book, *Handicrafts of the Southern Highlands: With an Account of the Rural Handicraft Movement in the United States and Suggestions for the Wider Use of Handicrafts in Adult Education and in Recreation.* Eaton was a New Deal liberal, linking craft to social reform, cultural pride, improved labor relations, and therapeutic uses. The book was the first text to document the region's craft schools, guilds, and individual and family endeavors. Its extensive bibliography is littered with government-issued reports on social conditions to which he alone had privileged access as the director of the Department of Surveys at the Russell Sage Foundation. While the foundation was based in New York, it had a history in the region. Nine years prior, it had also published John C. Campbell's *The Southern Highlander and His Homeland* (1921). Campbell was the husband of Olive Dame Campbell and the namesake of her aforementioned

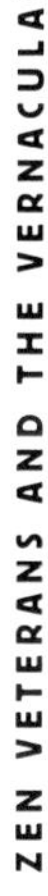

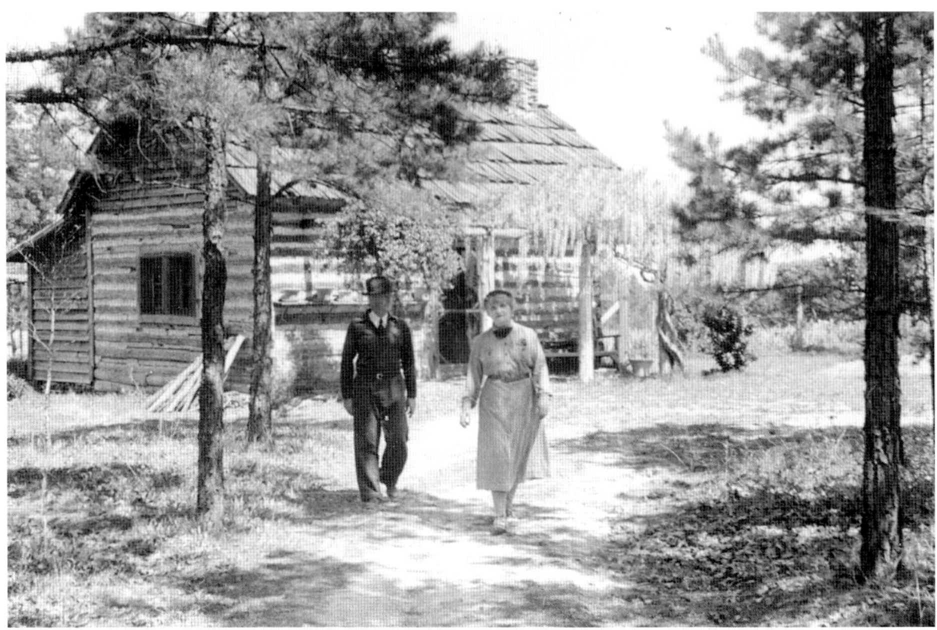

Figure 3.8 Jugtown Pottery (Moore County, NC), April 1938. Taken by Sharpe; identified as Jacques and Juliana Busbee, the founders of Jugtown Pottery, Conservation and Development. Courtesy of the State Archives of North Carolina.

school. Despite Eaton's rather bureaucratic approach, *Handicrafts of the Southern Highlands* also functioned as a detailed guide to the region's material culture, portraying craftsmen in their native environment. Through his extensive rural travels, Eaton enthusiastically profiled individual quilters, furniture makers, basket weavers, whittlers and carvers, makers of dolls, toys, and musical instruments, potters, and miscellaneous craftspeople who worked with leather, candlemaking, and the like. But the rural poor that Eaton was profiling—also as a means of social uplift—turned out to have a more complicated legacy.

To have been written up in Eaton's text was to achieve a certain modicum of fame, as it legitimated local people to themselves and to the wider world, promising future visits and sales by Northern tourists. As the proprietors of Jugtown Pottery, Jacques and Juliana Busbee (figure 3.8) were already quite famous regionally and among the potteries Eaton described in detail:

> Perhaps the product generally known to the public from the Piedmont section of North Carolina is the Jugtown Pottery, which has been revived and improved under the direction of Mr. and Mrs. Jacques Busbee, near Steeds, Montgomery County, North Carolina. These quaint and attractive articles have been described many times in newspapers and magazines and examples are to be found widely scattered throughout the country.

All of the pieces are turned on a primitive kick-wheel by workmen of the neighborhood and usually are in Carolina patterns and glazes. A few forms of exceptional interest to the student of ceramics and to artist and collector, inspired by old Chinese influences, have been developed by the Busbees; but it is the native pottery which is best known, and it is this traditional product which they have conserved, improved, and which they have made available through innumerable outlets the country over.[38]

While "traditional products" were most certainly the Busbees' economic mainstay, Eaton mentions, but quickly dismisses, their active pursuit of new, presumably more modern, forms, derived from "old Chinese influences." What this tells us is that the Busbees were less vernacular than Eaton would have his readers believe.

In fact, the Busbees were urban sophisticates and entrepreneurial middlemen, Raleigh-born, New York-based entrepreneurs who purchased a rural pottery, interested in reproducing the local salt-glazed stoneware they found so alluring. Jacques originally journeyed to Steeds, North Carolina, to seek out potters who would be interested in making such wares for sale in Juliana's Greenwich Village teahouse. For this purpose, Jacques Busbee opened Jugtown Pottery circa 1921. When Juliana's teashop closed around 1926, she joined her husband in Moore County, where they lived for the remainder of their lives. Neither Busbee actually threw pots on the wheel, instead employing local production potters. Busbee added additional so-called Oriental stoneware forms at the suggestion of Tiffany Studios in New York, with whom he had consulted. These wares were glazed with salt, or simple white, black, and green glazes, as well as one that Jacques named Chinese Blue.

Jugtown outlived the Busbees, becoming a mainstay pottery and tourist draw in the region known as Seagrove. Nancy Sweezy, former director of Jugtown from 1968 to 1980, writes, "Oriental shapes, produced with local materials and traditional turning and firing techniques, were assimilated into the local idiom and soon 'belonged' to Jugtown."[39] Its Eastern proclivities more than account for Juliana's query into Black Mountain's Pottery Seminar. On letterhead from the North Carolina State Art Society, in the capacity of vice-president, Juliana wrote, on September 5, 1952:

Gentlemen,

Your Pottery Seminar folder has just reached me, and naturally, am much excited and interested. Does it mean, actually, that Mr. Bernard Leach is to be there? And does it mean that one, this one, may come up for a day? We have been so thrilled with Mr. Leach's book on Pottery, as ecstatic about his pottery, that I would have to miss seeing him if he is to be with us. To say nothing of the others. The work here is the effort to

present the craft in the state — to hold to the N.C. tradition — when we stray from the local shapes, we go to the early Chinese for forms — the Han, T'ang and Sung dynasties. The settlers in this community came here around 1740 — and settled. In 1917, Jacques Busbee began this Jugtown interest, hoping to begin a new industry to our state. I wish that sometime you would come to see us in the back country of Moore County.
Very sincerely,
Juliana Busbee (Mrs. Jacques) Steeds, N.C.[40]

Based on her letter, it seems highly likely that Juliana Busbee attended a portion of the Pottery Seminar. But it is Louise Pitman's impeccable manners that confirm this. In her thank-you note of October 21, 1952, the then-president of the Southern Highland Craft Guild wrote "Those of us who attended yesterday not only enjoyed the demonstration but got a great deal from it and talking with the potters afterwards."[41] Like Juliana Busbee, Pitman was a New York transplant who arrived in Asheville in 1928 as a volunteer for the American Friends Service Committee, a Quaker social justice organization.

Pitman became active in the John C. Campbell Folk School, where Eaton met her during the late 1930s. He mentions her by name in his book:

> One of the most recent and active experimenters with natural dyes is the John C. Campbell Folk School, where much of the wool used in the weaving is colored by the old process. The dyeing at the school has been developed under the direction of Louise Pitman who has applied her dyes in three ways: to the raw wool, to the yarn before it is woven, and to the finished textile.[42]

Besides Eaton's tribute, Pitman was also photographed by Doris Ulmann (figure 3.9), a New York-based photographer working in the pictorialist tradition of photography popularized by Clarence White, creating painterly, softly lit portraiture.[43] Here, Louise Pitman is transformed into a local Appalachian woman, a native dyer in angelic white, standing over a hot outdoor kettle, humbly stirring her livelihood. But Ulmann and Pitman had more in common than the photograph suggests. Both were Columbia University graduates. Ulmann's photograph, then, furthered Louise Pitman's rural persona as she acquired regional acceptance. Like Jane Hartsook of Greenwich House Pottery, Pitman's legacy is the institutionalization of the organization itself: the Southern Highland Craft Guild is, at present, a highly influential regional membership-based association, dedicated to preserving the heritage of traditional craft throughout Appalachia. Its present-day Asheville campus supports three exhibition spaces, a craft shop, a library, and daily live craft demonstrations by local artisans.[44]

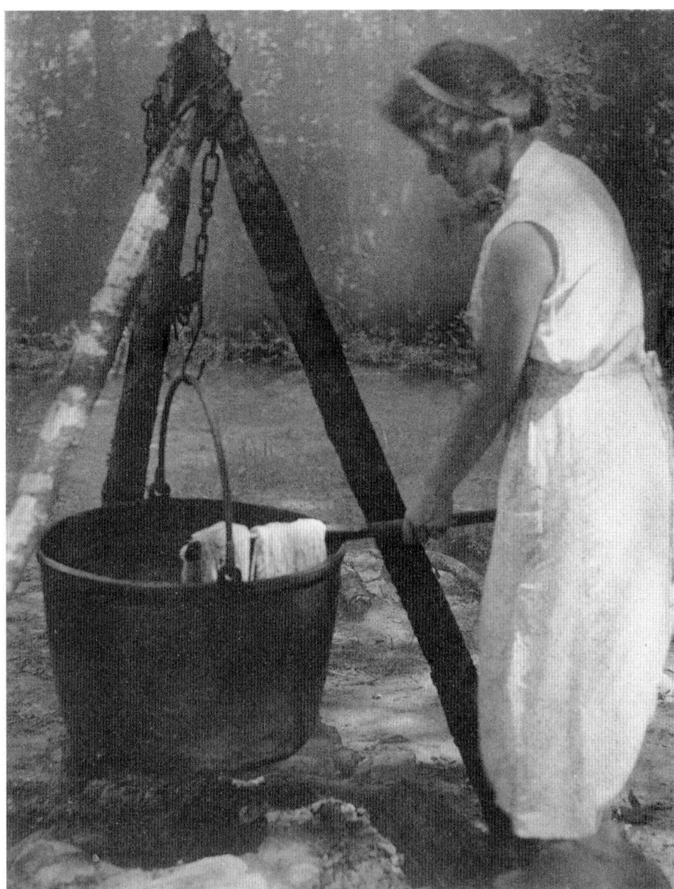

Figure 3.9 *Louise Pitman at the Dye Pot*, ca. 1933. Doris Ulmann (American, 1882–1934). Doris Ulman Photograph Collection. Image used with permission of the Hunter Library Digital Collections at Western Carolina University and the John C. Campbell Folk School.

The mostly Northern women fueling the Craft Revival in the American South were closer in spirit to Yamamoto than Yanagi, intent on offering manual craft skills as a recognizable path to a dependable wage and regional pride. Louise Pitman's performance of kettle dyeing was a performance not unlike Shoji Hamada's own dress and demeanor; both had rejected their elite educations and urbanity in favor of a rural lifestyle, with craft at the center. Their differences, however, were highly gendered: Hamada presided over a family-based production pottery that transformed him into a potter of international standing, with an active exhibition schedule, while Pitman assumed the persona of the craftswoman as a means of extending her reach and influence as a community arts organizer. Occurring contemporaneously, with distinct nationalist agendas, it can be said that Japanese peasant art and the Southern Craft Revival were similarly engaged with reviving traditional, hand-based craft processes, leading to cultural appreciation and economic renewal. Their combined emphasis on personal empowerment was

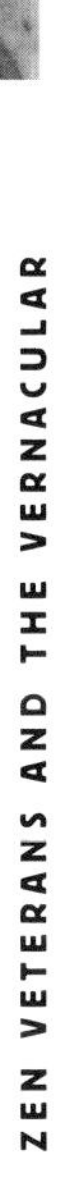

at distinct odds with the grand intellectual gestures associated with *mingei*: the no-self and the valorization of an idealized anonymous craftsman without any acknowledgement of his material needs in the modern world.

Beyond the Pottery Seminar correspondence, the other clear overlap between the insular Black Mountain community and the local Western North Carolina craftsmen was Edward L. DuPuy, the college's general handyman and janitor, who was actually a master cabinetmaker, woodworker, and amateur photographer. DuPuy is best known as Josef Albers's framer, constructing custom frames for Albers's paintings. However, he was also a member of the Southern Highland Craft Guild for over thirty years and a lifelong resident of the town of Black Mountain.

In 1967, DuPuy joined forces with a local journalist and published *Artisans of the Appalachians*, a celebratory book that combined brief oral histories with DuPuy's photographs of over sixty individual craftsmen, including Louise Pitman (figure 3.10). In the book, DuPuy commemorates her in a section distinct from the other craftsmen (organized by medium), designating her and six other women as "Teacher-Craftsman." DuPuy describes Pitman as a "Girl Friday" at the John C. Campbell Folk School, responsible for the business operations and sales at the school store. He reports that she became the director of the Southern Highland Handicraft Guild in 1951, a position she held for ten years. Though DuPuy credits her as "responsible for pulling the far-flung Guild together during those ten years," in the book, Pitman backs away from her accomplishments, saying only that she "made a lot of good friends."[45]

Compared with Eaton or Ulmann, DuPuy's characterization is the most profound: a moving tribute that acknowledges the debt that his own generation of native craftsmen felt toward the women who had established the region's craft schools, guilds, and training programs. Educated women, Yankees, or both, Pitman and her peers were skilled administrators whose roles in the Southern Highlands community called for them to adopt a traditional craft skill in order to gain local trust and recognition. And so they did, in the process becoming genteel Southern women, who kept their sophisticated backgrounds under wraps, even—or especially—when outsiders came calling.

Black Mountain College's longtime historian Mary Emma Harris was awarded seed money to begin documenting the oral history of the college in 1971. A North Carolina native, she was the first to ask DuPuy if there was a connection between the crafts taught at Black Mountain and the crafts in the immediate area.[46] DuPuy's reply:

> The pioneers didn't think of themselves as a hobbyist when they needed a bed. They made it because it was the only way of getting a bed, you see. And the same thing applied to other things that they needs for daily living—their pots, and their coverlets,

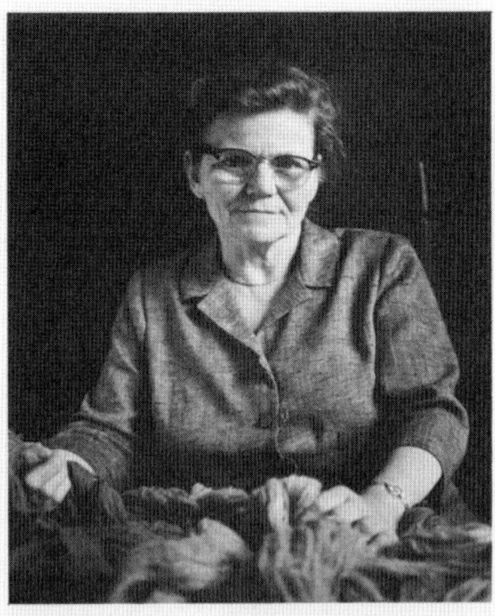

LOUISE PITMAN

Teacher-Craftsman

Asheville, North Carolina

Giving a year of your life to your country is not such a new idea. Louise says the American Friends Service Committee asked young people fresh out of college to do it back in 1928, and she was intrigued: "They offered things all around the world; all sorts of foreign assignments. Way down at the bottom of their list was a little school in southwestern North Carolina which was just getting started. Well, I met Olive Campbell and Marguerite Butler (now Mrs. George Bidstrup), who were organizing it, and I was impressed. And I took it. And first thing when I got to Brasstown, Marguerite said, 'We've got to pack up these brooms.' There were twelve of them, all different sizes and shapes, and if there was one thing I hated to do it was to do up packages and if I had known I would have to do up packages for the next fifty years—first at the School, and later with the Guild—I'd have gone back to New York!"

Soon after that, "The Guild got started," says Louise, "and the one great disappointment of my life was missing their first organzational meeting at Penland. It was Christmas; Marguerite and Olive Campbell went, but everyone else was down with the flu except me and I had to stay behind and see to things."

Louise's own craft was to be—somewhat to her surprise— vegetable dyeing: "The School had invited Wilmer Viner to come and teach it, and they said I was the one who would have to learn. And this is kind of interesting. Several years later the U.S. Department of Agriculture wrote the School that they were making a study of vegetable dyeing and didn't know where to turn for information. So I typed out the notes I had taken in the Folk School kitchen from Wilmer, and added what little I had learned, and they used it as a basis for their study . . . so Wilmer really started that."

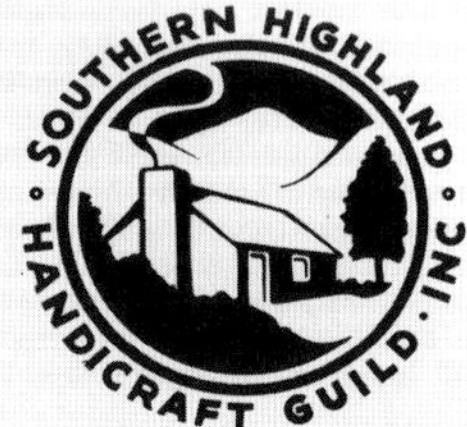

Louise was the Girl Friday at the John C. Campbell Folk School for seventeen years, as business manager and salesman for their growing crafts. It was good training for the job of Director of the Guild which she accepted in 1951: "I still tied up a lot of packages; wrapping things for exhibits all over the country; loading the truck with stuff for our out-of-town shops, and helping set up the Craftsman's Fairs each year." She drove all over the mountains of seven states keeping in touch with members: "It was a much smaller and more personal operation then than it is now. Thank the Lord I retired in time; I couldn't handle it now."

In response to someone's comment that she had been responsible for pulling the far-flung Guild together during those ten years, Louise says, only, "Well, I made a lot of good friends." Which is, perhaps, just what an impulse to give a year to your country will lead to, if not checked in time.

119

Figure 3.10 Louise Pitman, page from *Artisans of the Appalachians*, 1967. Photograph by Edward L. DuPuy; text by Emma Weaver.

and their brooms and whatnot. It was either make it or go without. . . . Of course today there are an entirely different set of circumstances. They [artists] turn them [crafts] into dollars simply because they are in demand as items of usefulness, beauty and items of appreciation. They [Black Mountain College] really didn't teach it as crafts. To some extent there was, undoubtedly. Now, I think they built one or two looms, two or three

> looms, and undoubtedly they used the design, the general design . . . and that had some bearing on that thing."[47]

This is perhaps the most damning response: that craft at Black Mountain was not really taught as craft, that it was something unrecognizable to DuPuy, a vernacular craftsmen, and, moreover, that this something "else"—design, or modern art—in turn did not recognize him either. Josef Albers had designed furniture at the Dessau Bauhaus in the late 1920s, working with bent plywood.[48] As well, the Alberses had a passion for indigenous objects, so theoretically their interest level in local craft should have been high. But designing furniture is not the same as building it. Molly Gregory, the college's resident carpenter, also built furniture that Albers had designed. Other than commissioning DuPuy's handmade frames, there is no record that the Alberses engaged with any other member of the Southern Highland Craft Guild, nor any record that either he or Anni knew of the guild's existence. Such a missed opportunity underscores the separate spheres of artistic production that were in effect during Josef Albers's leadership and which the Pottery Seminar sought to bridge or, more cynically, profit from.

The Pottery Seminar's Student-Veterans

Richards continued to promote the Pottery Seminar in New York. She secured its only press coverage and had planned to attend but, in the end, fell ill.[49] Through her efforts, *Ceramic Industry* magazine, of Chicago, agreed to send a freelancer, a former Navy journalist and photographer named Robert Diffendal, who was also an amateur ceramist, having studied at the School for American Craftsmen after his service. The magazine had no real stake in the themes of the seminar but was interested in using the material to sell more magazines, looking to acquire crowd photos that would enable people to identify themselves.[50] Unfortunately, due to extreme drunkenness, Diffendal was asked to leave the campus. Hazel Larsen Archer recounted this story to the magazine's editor, adding that

> We agreed that if he could go away for several days and return, completely sober, and remain so, that he could stay to do the story. He kept this agreement, returning, and spent two days gathering material for the story. I suspect, after some conversations here, that the drunkenness comes as an aftermath of war experiences. I believe Mr. Diffendal is trying hard to adjust himself. He is a likeable polite man when sober and I urge you, if he does submit material on the Seminar, to use it if you can.[51]

Diffendal's forgiven transgression on the route back to civilian life exemplifies the hope invested in the range of academic programs established for veterans.

Throughout the United States, the GI Bill was the impetus for university-based initiatives in manual arts and design, expanding college art departments at a rapid clip. Adding craft to university curricula helped to erode long-standing class barriers in postsecondary education. The highly skilled craft disciplines of weaving, wood, ceramics, and metals (American studio glass, which dates to 1962, did not yet exist) were routinely considered to be forms of vocational training rather than art education. Intent on fostering practical careers in industrial design and manufacturing, these programs were, by and large, aimed at student-veterans.

Founded by an educator rather than a patron, Black Mountain College was a fledgling institution without an endowment. Consequently, the World War II Serviceman's Readjustment Act of 1944, better known as the GI Bill, profoundly improved its shaky finances. Indeed, some of its most famous male students were veterans, including Kenneth Noland and Robert Rauschenberg. Black Mountain eagerly accepted the government subsidies attached to student-veterans, and it was only due to GI Bill tuition reimbursements that the college was able to pay its faculty a regular salary during the postwar period.[52]

Two months after the bill's passage, Black Mountain's *College Bulletin* opened with the declaration that "Black Mountain College has extended its curriculum and revised its calendar to meet the demands of the war."[53] This was an explicit attempt to lure recent GIs through wartime scholarships, an accelerated curriculum allowing students to finish in three, instead of four, years and expanded offerings in manual arts courses, which included woodworking, photography, and non–liberal arts offerings like first aid and surveying. Additionally, a quota of students would receive deferments through the Enlisted Reserve Corps while they completed college, to insure a future supply of officer candidates.[54]

Though women also participated in the war effort, men had many more opportunities through the GI Bill to obtain a secondary degree. This social disadvantage became less apparent after 1957, when the GI Bill subsidy was vastly curtailed. A cross section of the Pottery Seminar's participants demonstrates the gender divide. Karl Martz (1912–1997) and William Watson (b. 1914–?) were both professional potters who had already obtained MFA degrees before coming to Black Mountain. Having run his own pottery in Indiana, Martz founded the ceramic arts program at Indiana University in 1945, where he taught until his retirement in 1971 (figure 3.11).[55] With a large gap for war service, William Watson's résumé was more typical of the era. Thanks to the GI Bill, Watson completed graduate studies at Ohio State in 1946 and by the age of thirty-eight was already an associate professor of ceramics at Florida State College (now University) in Tallahassee. He had also spent a summer studying with Marguerite Wildenhain in Northern California. In 1954, two years after the seminar, he served as one of the regional judges for the Ceramic National competitions.

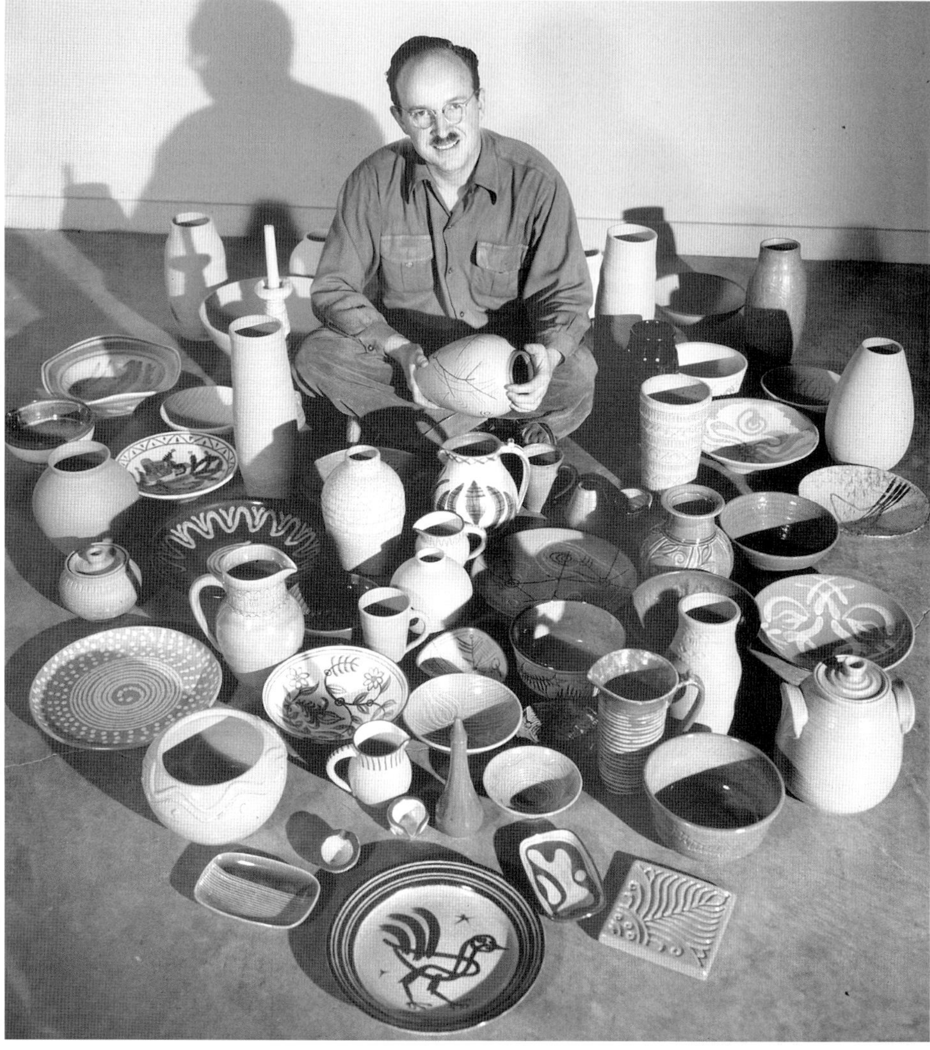

Figure 3.11 Karl Martz with his wares, ca. 1951. Hand-colored photograph. Courtesy of Eric Martz.

With a title and a more-or-less secure university position, Watson and his male colleagues were in far better positions professionally than the women of their generation, such as Lillian Boschen (b. 1919–?) and Janet Darnell (1918–1997), two highly skilled potters also in attendance, who, like Jane Hartsook, had to work harder to secure lesser opportunities. Like Watson, Boschen was also a war veteran, one of a small number of women who served in the US Navy, where she met fellow potter and teacher Frances Senska (1914–2009). Senska is best known for introducing Rudy Autio and Peter Voulkos—both student-veterans—to clay as undergraduates at Montana State in Bozeman. Senska was a functional potter whose own work was domestically scaled, such as in the lidded bowl (1988) of figure 3.14. She spent many summers studying at Pond Farm. While not widely known outside the West, she remains a widely admired and important figurehead regionally who succeeded not only in establishing a well-regarded art department at a rugged agricultural school but also in introducing a modernist approach to ceramics through her teaching and artistic practice.

With her G.I. Bill funds, Boschen had studied for an MFA alongside Voulkos at the California College of Arts and Crafts in Oakland, with Anthony Prieto (figure 3.12). She had also attended F. Carlton Ball's summer sessions at Mills College, where she had encountered Leach in 1950, writing, "When Mr. Leach conducted his workshop at Mills College in California, I was fortunate in attending and would most certainly look forward to being in his presence again."[56] Thanks to Senska's influence with Montana businessman Archie Bray, Boschen was appointed the first official resident director of the eponymous Archie Bray Foundation in Helena, Montana, a position she held for less than a year, from October 1951 until the spring of 1952. A pragmatist, Boschen made a prototype stoneware mug, introduced as a potential production line to generate income for the pottery (figure 3.13). Though Boschen completed her studies earlier than Voulkos, she was recommended only to fill the job until he finished graduate school. Voulkos and Autio had been the first potters to help Bray build the pottery during the summer of 1951. Rick Newby and Chere Jiusto write that Bray himself came to prefer Boschen for "Her attention to detail which Peter did not display, her willingness to do little things which Peter would look on as a nuisance and her willingness to follow my directions which I think Pete would not."[57] Regardless, Bray elected to retain Voulkos rather than Boschen, on the basis of his emerging national reputation and his charisma. This sort of casual sexism ultimately decimated Boschen's career, while Voulkos moved on rather quickly: he spent the summer of 1953 at Black Mountain and by 1954 had accepted a full-time teaching position at Otis College of Art in Los Angeles. Late in 1952, Boschen set up a small pottery of her own in Virginia City, Montana. No further exhibition record for her exists outside of Montana, under either Boschen or her married name, Tidball.[58]

Like Boschen, Janet Darnell also served in the war effort, but in a civilian capacity, as a welder in a Staten Island shipyard, a position that did not entitle her to GI Bill monies. Intent on a career as a sculptor, she worked on her own, living in New York and showing periodically in group exhibitions at 57th Street Galleries, but she sold nothing. She decided to try clay at a community pottery, the Inwood Pottery Studios, on 168th Street, and upon mastering basic skills, Darnell left New York City for the position of teaching pottery in the occupational therapy

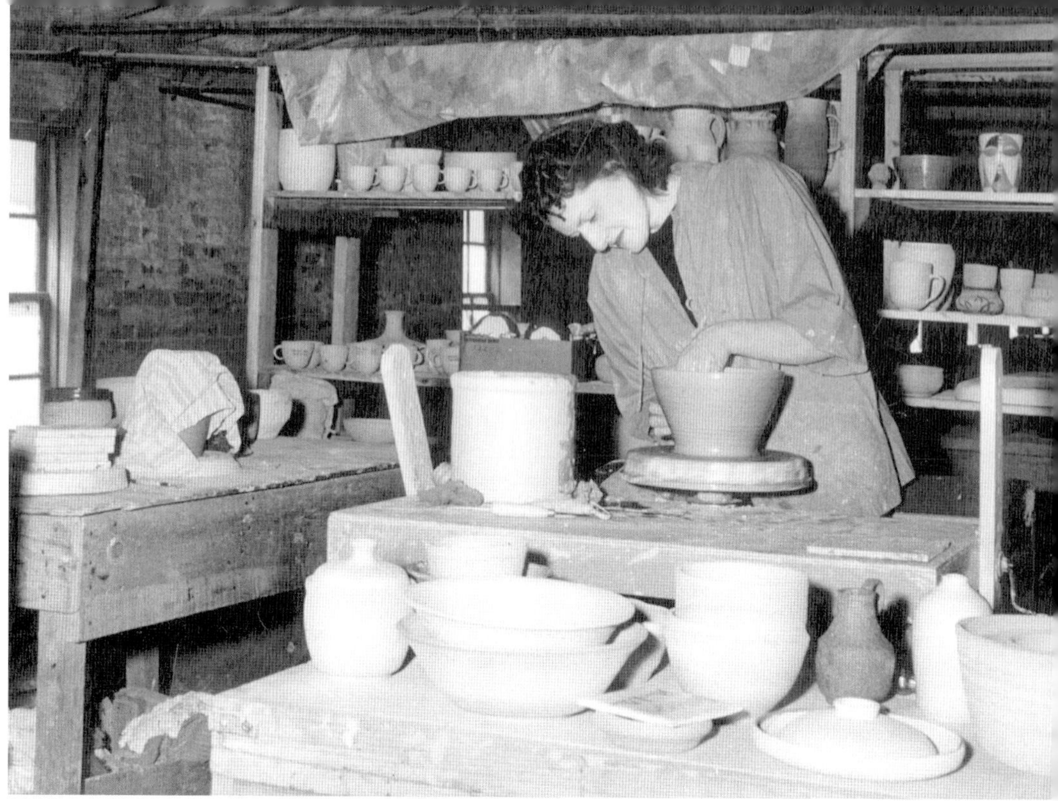

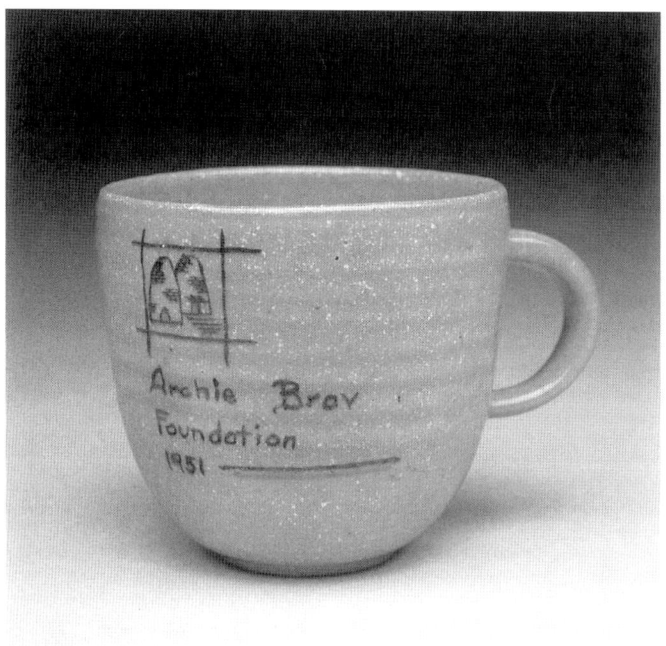

Figure 3.12 Lillian Boschen at Archie Bray Pottery, 1951. Collection of the Archie Bray Foundation Archives, Helena, Montana.

Figure 3.13 Archie Bray Foundation production ware cup, 1951. Glazed Stoneware, 3 x 4 x 3 inches. Collection of the Archie Bray Foundation, Helena, Montana.

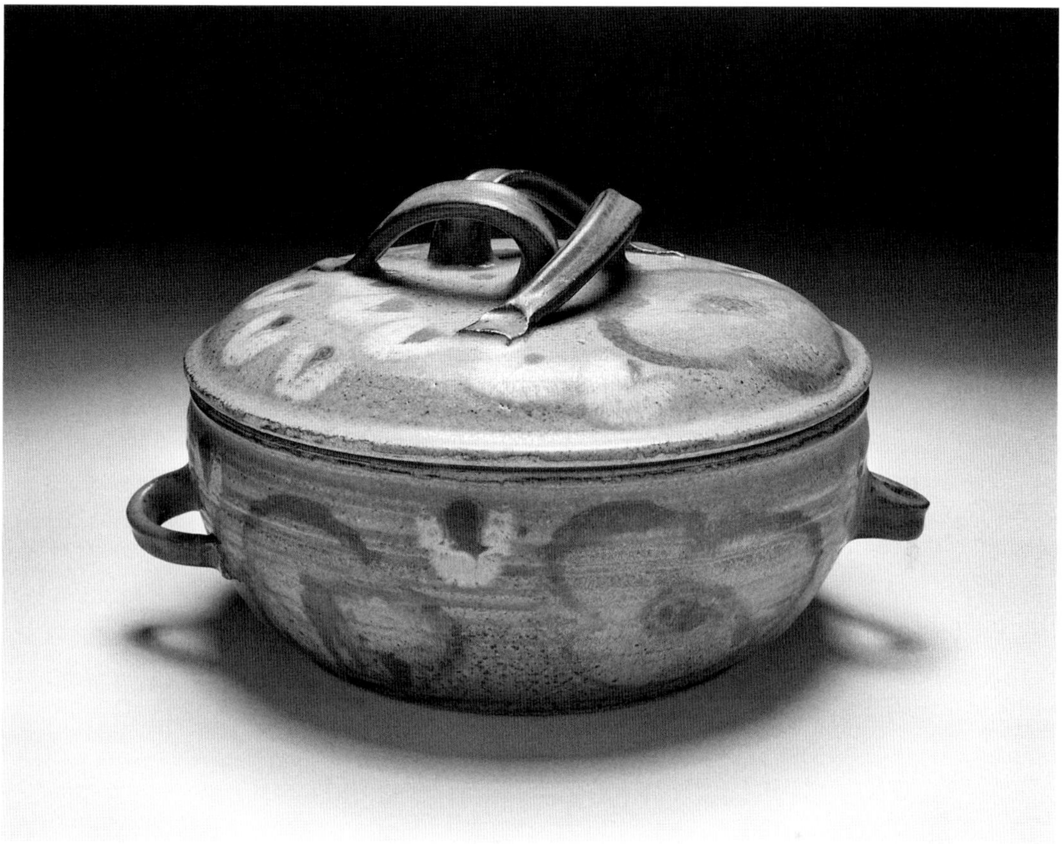

Figure 3.14 *Bowl*, 1988. Frances Senska (American, 1914–2009). Stoneware, 6 x 26.4 inches. Courtesy of the Nora Eccles Harrison Museum, Utah State University, Logan, Utah, Nora Eccles Harrison Ceramics Gift. Photo by Andrew McAllister.

department at Rockland State Hospital, a psychiatric facility in Rockland County, New York, about an hour outside Manhattan. In keeping with the era, Darnell was treated professionally as a therapist and caregiver, rather than as an artist. Of all the potters who attended at the Black Mountain Seminar, Darnell had the most direct contact with craft as a method of rehabilitation.

After her institutional experience, Darnell became the resident potter at Threefold Farm, in Spring Valley, New York, a self-sustaining anthroposophical community based on the teachings of Rudolf Steiner, with its own school and farm. Established in 1922, Threefold Farm was an Arcadian experiment, the first in America to utilize consciously organic farming methods such as composting.[59] From 1949 to 1954, Darnell built her own home and established a pottery, teaching classes to both adults and children. Darnell eventually became Bernard Leach's third wife, marrying him in 1956. But in her application to Black Mountain, she described herself as "the only potter in the area," writing:

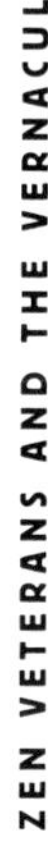

> It is my personal desire to attempt to bring a higher level of teaching and creative work into the field of adult education and accordingly, Mr. Leach's principles and Mr. Hamada's work are constantly in evidence in these classes. Because of living in an isolated area and being tied down with class schedules, I miss the normal exchange between working potters to a great extent and therefore look forward to such a seminar as this.[60]

A.J. Spencer, one of the few beginners who attended the Seminar, also alluded to the isolation he felt in his home state of Florida, writing, ". . . the opportunity for instruction from real craftsmen in this state is almost non-existent."[61] Spencer went so far as to differentiate between amateurs like himself and the "real craftsmen" he thought he might find at Black Mountain. Similarly, Darnell makes the case for the vitality of networks and an active discourse, espousing the individual potter's very real need for community and intellectual exchange.

Japanese Aesthetics and Zen Veterans

The Leach Pottery was located in St. Ives, an artist colony and resort town on the Cornish coast, where Bernard Leach was part of an avant-garde circle that included Barbara Hepworth, Patrick Heron, Ben Nicholson, and Russian émigré Naum Gabo. Leach favored high-fired stoneware for its associations with the wares of the Sung dynasty, as well as its expressive properties. Working in a limited color palette of yellow, black, brown, white, and blue, Leach's work is a fusion of Eastern and Western vernacular styles, using Oriental-inspired materials and brushwork to heighten the forms of vernacular English pottery, particularly rural, artisan-produced pitchers and jugs. The results were eclectic: solid, minimal forms decorated with simple calligraphic paintings of animals found in the English countryside such as roosters, goats, and owls. Leach himself preferred decoration to throwing, writing, "I am not, in my own opinion, a first-class thrower, and I have never achieved a complete mastery of throwing."[62] Yet Hamada praised Leach's aesthetic sensibility, writing, "His stance between East and West is a true balance, not a measured middle."[63]

Upon arriving at the Pottery Seminar, Leach refused to demonstrate his throwing technique, suspicious of the "unknown" and potentially poor quality of the local clays. While it has been suggested that Leach did not have the vast technical repertoire of either Hamada or Wildenhain, his negativity must have contributed to the Americans' already deeply ingrained inferiority complex, which was reflected as a constant preoccupation with materials, as though a better clay could somehow produce better pots. Ironically, Leach's preferences were anticipated by Karnes and Weinrib, who, as Black Mountain's resident potters, were familiar with his writings and also responsible for organizing the materials and equipment for the seminar. According to Karnes, all the clay was ordered and trucked in from elsewhere, because there was no stoneware—Leach's preferred clay body—in the area.[64]

At the seminar, Hamada, renowned for his spontaneity, quickly remedied Leach's faux pas and, with minimal fuss, began throwing. Karnes succinctly (and dryly) described the difference between the two:

> Leach was talking, philosophizing and everything, and Hamada just worked. He was wonderful . . . I think the important thing for me over the years was to not be afraid of being in front of an audience, because Hamada came to my mind. I could just sit there and do what I wanted too, and they [the audience] probably would like it.[65]

In addition to Karnes, Hamada proved inspiring to an entire generation of young American potters, in particular, Rudy Autio, Janet Darnell, Susan Peterson, and Peter Voulkos, all of whom saw him throw during the 1952 tour. Darnell described her experience of the Black Mountain Seminar and Hamada's influence, writing in 1975:

> During this two-week conference at Black Mountain College, there were many "firsts." The Oriental pots were just coming out of the basements of American museums and being exhibited for the first time since the long Pacific war. A few Japanese craft exhibitions were also coming over. Many of us had been studying them and saying "If only we could get their fine quality brushes, our decoration would be better" — but when we asked Hamada about his brush that was gliding over the pots making beautiful sensitive patterns, he said it was from the scruff of a local dog in Mashiko [Japan]! We then caught a stray, clipped pieces off his scruffy neck with nail scissors (and almost got bitten), bound the hairs together with sewing thread, tucked this tuft into a piece of hollow read [sic], and had a better brush than we had ever owned . . . . It was obvious how far away we were from the roots of things in our preoccupation with tools, equipment, and technical know-how. So I went to Japan to see more of what Hamada represented.[66]

While Rudy Autio stressed Hamada's—and Yanagi's—influence, he also took note of Leach's condescension:

> In retrospect, as I looked at Yanagi, he was a very gentle man with a lot of perception and he didn't look down on us, like perhaps Bernard Leach did a little. I never felt that about Hamada. Hamada tolerated us as being young people who were promising in the arts, and he was very kind to us, whereas Leach was a little impatient with us.[67]

In his article on the seminar, Robert Diffendal, on the other hand, described Wildenhain as being extremely vocal, initiating discussions on topics such as "honesty of design" and "pleasure in technique."[68] While the final program is lost, an archival typescript scheduled Wildenhain for two separate all-morning lectures titled "Aims and Limitations" and "Craft in the Machine Age." Additionally, during the final two days of the seminar (October 27 and 28) she spent six hours each day (9 am to noon, 2 pm to 5 pm) demonstrating a gamut of basic shapes and "the development of personal and expressive elements."[69] This allowed students a thorough grounding in her technique, marked by years of Bauhaus-enforced discipline, versus Hamada's noted spontaneity.

Distinguished potter Val Cushing, who first encountered Wildenhain at Alfred in the 1950s, described her as having "a charisma about her . . . she was very authoritative, very dogmatic, very much a person who had all the answers, but backing that up was, again, this incredible skill and technique that she had on the potter's wheel. And if you sat there, as I did, and watched her throw, it was like magic."[70] Powerful enough to achieve the same master status as her male counterparts, Wildenhain's persona could be considered as awe-inspiring as Hamada's.

At their personal potteries abroad, both Leach and Hamada threw wares communally alongside a staff of potter-laborers, which included their own sons. Such expressions of collective labor grew out of Yanagi's ideas, who praised what he called the "unknown craftsman," the anonymous, communal craftspeople of the preindustrial world, writing ". . . as in religion, a real salvation is found in the field of craft—one finally finds real self-affirmation in the abandonment of self.[71]" But the "unknown craftsman" ideal also held broad appeal for student-veterans, who had already absorbed the collective ethos of American military life. So it is not inconsequential that of the twenty-five studio potters in attendance, seven of the men were also war veterans, while at least two of the women had served the war effort.[72] While the seminar itself had an avant-garde slant, the idea of veterans throwing repetitively on the potter's wheel and listening to lectures on Zen philosophy resonates with the larger therapeutic practices detailed in chapter 1. Yet this situation is underscored by an irony best understood as an outcome of American wartime policies: the student-veterans eagerly studying Japanese ceramics on the GI Bill were the same young men who had fought and occupied overseas.

A number of American craftsmen had actually made pottery pilgrimages while serving on foreign shores, using precious leave time not for drinking and carousing but for their own edification, consciously seeking out potters such as Leach and Hamada. Among the potters who described such journeys were Dean Schwarz, Marguerite Wildenhain's last studio assistant, and Otto Heino, Second

World War veteran who taught at the Chouinard Art Institute in Los Angeles from 1959 to 1968 (see chapter 1). As he recalled:

> After I made another twenty-five missions, they [the Air Force] issued me a jeep and extra gas. I visited the museums and individual potters, silversmiths and painters for thirty days. Then I went back flying again. I'd visit craft shops—I visited Leach's pottery at that time [unspecified, ca. 1942]. They told me a lot of names to visit . . . . The silver didn't impress me because you'd pound it and there wasn't any action of a challenge. You could hit it and it would stay there and wouldn't move. But when I saw Leach's pottery where people were throwing, they had to master that clay or else the clay would take over. So I decided when I got back to the United States I would take up pottery. I took up pottery with the G.I. and Vivika [Heino, his first teacher, and later, artistic and life partner] was teaching.[73]

According to scholar Donald S. Lopez, Jr., "The hallmark of Buddhist thought is the doctrine of no-self."[74] In Buddhism, the autonomous self is considered to be the root cause of all human suffering. Rather, the path to enlightenment is the achievement of "no-self" through meditation, the utter absence of ego. In this way, Zen and *mingei* became casually interchangeable philosophies in the American context: the idea of "no-self" became (wrongly) synonymous with anonymity, and in turn, wheel-thrown ceramics became a therapeutic extension of Zen. Yet both philosophies offered compelling alternatives to the existential crisis of self-expression, which consumed the United States in the years following World War II. Heino's attraction to the mastery he witnessed at Leach's pottery was a way of foregoing individual expression in favor of the collective task of conquering clay's formlessness.

Active duty was a life predicated on extreme uniformity, down to the most mundane details of daily experience such as dressing, eating, and sleeping. In everyday actions, the individual receded in favor of standardization, conforming to a routine that became nearly mechanized in practice. This was a useful quality in the practice of wheel-thrown pottery, in which the goal is to be "at one" with the clay, to the point where the hands' actions become involuntary, or intuitive, working in tandem with the conscious mind. Such mind-body practices came to be idealized in the West as "Zen," a spiritual practice assumed to be the same thing as Zen Buddhism.

Anthropologist Brian Moeran argues that in Japan, *mingei* came to be seen as highly politicized, indicative of a new wave of nationalism.[75] Design historian Yuko Kikuchi expands on this, arguing that Yanagi's *mingei* was resolutely imperialist, defining itself as not only Japanese material culture but rather as *all* the handicrafts of Japanese-controlled populations, including Korea, Taiwan, Manchuria,

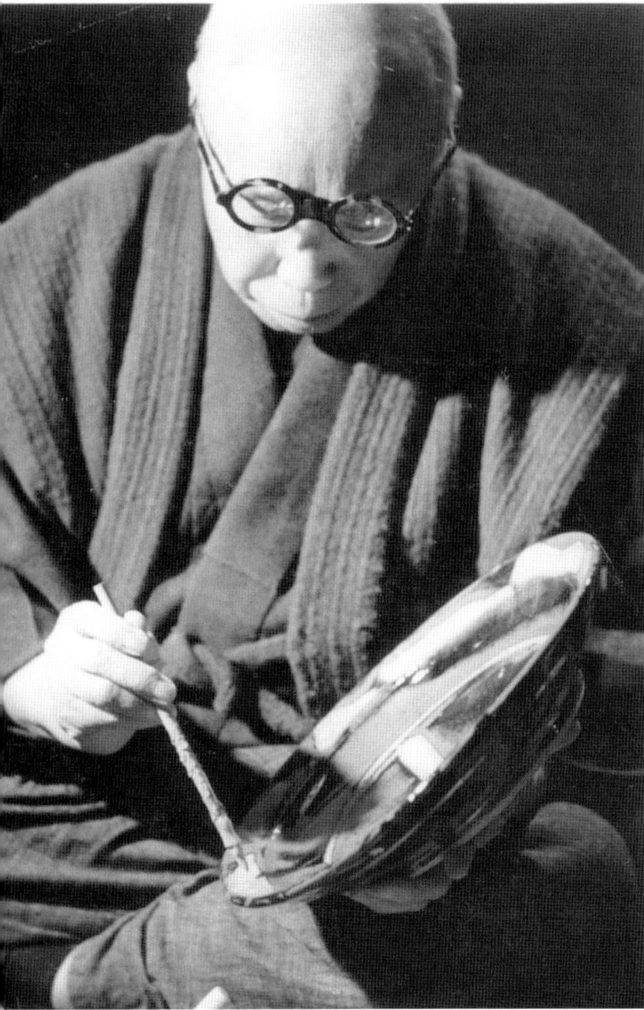
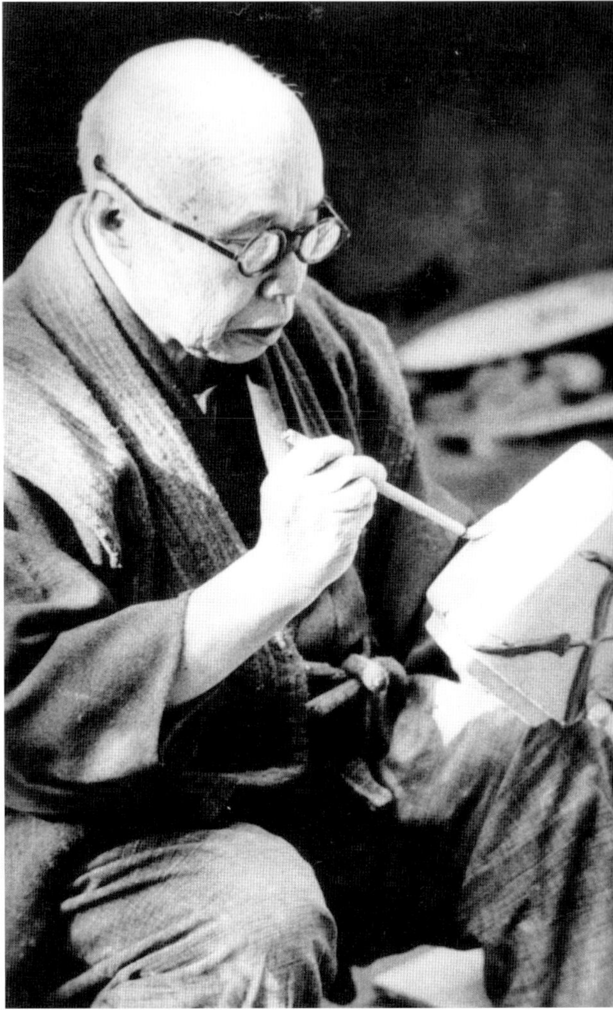

Figure 3.15 Two archival images of Shoji Hamada, ca. 1972. Unpublished photos by Susan Peterson, in preparation for *Shoji Hamada: A Potter's Way and Work* (1974). Susan Harnly Peterson Archives, Ceramics Research Center, ASU Art Museum, Arizona State University, Tempe. Courtesy of Jill Peterson Hoddick, Jan Peterson, and Taag Peterson/the Estate of Susan Peterson.

and Okinawa. Such layers of colonization are crucial for recovering a fuller picture of the era. While Japan was, for centuries, the dominant power in East Asia, in the American context of the Black Mountain Pottery Seminar, Japan was perceived as weak, a former enemy that had been dramatically humbled. In turn, humility became an important theme of the seminar itself, embodied by Hamada's silent presence (figure 3.15). In her 1974 monograph, Susan Peterson characterized Hamada's appearance as "similar to the look of rural Japan, soft and mellow . . . He wears a treasured old kimono vest, woven years ago by a friend, which tops the traditional country style trousers, which are cut fully, tied at the waist, and tight at the ankles."[76] An academically trained ceramic engineer well versed in glaze chemistry and kiln technologies, Hamada nonetheless performed the role of the potter-laborer. His everyday attire was a conscious display of agrarian values: modesty but also, more importantly, a distinct disidentification with Japan as a sophisticated, modern nation or an instigator of aggressive warfare.

Through Hamada's large technical repertoire, the artist revived ancient forms and employed a variety of complex glaze techniques, such as using wax resist before dipping in limestone or iron glazes. Adhering to the principle of the "unknown craftsman," Hamada did not sign his pots, but neither are his works anonymous. Rather, they retain a signature look, decorated with the artist's spare calligraphic brushwork, often in the form of a recurring motif, a broken bamboo stem rippled by wind (figure 3.16).[77] His forms embody an earthy dynamism best achieved through risk: the split-second decisions made on the wheel. Chance was another strong component of the *mingei* sensibility, though it was never called that. After her visit to Japan in 1954, Darnell elaborated on Hamada's "spontaneous vitality":

> He is striving for the spirit of the form in clay and his working method is always as we observed on his tour in America: the pot comes up and at the first spontaneous burst of life he stops working it. It may not be quite smooth, even or centered but these factors become secondary and he does not sacrifice spontaneous vitality of form to a mechanical slickness and perfection.[78]

Through her careful description, Darnell theorizes Hamada's crucial investment in chance. For Hamada, chance is both an endpoint to the process of making and a refutation of perfection: once the form stops being *live*, the work is considered finished, regardless of any asymmetry or flaw. Yanagi also described the necessity of imperfection, or, as he wrote, "The love of the irregular is a sign of the basic quest for freedom."[79]

Such conceptions of clay's vitality and the search for personal liberation and spiritual fulfillment through functional pottery are iterations of the ideas once espoused through the World War II-era therapeutic discourses of craft. Yet at Black Mountain, they were reframed as Zen. The process of throwing pottery as a gesture of healing, soothing the trauma borne of warfare and violence, became subsumed by a discourse of Eastern spiritual practices and the "no-self," sacrificing the ego by engaging the philosophical precepts of Zen. In his writings and his Black Mountain lectures, Yanagi privileged chance and the beauty of imperfection as aesthetic ideals meant to restore spiritual unity and cleanse the mind of its need to control consciousness. According to Darnell, his lectures were so well received that students requested copies.[80]

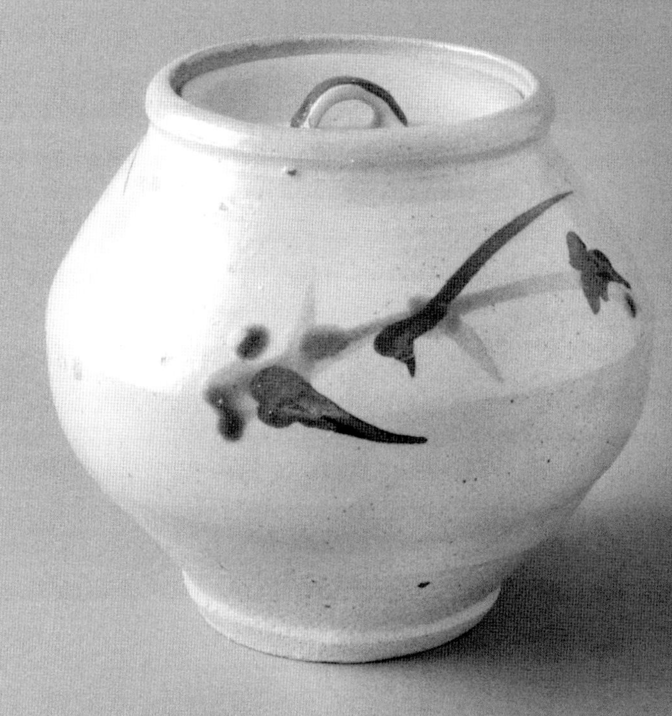

Figure 3.16 *Covered Jar,* 1945. Shoji Hamada (Japanese, 1894–1978). Image courtesy of the Pucker Gallery.

At Black Mountain, working on the floor with his hand wheel placed before him, Hamada became the living example of Yanagi's principles. Using the primitive posture of the anonymous, preindustrial artisan, he sat on the floor cross legged, a position "impossible for most Westerners," in the words of Peterson's racially tinged declaration.[81] With only his arms and hands in constant motion, he held his body almost completely still and worked noiselessly. As Peterson described it: "As he pulls up and opens the clay, shapes seem to arrive effortlessly. His hands, right one inside, left one outside, are almost parallel with his own middle. The clay cylinder goes up fast, then in, and the profile is drawn out into fullness."[82] But while Hamada would perform in exquisite silence, Leach stood alongside, a mediator providing commentary, adding a layer of Western-style interpretation. Though Hamada spoke (and wrote) English fluently, he excelled at the role of the stoic Zen master, allowing Leach to "translate" on his behalf.[83] Bald and monk-like in his intensity, Hamada must have seemed like a present-day Buddha. Certainly, his simplicity was electrifying. Through the live pottery demonstration, Hamada restored the fleeting, experiential quality to a process—throwing at the wheel—that had the limitless potential to be authentically new each time. Such presentness had a shaman-like effect on the students, introducing a style of performance previously unseen in American ceramics. Black Mountain's student-veterans became, in effect, Zen veterans, and the therapeutic and social welfare dimensions of craft, forever associated with women, were readily cast off in favor of a compelling (male) philosophy that had real social import and cultural cache; on a campus celebrated for its avant-gardism, Zen helped to elevated ceramics to the realm of a fine art that existed in tandem to painting, sculpture, or the Happening.

The potter Rudy Autio stressed the tremendous impact that Hamada's technique made on him:

> It was the economy of everything he did. The feel he had for clay, which can't really be described . . . the mind and idea and the material was all together . . . . That was a real revelation to me, just to see it. I think watching Hamada meant more to me than hearing Leach explain the philosophy of ceramics. . . . He didn't have to speak English to be able to tell me what he was doing, you see.[84]

Hamada's "technique," then, was strategic and entirely self-aware: a process of refusal through silence, so that, accruing mastery through silence, he embodied the omniscient Zen master feigning incomprehension, performing the Zen koan, a riddle or parable said to enhance spiritual awareness. By fooling his audience, Hamada perpetuated a blankness that, like a screen, allowed his audience to conceive and project their own interpretations.

Throughout the 1950s, America's engagement with Zen was prolonged. Zen permeated nearly every facet of avant-garde production from Beat literature, atonal music, and abstract expressionist painting to modernist pottery. At the seminar, Hamada's casual handling of clay, combined with Yanagi's lectures on the handmade—what he called "speech without words of Zen"[85]—proved to be a seductive combination and reflect the earliest instantiation of Zen philosophy being disseminated to an American craft audience. This is particularly important as John Cage has been singularly historicized as Black Mountain College's primary interlocutor on Zen aesthetics.[86]

In the 1950s, it seemed possible that Eastern thought could cure the West of its all-consuming belief in free will and individual thought—in effect, to solve the problems inherent to Western modernity. Or, put another way, Zen had the potential to short-circuit abstract expressionism's abiding self-interest in the individual interlocutor. Caroline Jones has already demonstrated Cage's centrality in dismantling what she terms the "abstract expressionist ego," identifying his role as "Zen master" within two different camps: the abstract expressionists themselves and the generation of 1960s process artists that come after, such as Robert Smithson and Robert Morris.[87]

In Cage's *Silent Piece* (1951), as in Hamada's silent pottery demonstrations, experience and expression were indeed merged: presence was embodied by the performance of a work of art *in process*, whether a composition or a vessel, the continuous making of an open form, or *live form*. The indeterminacy present during the act of throwing forms on the pottery wheel is akin to the indeterminacy of the Cagean performance and not unlike the abstract expressionist canvas, where thumb or fingerprints (Pollock especially) sometimes appear, also registering the momentum of the expressive gesture. The meditative quality of handiwork, combined with the rhythmic rotation of the potter's wheel, was a powerful repose of the essence of Zen contemplation.

Highly educated and fluent in English, Yanagi had had been the student of a trio of scholars, botanist Tanosuke Hattori and the philosophers Kitaro Nishida and Daisetz T. Suzuki, who had collectively introduced him to Western humanism in the natural sciences and literature.[88] As the foremost Buddhist scholar working transnationally, Suzuki would come to have a great deal of influence in the West, as the source of Zen in 1950s America. Like Yanagi, Cage also considered himself a student of Daisetz T. Suzuki, though he could probably be more correctly construed as a disciple. In 1950, at the age of 80, Suzuki was invited by the Rockefeller Foundation to do a lecture tour of American universities, finally settling at Columbia, where he taught intermittently between 1952 and 1958, making himself available to seekers such as Cage. Other art world luminaries who attended Suzuki's lectures included Philip Guston, Abram Lassaw, Ad Reinhardt, and Betty Parsons.[89]

Yanagi himself also became a significant figurehead abroad. As a young man, he quickly established himself as an important tastemaker and cultural critic in Tokyo, as the founding member of the avant-garde journal *Shirakaba* (White Birch), which was key in introducing principles of Western modernism to Japan through the arts and literature. His first book, a study on William Blake, the first ever in Japan, was published in 1914.[90] In 1927, Yanagi published his essay *Kogei no Michi* (The Way of Crafts) and organized the first exhibition of Japanese folkcrafts, as well as a guild of craftsmen, that same year in Tokyo. In 1929, he presented a lecture series in English titled "Criterion of Beauty in Japan" at Harvard University, which crystallized the ethnocentricism of his endeavor, distinguishing *mingei* as inherently Japanese. This culminated in his establishment of the Association of Japanese Folkcrafts (1934) and the Japan Folk Crafts Museum (1936), both in Tokyo.[91]

Yet the popular American conception of Zen was a distortion of classical Zen, a ritualized sect of Buddhist monastic practice that, like other forms of the religion, traces its roots to the Indian monk Bodhidharma, who traveled to China in 520 AD, achieved enlightenment (*satori*) through meditation, and incorporated many Taoist, and even Confucian philosophies into Chinese Ch'an, or what became known, in twelfth-century Japan, as Zen. In that era, Zen was adopted by the ruling class and became popular with its military elite. However, as Robert Scharf has argued, Zen is the only sect of Buddhism that was brought to the West, rather than "discovered" by it:

> Zen was introduced to Western scholarship not through the efforts of Western orientalists, but rather through the activities of an elite circle of internationally minded Japanese intellectuals and globe-trotting Zen priests, whose missionary zeal was often second only to their vexed fascination with Western culture.[92]

While Scharf's essay traces the widespread influence of Japanese philosopher D. T. Suzuki throughout America, his characterizations of East-West reciprocity are useful correlatives for the Black Mountain Pottery Seminar: Bernard Leach could most definitely be characterized as a "Western orientalist," while Hamada and Yanagi were "internationally minded Japanese intellectuals." In contrast to Hamada's careful projection of the laborer-potter through his peasant garb, the bespectacled Yanagi was the consummate Western intellectual, always photographed in a pressed three-piece suit and tie. Together, Leach and Yanagi had "discovered" ancient Korean and Chinese ceramics made by anonymous artisans, in particular, ceramics from the T'ang (618–906 AD) and Sung (960–1279 AD) dynasties, praising them for their humble, pastoral origins.

As an aftereffect of the Pottery Seminar, *mingei* became indistinguishable from the American-type Zen introduced by Yanagi, espoused by Leach, and silently enacted by Hamada. The stridency of Wildenhain's European modernism became a secondary strain of thought, predicated on the edict of individualism rooted in Western humanism. This is the inverse of Zen Buddhism, where enlightenment (*satori*) is attained through the absence of ego. In reality, however, both strains spread rapidly after the seminar, taken back to college classrooms, workshops, and community centers throughout the United States, where most of the attendees were already professional potters and instructors.

The Pottery Seminar represents a decisive turn away from the protofeminist strains of craft practice that stretched across three decades, the 1930s through the 1950s, ceramics as either a therapeutic process or a tool of social welfare. These were discarded in favor of an importation, so-called Zen pottery, that stirred a younger generation of students, many of whom had led the war effort, and went on to disseminate a version of the performance-intensive ceramics they had experienced at Black Mountain, filtered through their own aesthetic interests and training. Zen, rather than the therapeutic, proved better suited to the discipline's interest in promoting its vanguard sensibilities. These tensions are embodied by the legacy of M. C. Richards, whose silent influence undergirds the Pottery Seminar and whose own writing and thinking, examined in the next chapter, became a site for the reconsideration of ceramics as a live, process-based art form that had the potential to be both therapeutic *and* radical.

M. C. RICHARDS'S VANISHING POINT

CHAPTER FOUR

In 1972, the American artist Mary Beth Edelson appropriated one of the masterworks of Western art, Leonardo da Vinci's *Last Supper* (1498), as an act of feminist defiance. The resulting collage, *Some Living Women Artists/Last Supper* (1972) (figure 4.1), is a work in which da Vinci's religious iconography is replaced with a matrilineal clan of twentieth-century American women artists. All the male heads of the apostles are transformed into artists who had achieved professional success; the bodies remain as they were. Such an uneasy arrangement is expressed through a hybrid title that conveys the ambivalence of shared cultural power.

Through photomontage, Jesus and his twelve disciples are transformed into a secular dinner party of protofeminist women, eight years before Judy Chicago's large scale sculpture installation *The Dinner Party* (1979) had been either conceived or celebrated. The appearance of photography within a much older genre, fresco painting—the reproduction of which was taken from an art history textbook and blown up using early photocopy technology—exemplifies the new media concept of *remediation*, in which a new technology reanimates, or offers an adaptation or accommodation of, its predecessor.[1] Edelson's resulting lithograph is an offset print that could be easily reproduced and circulated like a poster, offering the widest possible distribution. Da Vinci was representative of the sort of dead white male artist dismissed by feminists as irrelevant. Instead, the artist translates *Last Supper* into a visual vocabulary instantly legible and, in turn, widely celebrated by her own contemporaries.[2]

The women offered seats at Edelson's secular supper were known as trailblazers but not necessarily feminists. Yet their artistic careers were marked by grit

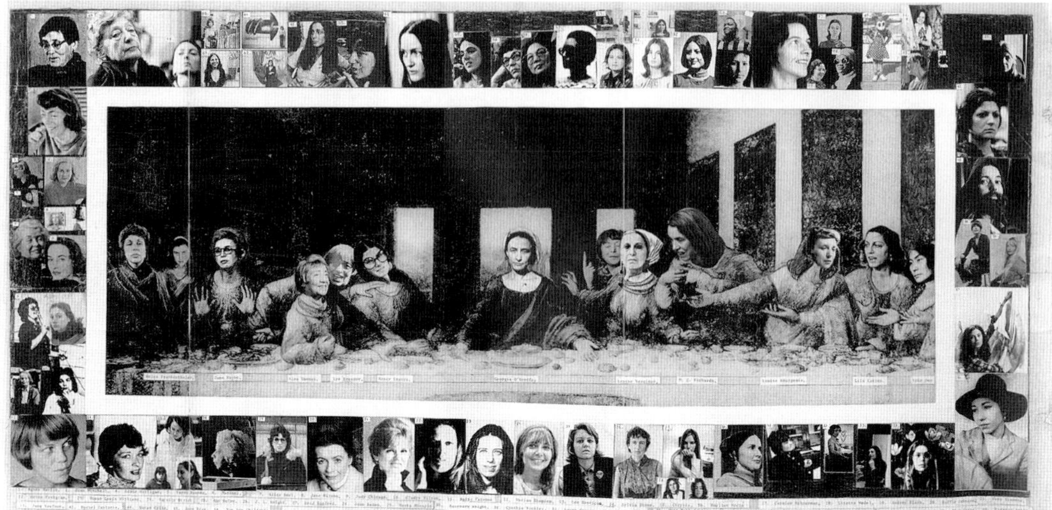

Figure 4.1 *Some Living American Artists/Last Supper*, 1972. Mary Beth Edelson (American, b. 1934). Original layout for poster production, photographs, pencil, ink, china marker, and reproductions. Courtesy of the artist.

and perseverance, serving as a model to second-wave feminist artists—Edelson's own generation—as they struggled for parity and recognition at the dawning of the 1970s. While famed in the art world, the artists were hardly household names: Lynda Benglis, Louise Bourgeois, Helen Frankenthaler, Nancy Graves, Lila Katzen, Elaine de Kooning, Lee Krasner, Louise Nevelson, Georgia O'Keeffe, Yoko Ono, M. C. Richards, Alma Thomas, and June Wayne. But they would have been easily recognizable as predecessors to an audience of like-minded women, other feminists with whom Edelson established a community as one of the founding organizers of the inaugural Conference for Women in the Visual Arts (CWVA), held at the Corcoran Gallery in Washington, DC, in 1972. In this sense, Edelson remediates the concept of the early modern master—da Vinci's most agreed upon and celebrated attribute.

Conveying a combination of humor and homage, the elder of the group, Georgia O'Keeffe, is depicted as Jesus. In her scholarship on O'Keeffe, Anne Wagner has also noted Edelson's tribute, but she demotes it to the realm of the reversal, women in exchange for men:

> We have failed to find the terms in which to *see* women's art, failed to point to it in ways that make enough cultural or aesthetic sense. Too often the strategy is merely a reversal: for what was once male, read female; where once was Jesus is now O'Keeffe. And in place of Leonardo, who?[3]

But Wagner has it wrong: Edelson is not performing a simple substitution. One of the most distinguished and celebrated features of the fresco itself is its use of perspectival form, in which the picture plane projects a spatial continuum, leading to a central vanishing point. Edelson's collage explores the metaphoric potential of the vanishing point as a means of signposting women's invisibility in the history of Western art, as examined through the prism/prison of modernism. What does it mean to vanish in one's own lifetime, even as one is still actively living and producing? It is this threshold of disappearance that best expresses the gender asymmetry of endurance; the prospects for male artists improve as they peak artistically and their exhibition record accrues. The burden of disappearance lies with women.

The potter and poet M. C. (Mary Caroline) Richards has largely vanished from the history of postwar American art. In 1964, Richards published a book that became her most important achievement, *Centering: On Pottery, Poetry, and the Person*. But she is best known as a shadowy figure in someone else's art work: the nameless poet reading poetry from atop a ladder in John Cage's *Theater Piece #1* (1952), widely designated the first postwar "Happening" in modern art. As the least recognizable member of Edelson's entire assemblage, she is given one of the most prominent seats at the table. *Some Women Artists/Last Supper* establishes Richards as an underappreciated elite, belonging to a pioneering generation who had all weathered enormous struggles to singularly carve out space in a culture that did not readily accommodate female ambition. Some, like Louise Nevelson and Georgia O'Keeffe, rejected traditional marriages and motherhood. Others had careers that were at times overshadowed by their attachments to legendary men: Louise Bourgeois (Robert Goldwater), Helen Frankenthaler (Clement Greenberg, Robert Motherwell), Elaine de Kooning (Willem de Kooning), Lee Krasner (Jackson Pollock), Nancy Graves (Richard Serra), Yoko Ono (John Lennon), M. C. Richards (David Tudor), and O'Keeffe as well, in the early part of her career (Alfred Stieglitz). Still others were nearly forgotten: Lila Katzen was a sculptor of large-scale metal abstractions active in New York in the 1960s, and Alma Thomas, the only African American artist represented, was an abstract painter attached to the Washington, DC, color school who had made her living as a schoolteacher. She came to prominence too late to enjoy her first posthumous retrospective in 1981.

Set in three-quarter perspective, Richards is a stand-in for the apostle Philip. Reflecting back on this decision, Mary Beth Edelson writes:

> I decided early on, in thinking about that first collage poster, not to research who would be appropriate for what position given that there is so much controversy about the identity

> of several of the figures in the original Last Supper, and a dialog[ue] about that identity was not the intention of this art work in any case. The intention was to challenge patriarchal religions for cutting women out of positions of authority and power—no matter who, what or when—we were not invited, included or welcomed—period. As far as selecting her image for that position is concerned, my sense of her was that she was a self-contained person who had a good grip on her own power and was self-assured. The photo that I had of her conveyed that. In addition, as a purely artist decision her image worked well with the over all intention of the collage poster, to show women in a strong positive light.[4]

Richards was a powerful and catalytic figure. As the sole woman in her artistic milieu, she was one of the primary, but unacknowledged, intellectual forces of the burgeoning 1950s queer aesthetic, though she herself was not queer. This was accomplished through her role as a translator that went well beyond the transmission of language and text. She was a crucial interlocutor between disparate artistic media and its networks of artists and thinkers. Further, she shaped the role that translation itself plays as a form of remediation, or what Jay David Bolter and Richard Grusin describe as "the transparent interface that stands in an immediate relationship to the contents of that medium."[5] For much of her varied career path, Richards was consumed with the transformative potential that the creative act, as a *live form*, held. Working across literature, theater, music, and the visual arts, in the multivalent capacities of professor, translator, performer, potter, poet, and philosopher, she readily embraced an interdisciplinary practice before such a term had come into full usage, let alone presence, inside the academy. But the immediacy and broad scope of her diverse set of practices was not well received in her own lifetime, nor has it been revised posthumously as a distinctive contribution. Richards's lived experience destabilizes the accepted postwar narrative in modern art by making apparent the still unacknowledged masculinist ideologies that compounded and ultimately defined many of her choices, including her eventual self-exile from New York during the late 1960s and the resultant loss of one of the avant-garde's intellectual lights.

In particular, I will examine Richards's erasure as a collaborator with and, indeed, influence upon John Cage, whose legacy has eclipsed her own. Further, I map Richards's career post-Black Mountain and her contributions to the studio craft movement, which were borne of her frustrations with the insularity of New York's models of artistic collaboration. Celebrated as a teacher and thinker but disdained for her communally-engaged artistic pursuits, M. C. Richards's career functions as a counterarchive, or what anthropologist Ann Laura Stoler has called a "disabled history," emblematic of the discontinuous histories, untranslated documents, and incomplete narratives that populate colonial archives.[6] In translation, Richards's paradoxical *oeuvre* casts a shadow of difference over the generation of her all-male peers that were richly rewarded for their contributions to American arts and letters.

The Pottery Happening, 1958

In 1958, Richards created a one-evening event in New York's East Village titled *Clay Things to Touch, to Plant in, to Hang Up, to Cook in, to Look at, to Put Ashes in, to Wear, and for Celebration* (figure 4.2). The documentation for this event is a single poster in the Mary Caroline Richards papers at the Getty Research Institute.[7] While clay items were sold during the opening, there is no evidence that there was actually anything left in the gallery after the Friday evening event.

If it weren't for the poster's date (April 25, 1958) and the event's attendees, which likely included composer John Cage, dancer and choreographer Merce Cunningham, musician David Tudor, and others from Richards's inner circle, this hand-drawn poster of graphite block letters would merely be a merry bit of ephemera—an example of late-1950s New York avant-garde detritus or everyday whimsy. The early date of *Clay Things* has the potential to alter the traditional narrative of postwar experimental performance art. Created one year before Allan Kaprow's *18 Happenings in 6 Parts* (1959) and three years before Claes Oldenburg's sculptural environment *The Store* (1961), *Clay Things* moves Richards to the center, rather than the periphery, of experimental practice. Like Kaprow's and Oldenburg's events, Richards's was also held in a space on Manhattan's Lower East Side, in this case at the Nonagon Gallery, a cross-disciplinary site where Charles Mingus would record a live album the following winter.[8]

Richards taught at Black Mountain College from 1945 to 1951. Her experiences there inaugurated a lifelong commitment to communal living. Following her departure, she became the epicenter of the experimental Gate Hill Cooperative, an artist's community known as Stony Point, in Rockland County, New York, which was established in 1954 by four couples, all Black Mountain refugees: Richards and her partner, David Tudor; the potters Karen Karnes and David Weinrib; the architect Paul Williams and his wife Vera, a writer; and John Cage and Merce Cunningham. Richards had high hopes for the community; she was interested in establishing what she called a "weekend Black Mountain," where, as she wrote, "People from the [New York] city could come out Friday night or Saturday, have 2 days. Everybody here has something to teach, if they want it."[9] While Paul Williams was the sole source of funding for this venture, he indicated to Richards that the community would not occur without her approval, or her "go ahead."[10] Eager to try again, she detailed her plans in numerous letters and proposals, trying to entice many of her former colleagues, including Charles Olson and Cage, into

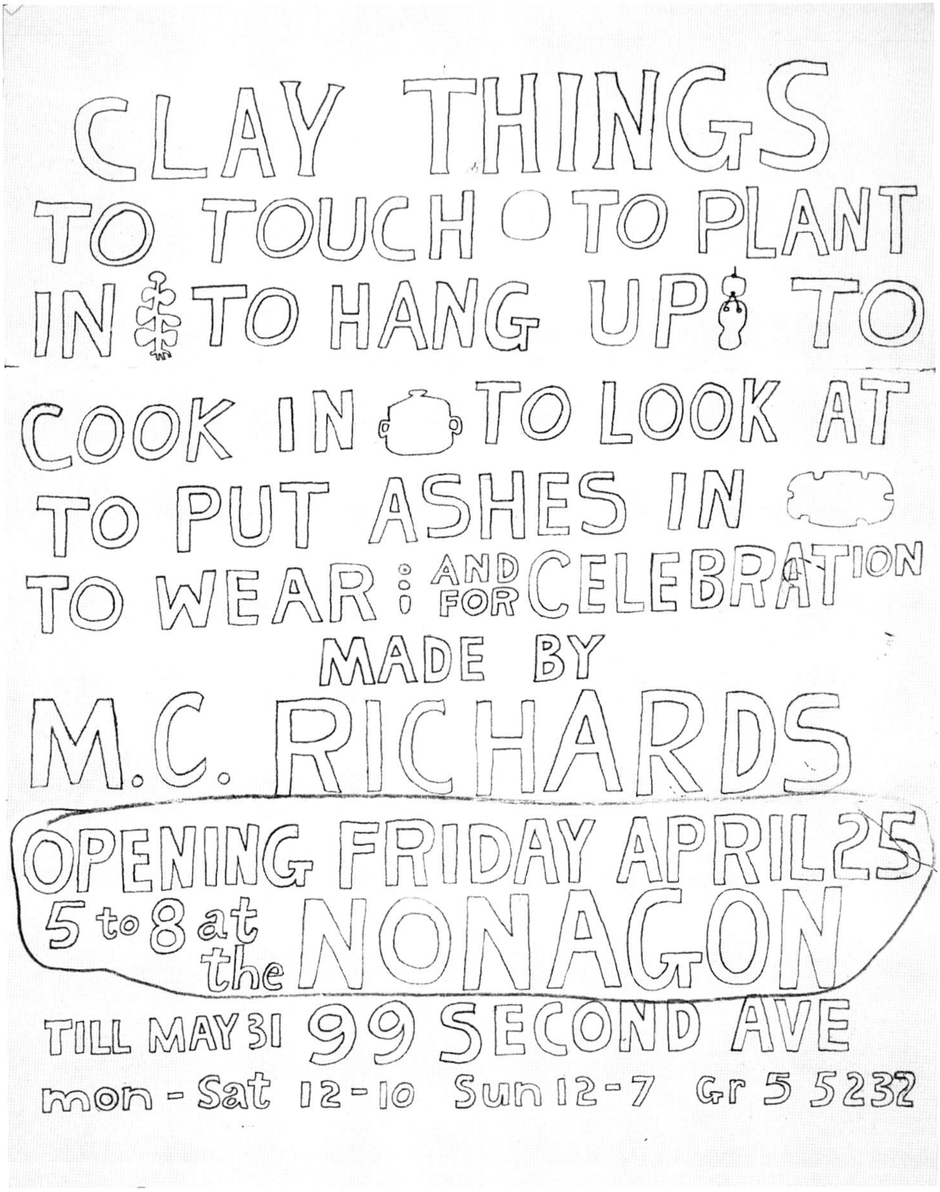

Figure 4.2 Flier for *Clay Things to Touch*, 1958. Collection of Mary Caroline Richards Papers, the Getty Research Institute, Los Angeles (960036 ADD2). Courtesy of the Estate of M. C. Richards.

teaching at this new experimental venue. But this did not come to fruition. Cage toyed with and eventually rejected her overtures, while Olson decided against joining the community.[11]

Written as a series of commanding infinitives: "to touch," "to plant in," "to hang up," "to cook in," "to look at," the directional format of Richards's poster prefigures many of Kaprow's instructional pieces of the early 1960s and even those of the core Fluxus group, most notably Yoko Ono's series *Instruction Paintings* (1961–62) and Alison Knowles's *Make a Salad* (1962). With pots at the center of the event, Richards exaggerates their usefulness, a comment upon the historic, nearly excessive utilitarianism of the ceramic object (as planter, as teapot, as casserole). Yet the categories break down in her litany of uses: "to put ashes in" seems dour, signaling death, until it is paired with the spirited proto-Fluxus silliness of "to wear," as though her "clay things" could be a jangle of ceramic bangles—purely frivolous, prone to cracking and breakage. Rooted in verbs of action, the infinitives careen from their path of stability and are replaced by the final pronouncement, "and for celebration," which commandeers revelry, merriment, and good fun.

There is a potlatch quality to Richards's event, one sparked by a go-for-broke approach. The process of making and storing heavy ceramic goods requires months of careful planning, culminating in a festivity that results in a generous and complete distribution of goods, during the opening celebration. Taking place in a space that sold works of art, Richards's event foreshadowed Oldenburg's *The Store* quite literally: both functioned nominally as gallery exhibitions open select times and days over the duration of a month.[12] Yet the 1958 opening night itself was just one in a series of events Richards and her friends would traverse the city to attend that evening. As she wrote afterward, on the back of the poster:

> Terrific opening. Scads of people. Good punch. I was in a dazzle of joy and excitement. I'm not sure how many "pots" got sold, but as far as I'm concerned, it was all a smashing success! More fun tonight: going to Jim Herlihy's Broadway success *Blue Denim*, then to an opening of an exhibit of John's scores, WOWEE, big whirl![13]

Her account references the overlapping events, interconnected people, and avant-garde worlds that Richards embodied as a pivotal figure in late 1950s New York. "John" was John Cage, and his event scores were as hermetic as James Leo Herlihy's Broadway play *Blue Denim* was populist. Cage's notations had limited appeal beyond his a small circle of musicians and visual artists, while Herlihy's drama was a coming-of-age narrative about American teenagers, Midwestern sexual mores, and an out-of-wedlock pregnancy. Yet Cage and Herlihy had something in common. Each represented successive generations (Cage being the elder) of New York's homosexual avant-garde, and both had worked closely with Richards at Black Mountain.

A self-described "teacher, potter, poet,"[14] Richards was awarded her doctorate in English in 1942 from the University of California, Berkeley, at a time when women were routinely discouraged from pursuing graduate studies. At Black Mountain, Richards was instrumental in initiating the college's experimental theatrical productions and publications. She translated Erik Satie and Jean Cocteau for collective theater performances, directed several joint student-faculty productions, and, most notably, began translating Antonin Artaud's *The Theater and Its Double* in 1951, one of the formative texts of twentieth-century experimental theater. Her work was published by Grove Press in 1958 as the first English-language translation of Artaud's collection of manifestos. Richards also initiated the Black Mountain Press, used for the college's literary publications, which later morphed into the *Black Mountain Review*, the literary journal that reinforced the college's reputation as a haven of experimental poetry.

Richards's own intellectual labor, however, has been subsumed by her proximity to the men of her circle. She is best known as one of Cage's collaborators. A now-apocryphal work widely credited as the first Happening, *Theater Piece # 1* (1952), was never documented visually, and no score has been definitively identified. Various versions of its retelling do not credit Richards at all, and she is sometimes excised completely.[15] This genealogy of early performance is destabilized by *Clay Things*, which prompts larger questions regarding the nature of influence and obscurity and reveals how avant-garde mythologies are perpetuated by casual retellings rather than historically accurate accounts. It seems that Richards, not Kaprow, organized what is arguably the *second* documented Happening. Moreover, its medium—ceramics—appears antithetical to the anti-object stance of postwar performance.

Such a reorientation of postwar history is tantalizing, tantamount to assigning an often-overlooked but important collaborator a centrality that would make many art historians uncomfortable—not least because of the dearth of substantial information surrounding the nature of the event and the artist's intent and the tendency of artists and historians alike toward documentary and nostalgic recollections of art-historical moments that prove elusive materially or conceptually, through, for instance, oral histories recorded years after the fact. Aside from Richards's own firsthand account on the back of the poster, there is an utter absence of any other eyewitness accounts.[16] As Judith Rodenbeck points out, the fragile history of the 1960s-era Happening is mired in a profound struggle regarding the location of the "work," whether its photographic or material documentation, such as texts or oral recollections, can ever really constitute a substantive accounting of the work, since there is no indexical relationship between time-based art and its historical afterimage.[17]

Given that *Clay Things* coincided with the publication of Richards's translation of *The Theater and Its Double*, the text itself might have been a source of inspiration for the event. *The Theater and Its Double* is rooted in the notion of a radical theater that advocated for the artist as a "living instrument," that is, the actor who did not repeat gestures or become paralyzed by affect, thereby invigorating the audience. Richards's pots, with their myriad utilitarian uses, may be read here as literal "living instruments" imbued with not only their literal exchange value but also a sense of constant re-performance. In this way, the performance is transferred to the objects themselves, to be reenacted continuously in everyday life as a *live form*.

Richards herself approached ceramics in this manner, working toward spiritual engagement and communalism in a medium that was often plagued by technical concerns and individual achievement. She instead embraced the role of the enlightened amateur, making and then giving away (or selling for very little) most of her pots, utilizing their production as the basis for an event itself. In guiding students toward non-object, process-based production, such as returning unfired clay to the ground after making a collective pot, Richards treated pottery as a medium that offered a participatory experience, one in which the object itself was secondary to its process. In this way, *Clay Things* anticipates the collectivity and vitality of socially engaged artistic practice, fifty years later, seen, for example, in the compendium of civic and social practice work in large-scale recent exhibitions such as *Living As Form: Socially Engaged Art* (Essex Market, Creative Time, New York, 2011) and *Ripple Effect: Currents of Socially Engaged Art* (Art Museum of the Americas, Washington, DC, 2013).[18]

Many of Richards's own ceramic forms engage with ideas of awareness, epiphany, and struggle, working through serial, repetitious experiments that resulted in misshapen, often unglazed forms: the same sort of messy formlessness that Paul Mathieu was critiquing in the *Dirt on Delight* exhibition, albeit nearly sixty years later. This has been the overarching critique of Richards's work: that it is amateur, what today would be known as sloppy craft.

Like Marguerite Wildenhain, Richards contributed to the early 1960s experiments in the dematerialization of the art object. But Richards's pedagogical process was very different than Wildenhain's. She did not teach the fundamentals of wheel thrown ceramics. She purposefully utilized the medium's most basic building blocks, coil and pinch pots, for the entirety of her *oeuvre*, such as in *Orange Pinch Pot* (1990s) (figure 4.3), a late work, reminiscent of the beginning forms that emerge from children's art classes. In choosing to shape vessels by hand, she produced off-kilter wares that reveal their own subjectivity, literally produced from the privileging of instinctual feeling.

Richards's key contribution is her emphasis of process over skill. A dynamic speaker and performer, she was highly skilled at guiding groups of students in the tactility of touching and shaping, speaking through form to creating a heightened awareness of the connections between materiality and feeling one's way in the world. Richards returned clay to its roots as a pliant and connective material, as a way of joining students to each other. Her role as a functional potter can

Figure 4.3 Orange pinch pot, ca. 1990s. M. C. Richards (American, 1916–1999). Earthenware. Collection of the Black Mountain College Museum + Arts Center, Asheville, North Carolina. Photo by Alice Sebrell. Courtesy of the Estate of M. C. Richards.

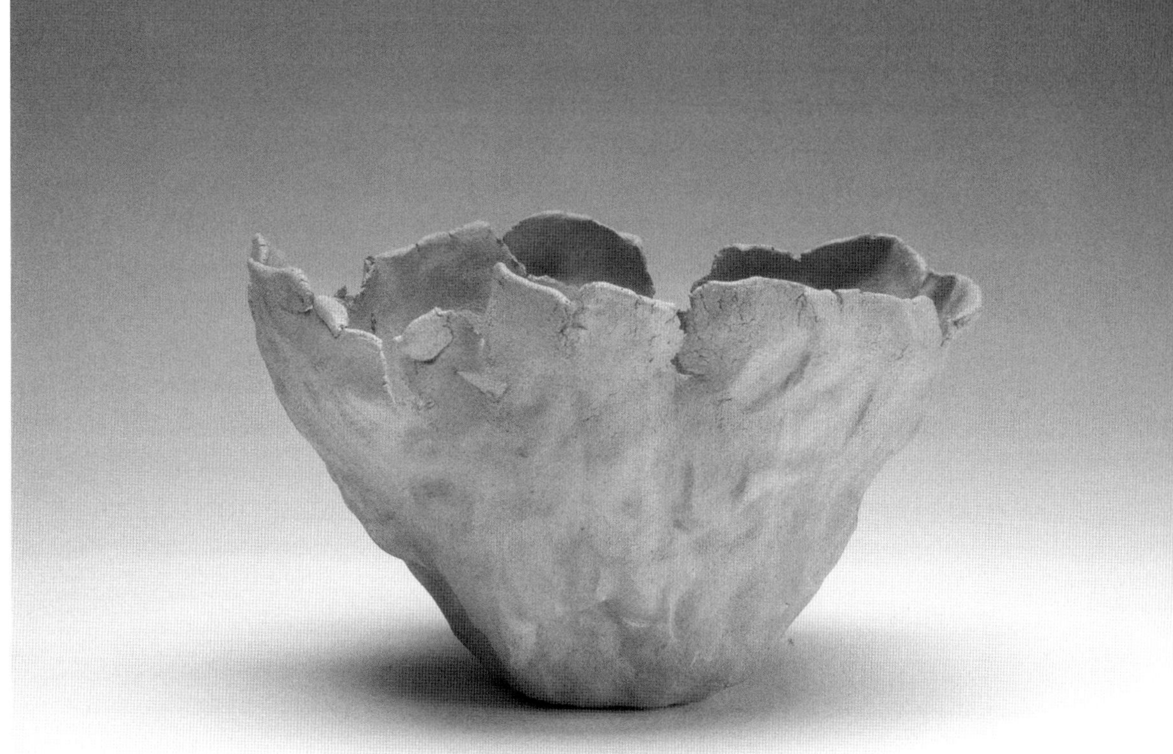

therefore be read as an early and potent antecedent to the current social practice projects occurring in the twenty-first century

Richards's conceptualization of ceramics as a fully communal and performance-based practice drives a wedge between the functional versus sculptural debates, in that she embraced functionalism in order to focus centrally on process and, in doing so, wrote an influential metaphysical treatise on pottery. In *Centering*, she defined her conviction in creative work as "... a training of each individual's perception according to the level on which he is alive and awake."[19] This awakening is a coming to form, a literal aliveness that also functions as a means of translating the capacity for analysis and redirecting it as an intuitive perceptual and cognitive form of making.

Centering's origination came, in part, from Richards's long struggle to find meaningful collaboration and artistic exchange within her peer group. But her conviction in the transformative social dimension of pottery eventually necessitated her own break from the avant-garde community she inhabited. For this reason, a history of her career and networks is crucial to recovering the fuller picture of her radical approach toward ceramics.

The Pot Shop

It took until May of 1962 for Black Mountain College finally to die in true pauper fashion, selling off part and parcel its material possessions. Its microscopes, its livestock, its books went to local schools and farmers, and its land, an idyllic spot on Lake Eden, went to the Christian summer boys camp, Camp Rockmont, that still inhabits it today. Charles Olson remained behind to attend to the bureaucratic responsibilities he had neglected during his four years as rector, settling debts and negotiating lawsuits involving former faculty members, many of whom were ultimately paid in sides of beef.

In June of 1957, Olson complained of this task in a letter to Richards:

> My dear darling MC—
>
> . . . Taking an ungodly amount of time to do it up right, but I am now down to the last furniture, the beef job + the Pot Shop (of which Turner finally baled out, alas!) I suppose the final details (like files [and] things) will take longer than I think . . . But I count the days and the land report is [in]. I have a PASSPORT SO? WO IST DER FUTURE?[20]

For Richards, the future was craft. By 1962, the year Black Mountain died, when the books finally reached a "zero" balance, Black Mountain College ceased to exist. Its failure to sustain itself was a bruise on the psyche of the American avant-garde, but the college's legacy would continue to loom large. Richards was in the early stages of writing the book that came to be known as *Centering: In Pottery, Poetry, and the Person*, published in 1964, which became an instant classic among potters and craftspeople, selling over 120,000 copies and in continuous print since. Richards herself struggled with her own disillusionment regarding the decline of the college, and her book *Centering* was born of this trauma, in which she wrote:

> I began consciously to turn to toward the wisdom of pottery at a time of breakdown of other values. It was a major life-shock for me to discover that the community I had chosen, that of artists and teachers . . . was as flawed as the rest of society . . . I faced in disbelief and rage and despair the reality that Higher Education had not equipped me nor my colleagues with human insight nor a loving will. Knowledge had not ennobled our behavior.[21]

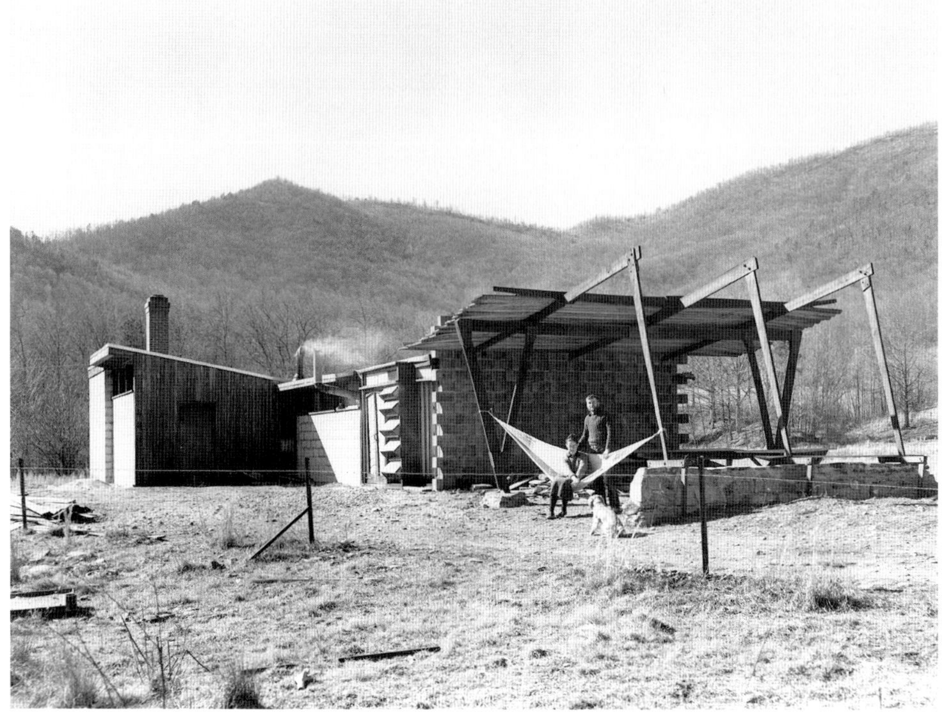

Figure 4.4 Photograph of David Weinrib and Karen Karnes, exterior of the pot shop, ca. 1953. Designed by Robert Turner. Organizational Records, Black Mountain College Records, 1933–1956. Courtesy of the Western Regional Archives, State Archives of North Carolina.

Richards had thrown her first pots in Black Mountain's Pot Shop (figure 4.4), aided by Robert Turner and Peter Voulkos. Karen Karnes and David Weinrib had hired Voulkos for the 1953 summer session, when Richards returned as a student instead of as faculty.[22]

No doubt this was surprising to many, as she had resigned two years prior. So returning to Black Mountain was a deliberate and humbling decision, reflected in an early platter she made in 1953, during her only summer as a student in the Pot Shop. A square with rounded corners in earth tones, *Leaf Platter* (1953) (figure 4.5) could even be described as earnest, unashamed of its own clumsiness. The glaze is applied unevenly, with a thick brushstroke, and the leaf design is generic, without any of the detailed incising found in Wildenhain's works. Such carving and rendering skills were too difficult for a beginner. As Richards explained:

I permitted myself a kind of freedom in the use of clay which I would not have known how to find in the VERBAL world . . . As soon as we find ourselves spellbound by order

Figure 4.5 *Leaf Platter*, 1953. M. C. Richards (American, 1916–1999). Stoneware. Collection of the Black Mountain College Museum + Arts Center, Asheville, North Carolina. Photo by Alice Sebrell. Courtesy of the Estate of M. C. Richards.

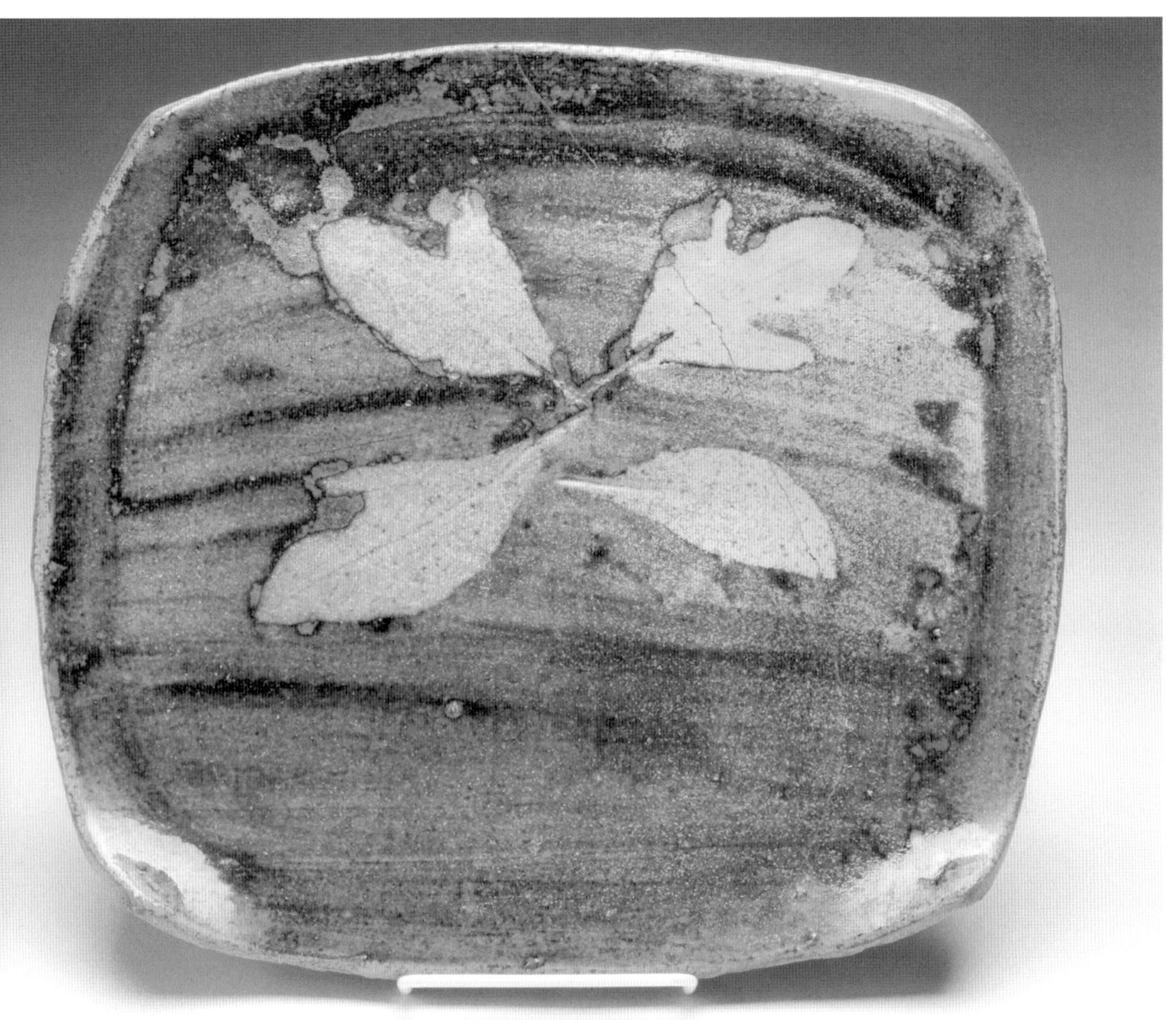

> and our ability to control our mediums and our tools, to do exactly as we want, we must do the opposite as well.[23]

She credited Voulkos, eight years her junior, as being one of her most influential teachers.[24] The admiration was mutual. Voulkos's first trip to New York City was with Richards, after the summer session. For the entire month of September 1953, Richards housed him in the East Village apartment she shared with David Tudor. The trip proved to be a life-changing experience for Voulkos. Richards was his guide while he drank in the New York avant-garde, meeting Franz Kline and Willem de Kooning at the Cedar Bar, visiting John Cage, and seeing contemporary art in New York galleries for the first time. Through Richards's networks, Voulkos was ignited by New York's raw and seductive creative energy, moved to abandon the potter's wheel when he returned to Los Angeles to teach at Otis in the fall.

The 1950s and 1960s were a heady time of heated debates over issues of self-expression and the responsibilities of the potter. At issue was whether or not potters could responsibly negate traditional forms in favor of eclectic, non-utilitarian ceramic sculpture, such as Voulkos's *Untitled Plate* (1973) (figure 4.6), its form violated with holes. Richards's two teachers, Turner and Voulkos, represented opposite ends of the spectrum: Turner made a combination of thrown and mold made pottery, simple, elegant wares, while Voulkos hand built with slabs of clay, steadily increasing the size and antifunctional stance of his work to increase its stature within the art world. Richards seemed to draw deeply from the lessons of both, admiring Voulkos for his fearless aptitude for making and unmaking form, for working the clay until it collapsed, even emulating his style at times, such as in an untitled work from the 1960s, in which she also utilized slab and wheel construction, layering a cacophony of shapes (plate 5). But ultimately, she preferred Turner's consistency, writing years later, in 1977:

> Turner was the only potter I ever knew who would throw a pot on the wheel, study it slowly, put it in the damp box, and then back on the wheel the next day to throw it some more. He took his time. This has been an important guiding image for me through the years: to take time, making the contact deep and personal, low key, attuned, listening to the breathing of the form, to what the clay is saying.[25]

Centering: In Pottery, Poetry, and the Person (1964) takes its title from the action of centering clay in the middle of the pottery wheel, describing the crucial moment that enables the transformation of raw material into vessel. Without a proper center, the clay form is permanently off-center, unable to grow into a stable form. All thrown vessels, even those that are intended to counter gravity, still begin with a centered lump of clay.

Through her initial organization of the Pottery Seminar, Richards developed an interest in Zen. In 1957, a compilation of Zen koans was published under the title *Zen Flesh Zen Bones*, which had a section called "Centering." *Centering* likely acquired its title based upon it, inspired by brief directives such as this: "11. Place your whole attention in the nerve, delicate as the lotus thread, in the center of your spinal column. In such *be transformed*."[26] But *Centering* is less of its medium than it initially seems. It does not strictly advocate for a renewed interest in the handmade, nor does it spend much time actually talking about

Figure 4.6 *Untitled Plate*, 1973. Peter Voulkos (American, 1924–2002). Stoneware and porcelain. The Museum of Fine Arts, Houston, the Leatrice S. and Melvin B. Eagle Collection, museum purchase funded by the Caroline Wiess Law Accessions Endowment Fund. Courtesy of the Voulkos Family Trust.

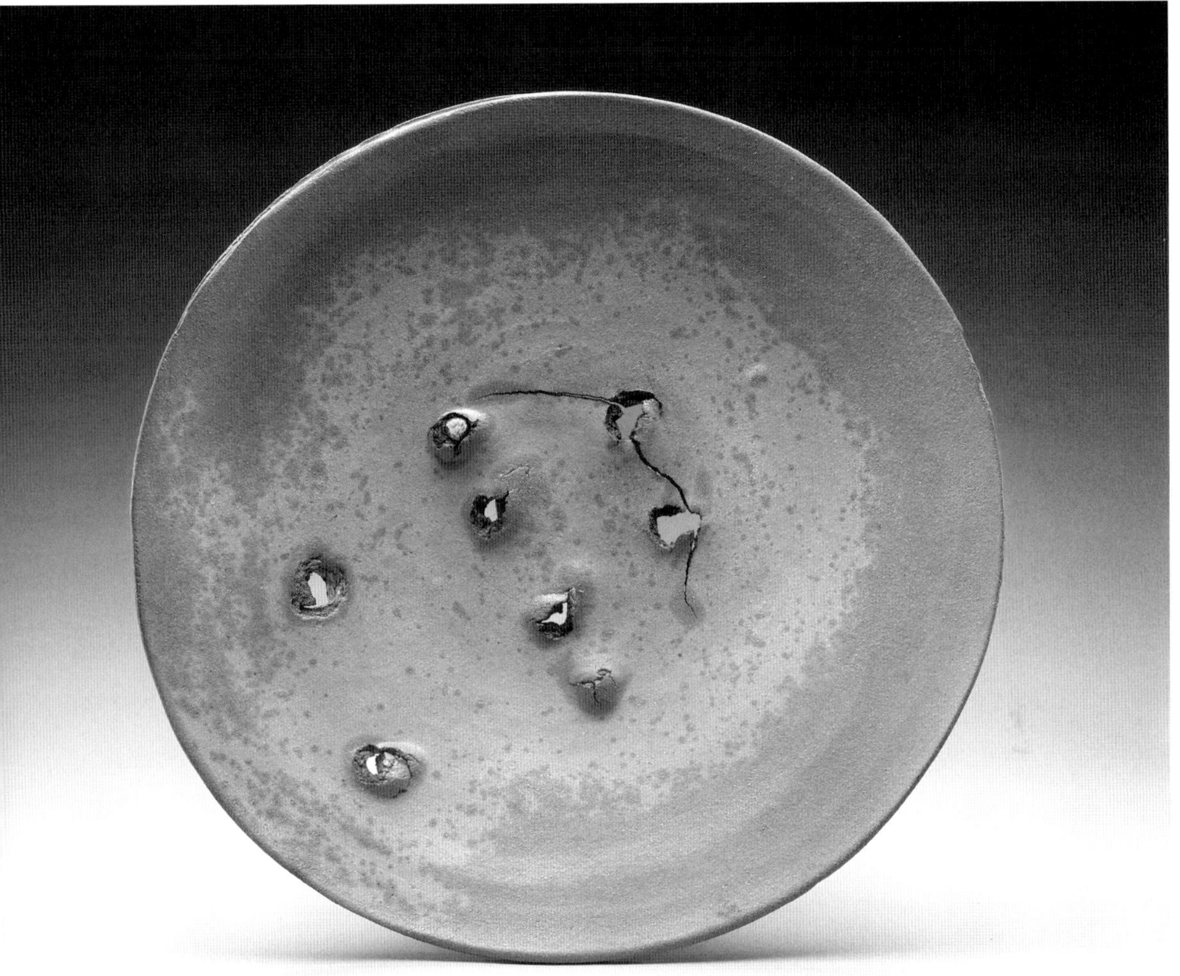

pottery or its role in society. Alongside Wildenhain, Richards was the only other woman in the early 1960s writing on pottery, though her book is notably absent of technique. It is unlike any other potter's book of the era in that it is a purely philosophical text, using clay as a metaphor for unmediated contact with the raw materials of the self. Richards internalizes *live form*, envisioning the self as continuously emergent, a privately evolving entity that marks the public sphere in productive, meaningful ways rather than simplistic declarations of the artistic ego. Richards believed the experience of form to be a transformative experience.

Enlisting art in the transformation of public life became one of the goals of Left intellectuals such as philosopher Herbert Marcuse, literary critic Dwight Macdonald, and psychoanalyst Paul Goodman, who also had Black Mountain ties.[27] Their work during the 1940s and 1950s had a large impact on the social and intellectual climate of the 1960s, whose activist-intellectuals, like Abbie Hoffman and Allan Ginsberg, became known as the New Left. *Centering* is in dialogue with this larger humanist tradition, published the same year as Marcuse's *One-Dimensional Man* and Marshall McLuhan's *Understanding Media: The Extensions of Man* (both 1964) and two years prior to Norman O. Brown's *Love's Body* and Phillip Reiff's *Triumph of the Therapeutic: Uses of Faith After Freud* (both 1966).[28]

A spiritual treatise on craftsmanship, *Centering* offers the self as an object of study, reflected through a discussion of form as a metaphor for unity. Richards's text is distinctly of its era: its analytic shamanism is particular to the 1960s, when humanities professors like Brown and Marcuse (so-called "father of the New Left") came to be revered by the burgeoning counterculture and radical student groups, such as Students for a Democratic Society. (It is Marcuse, the youngest member of the Frankfurt School, making peace with those ghosts of the Old World, Freud and Marx, who shooed hippie-pilgrims away from his door and yet is credited with coining the anti–Vietnam War peace slogan "make love, not war.") During this moment, pedagogy itself became a performance, an unintended consequence of finding the *right* audience—the disaffected college student, who is affirmed in his (in the 1960s, it could only be *his*) desires, which culminated in his conflict with the repressions of society. Such male repressions are specters in Richards own career, continually haunting her entwined personal and professional life.

Centering's closest kin is McLuhan's *Understanding Media: The Extensions of Man*, published concurrently through a mainstream versus a university press (McGraw-Hill vs. Wesleyan) and subsequently to much wider acclaim. But the similarities between the formation of craft and media discourses at this time are striking. Written by thinkers from similar intellectual backgrounds (both earned doctorates in English literature in 1942), the books are parallel texts with inverse premises. Both attempt a radical reformulation of individual consciousness, and both owe much of their shape and character to the broader sociopolitical backdrop of the human potential movement of the 1960s.

While Richards and McLuhan were part of the Left generation, their intellectual peak came later, when the passionate stance of their respective texts coincided with the urgency of the 1960s. In keeping with New Left values, both Richards and McLuhan urge a rejection of commodity culture and attempt to integrate the individual within the larger social body. Both argue for the social significance of emotion, the authenticity of human relations and relationships, and call for social values that incorporate wholeness in living. While McLuhan argues that industrialization has permanently altered and irrevocably fragmented the individual, Richards advocates for a person intact despite the shattering effects of consumerism and capitalism.

Together, Richards and McLuhan employ a language of physiological concepts. Below, both authors invoke homeostasis and equilibrium:

> MCR: When man intervenes and disrupts the **equilibrium**, nature tends to restore it. But there is a limit beyond which restoration is not possible. . . . Any situation in a state of permanent distortion naturally tends toward disruption and rupture. . . . This law seems to me also clearly at work in a person's evolution toward wholeness. Even the body seems to operate under this law. Physiologists call it "**homeostasis**," perhaps it is the body's natural centering process, a tendency that is always operative within it. On the other hand, substance is always "feeling" the pressures upon it, whether we are aware of them or not. In a condition of "fatigue," human beings like metals may break down under the effect of a slight but alternating stress.[29]
>
> MM: The natural tendency of the enlarged community of the city is to increase the intensity and accelerate functions of every sort, whether of speech, or crafts, or currency and exchange . . . The city, as a form of the body politic, responds to new pressures and irritations by resourceful new extensions—always in the effort to exert staying power, constancy, **equilibrium, and *homeostasis*** . . . . The alarm of the village, followed by the resistance of the city, expanded into the exhaustion and inertia of empire.[30]

Both writers resort to a preconditioned idea of "naturalism," where the primacy of human nature—found for Richards in the body's sympathetic responses, for McLuhan in social instinct—is privileged above intelligence. McLuhan utilizes the metaphor of the body to describe the inanimate social phenomena of the teeming metropolis. Richards applies the same metaphor to the intimacy of the personal sphere and the demands made upon the individual. Both writers chart a generalized sense of anxiety and peril, ultimately resulting in fruitless despair that can only be combated by the formation of either an internalized calm (Richards) or exterior control (McLuhan). McLuhan describes a series of human-coordinated responses in an effort to gain stability within a larger social group, while Richards maintains that equilibrium is a state of balance achieved through either the natural world or the artistic impulse.

McLuhan's text is structured as a series of brief essays on a variety of information age topics, ranging from gadgets to automobiles. As somewhat incomplete sketches, these outlines map out his interest in the inherent impulsiveness and lack of consciousness in a consumer-driven society. Most importantly, he theorizes the idea of the medium as a physical extension, or "self-amputation," of the physical body. The application of human physiology to interpret man's object-based desires, diversions, and modern inventions was a way to annex the very mediums, namely television and advertising, that seemed so far removed from the haptic and emotional life of the body. McLuhan's attempt to analyze their usage as a means of social control is his most enduring legacy.

Richards's book is a manifesto on wholeness, striving to give form and moral purpose to what she terms "metaphysical participation."[31] Unlike McLuhan, she is not interested in analyzing the political and social organizations that came to dominate postwar America. Instead, she looks to radical pedagogy and education as a means of reenvisioning the social life of the nation. She refers often to her own experiences at Reed College and Black Mountain College, as well as to other utopian and socialist notions of collective living, including those of John Dewey, Paolo Soleri, and Rudolph Steiner.

Richards's book is less disjointed than McLuhan's, structured around the vital action of centering clay as the central metaphor throughout her text, defining it as "the speech between the hand and the clay."[32] Centering is thus interpreted as a speech act, in the tradition of J. L. Austin, in which speech itself constitutes an action not unlike *live form*, but closer, in fact, to Walter Ong's *Presence of the Word* (1967), in which speech is in collusion with touch through its auditory properties, a dependency fostered through the Judeo-Christian tradition but out of favor in the modern secular world. A Jesuit priest and literary scholar, Ong was one of McLuhan's early mentors when the two taught together at Saint Louis University in the early 1940s.[33] In an undated letter to Cage, Richards reports having been at a dinner party in New York with Ong. Uncannily, she also thanks him for his Christmas gift, which, lo and behold, was *Understanding Media*, which she seems to have appreciated, writing, "It certainly fit right in with the thoughts I had had for my speech about the nature of book culture and its limitations, and there are several things I have been meditating on since reading it and which serve to strengthen my anti-book mood. It is not a good mood to be in if one is thinking about making a book."[34]

Spirituality permeates both McLuhan's and Richards's books as the authors initiate a process of diagnosis that leads to a kind of awakening or conversion, something both authors experienced personally. Raised Protestant, McLuhan converted to Catholicism in 1937, attending Mass daily throughout his adult life, while Richards joined the Theosophical Society sometime around 1966. A strong strain of lament courses through both texts, a sorrow for the materialism, affluence, distraction, and selfishness that had evolved as social values during the postwar era. McLuhan's prose style is quasi-mystical in its uncanny predictions and glib explanations, while Richards's approach, like Ong's, is more theologically driven. During the last fifteen years of her life, she lived in rural Pennsylvania, in a handicraft- and agriculturally based anthroposophist community known as Camp Hill Village, one of eighty worldwide, where mentally handicapped people lived, farmed, and produced crafts alongside their teachers, all of whom were volunteers, including Richards herself.

In their near-fanaticism regarding the capacity of language to restore lost spiritual values, both McLuhan and Richards owe a great deal to the New Critics, embodied by Cleanth Brooks's 1947 classic *The Well-Wrought Urn: Studies in the Structure of Poetry*. Beginning with this title and continuing in a later volume, *A Shaping Joy: Studies in the Writer's Craft* (1972), Brooks created an evocative and arguably conscious hybridity through his use of the potter's craft to suggest a well-written poem. New Critical formalism envisioned poetry as the preeminent literary form, in which the precision and specificity of language asserted a structural unity, thereby achieving resolution and harmony. Crucial to the understanding of poetic structure was the act of interpretation. This was achieved through close reading, with an emphasis on craftsmanship. Brooks was not a Left intellectual but rather from the opposite end of the ideological spectrum, a core member of both the Fugitives and its latter incarnation, the New Critics, a dominant force in midcentury American literature.[35] Though Richards's book appears later than the core of New Critical writings, her ideas undoubtedly sprang from this earlier era, as this was foregrounded in her academic training in English literature.

At the Center: Conflict and Collapse at Black Mountain College

From 1943 until 1945, Richards taught English at the University of Chicago before turning to alternative models of education. Along with her then-husband, the philosophy professor Bill Levi, the couple joined the faculty at Black Mountain. She presided at its helm as chairman during a time of intense intergenerational strife between its factions.

A key pedagogical figure at Black Mountain, Richards was a beloved faculty member, and Mary Phelan Bowles, a student during the late 1940s, described her as

> . . . a bridge between the intellectuals and the artists, perhaps. She was married to Bill who sort of represented the intellectuals. At the time I was there, there was something of a split . . . a tension, or something. The artists represented a certain approach. The intellectuals represented another kind of approach. And M.C. was really sort of part of

> both worlds. She was married to Bill Levi, who was the most intellectual kind of man. And she, herself, understood that kind of language and was excellent at it. But by nature she was more like the artists in that she was very spontaneous and full of feeling. . . . And I think Bill's approach to life kind of left her cold.[36]

The split to which Bowles is referring was the battle between the college's long-term professors and the new, younger faculty, led by Levi. On the surface, there was a basic feud over educational principles, whether or not the college could fully sustain a broad range of humanities offerings without hiring additional faculty. Josef Albers favored pushing forward an arts-only curriculum, renaming the school Black Mountain College of the Arts. However, underlying this disparity were larger issues. The architect Paul Williams, a student turned faculty member, characterized it as a "classic battle between the old timers and the newcomers."[37] The "old timers" were the physics and mathematics professor Theodore Dreier and his wife, Bobbie, and the artists Josef and Anni Albers. Classic though it was, this battle deeply ruptured Black Mountain's creative community and had devastating consequences, resulting in a mass exodus of three couples: the Dreiers and the Alberses in 1949 and the very public dissolution of the Levi-Richards marriage.

Perhaps the most obvious friction was caused by the discomfort of forced intimacy: when the Dreiers went on sabbatical during the 1947/48 school year, the Levis moved into their home, sharing one half of a single house with a common living room, putting them in direct proximity to their neighbors, the Alberses.[38] Seated together, but apart, a 1948 photograph (figure 4.7) shows Richards and Levi smoking languidly on either side of the couch of their shared domestic space. Such a stance foreshadows the difficulties that were to come, but the focal point of the photograph is Josef Albers's lithograph, which remakes the artist's very formal geometric shapes into a pair of eyes. Slightly off-kilter, the work hangs above their heads like a lofty intrusion. The couple is unable to escape the presence of Albers. Or rather, Albers himself was unable to escape Levi, a constant public irritant. Without a private space to retreat into, it seems likely that Black Mountain's crisis escalated disproportionately, burning larger and hotter than it might have without such bizarrely cloistered living arrangements.

In the summer of 1948, Richards sent a biting letter to the Dreiers, a testament to the clear generation gap. Emboldened perhaps by their physical distance, she accuses not only the Dreiers but the Alberses, too, of stultifying the college's legacy through a distastefully conservative agenda meant to bring security and comfort to an aging faculty (the Dreiers were in their fifties, the Alberses in their sixties), rather than making good on the radical educational initiatives for which the college was known:

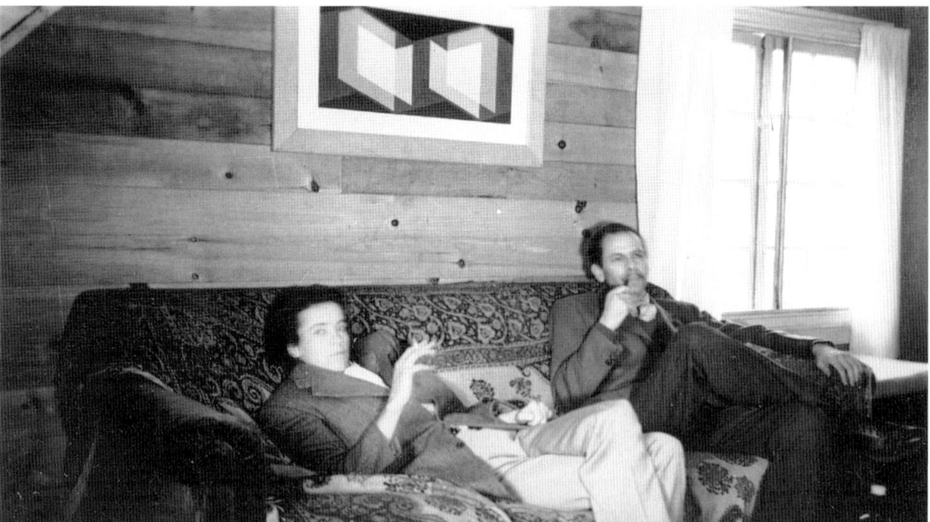

Figure 4.7 M. C. Richards and Bill Levi at home, Black Mountain College, ca. 1948. Collection of Mary Caroline Richards Papers, the Getty Research Institute, Los Angeles (960036). Courtesy of the Estate of M. C. Richards.

. . . But perhaps the fact is Black Mountain is not an adventure any more. Maybe it has outgrown its experimental phase and is now on the threshold of, shall we say, maturity? If so, so; but for heaven's sake, let's be clear about it and not get people into the place under the delusion that they are going to be shaping the future . . .

Now I should think that, unless you consider Black Mountain your capital investment, you should be very careful not to imperil its youthfulness by a desire to settle down at last. I certainly do hope that when I grow older, if I see myself relapsing into an anxiety about spending my old age in the work-house, I will have the generosity to withdraw from company whose liveliness might be dampened by my anxiety. If what you and Albers want is a secure old age, I should think there would be many better places of securing such than Black Mountain College. I have already in my travels observed numerous opportunities accompanied by respect and high income for the aged, the weary, and the dead. Maybe I am speaking too strongly for your stomach, but I do feel very strongly that you should see very clearly the effect you are having on the college and that you should be sure it's the one you want and that then there should be an end to any double-talk about experimentation or community or self-determination . . .

I hope you will get some idea of the worries I have on my mind about BM and you-all. I wish you had a little more candor and charity, and a little less sanctimony. I probably wouldn't like you as well as I do and feel as free as I obviously do to tell you how I feel about things if I didn't think you possessed the ordinary human frailties. So don't point the finger of scorn at me and others, but try to understand that we love BMC too and have invested what we possess of hope and faith and love in its processes.[39]

Despite Richards's heartfelt and cherished principles, such charges were deeply offensive: first and foremost because she and her husband were in the Dreiers' personal living space, and secondly because she and her thirtysomething postwar colleagues had not been present for the hardships during the Depression and the shortages—including food and lumber—of the war years. Unlike the Alberses, who were childless, the Dreiers were differently enmeshed in their surroundings, having lost a child during their tenure. Their nine-year-old son Mark was buried on the property, killed instantly when the college's truck accidentally overturned in 1941. One of the college's landmark buildings, Quiet House, was erected as a memorial, made of found stones and wood.[40]

Such a letter fulfilled the philosopher Norman O. Brown's bemused, years-later pronouncement "M.C., you wild and lawless character," which expressed affection tempered perhaps with concern.[41] While clear-eyed, maybe even courageous, her letter breached all decorum, simultaneously showcasing and compromising her own integrity as one of the leaders of the community. However, it is Levi who has been most frequently viewed as the instigator; his bullying and brusque demeanor at faculty meetings was notorious within the community at large and included several physical altercations.

At some point, the couple was depicted in caricature by an unknown student: Richards is thoughtfully arch, smoking a cigarette, one eyebrow raised, while Levi sneers midmovement, a gesture that declares his bristling impatience (figures 4.8 and 4.9). While she may well have disagreed with his public behavior, Richards's marriage left her on the receiving end of external criticism intended for Levi, though in hindsight, there were many sympathizers. The ceramic sculptor Stephen de Staebler, who attended Black Mountain the summer of 1951, recalled that

> the environment at Black Mountain took a tremendous toll on people. Their private lives were just too, too exposed, . . . I mean, I'm not surprised it folded because maybe it was ostensibly for financial reasons, but I would think there would be a lot of psychological reasons why leadership would be a tremendously difficult thing to pull off where the community that stayed around longest just by kind of proprietary interest would almost be the dominant ruling force.[42]

Fall 1948 progressed: Levi negotiated a year-long sabbatical for himself and Richards authorized by Albers as a compensation for the considerable time and monies he and Richards had spent interviewing candidates in Chicago for new faculty positions. But relations worsened, rather than improved; the 1948/49

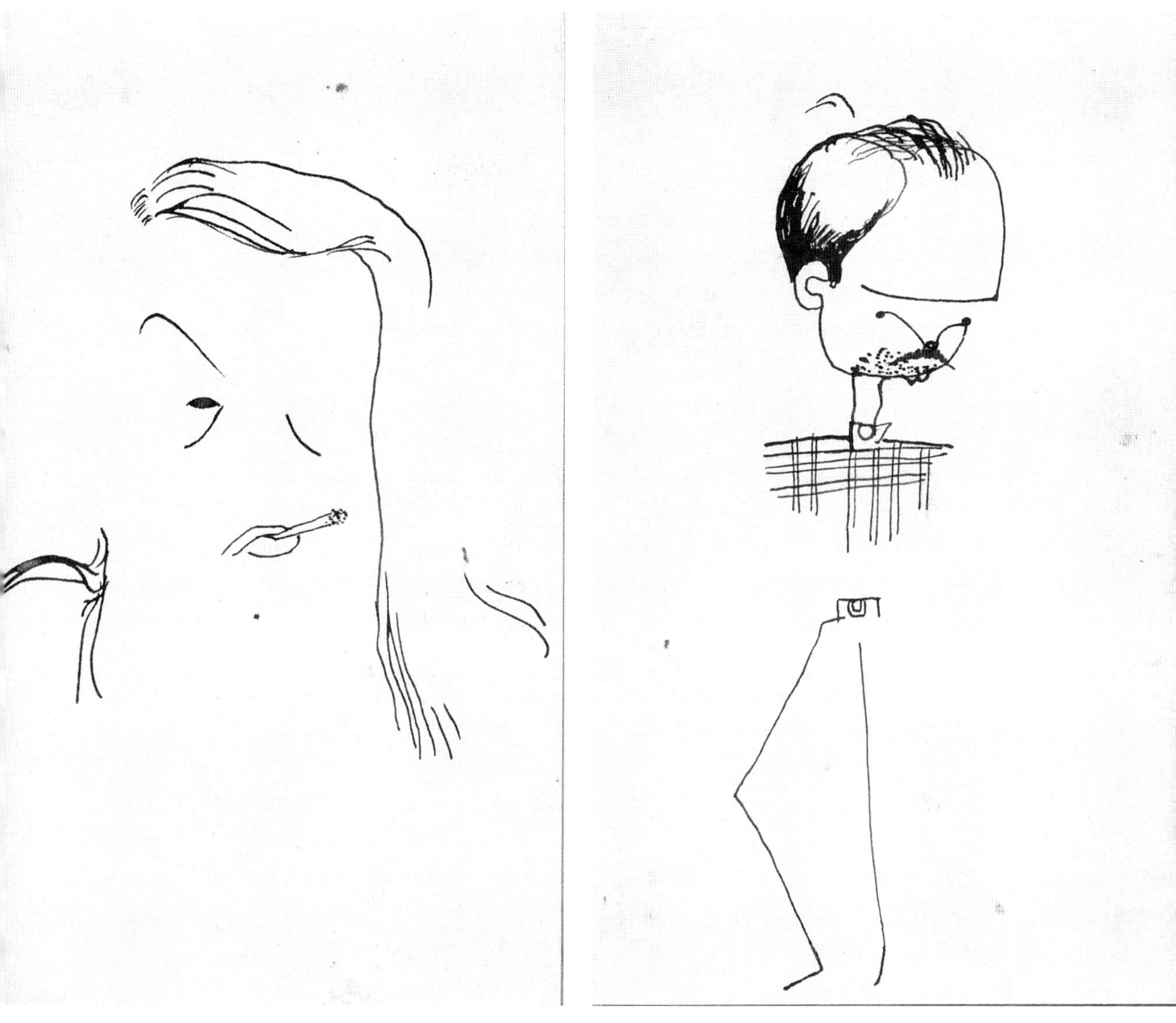

Figures 4.8 and 4.9 Unknown artist, sketchbook, Black Mountain College, ca. 1945–1949, Collection of Mary Caroline Richards Papers, the Getty Research Institute, Los Angeles (960036). Courtesy of the Estate of M. C. Richards.

academic year proved nearly catastrophic, a series of embittered negotiations by all parties toward a complete overhaul of the college's administration, which eventually culminated in all three couples' departures: the Dreiers abruptly resigned midsemester, the Alberses left at the end of 1949, Levi left in 1950, and by 1951, Richards herself had resigned from Black Mountain.[43] Recounting the events nearly two decades later, she described Levi's behavior as typical of ". . . men who have a feeling for this kind of politicking in an academic community . . . .

They love it, they feed on it . . . importing patterns of warfare and politicking and gaining supremacy."[44] But neither was she blameless in the debacle. If Levi was the petty warrior, Richards was his erudite orator, which is perhaps the best indication of her participation in the coup, translating the concerns of the rebellious youth into a tongue lashing directed at the community's elders.

The letter came back to haunt her years later, when, in 1961, she approached both couples about her desire to write a history of Black Mountain College. No doubt she was looking for more than reconciliation; she was, in effect, asking for their blessing and their consent. This went beyond mere cooperation; agreeing to extensive interviews would have meant working through a painful shared history and, finally, allowing Richards, one of the perceived instigators, to produce a judicious version of events in the retelling. "I hope you will invite me to visit you," she implored. But both parties refused, offering nearly identical replies: that Richards had only known Black Mountain in its decline, not during the interesting years of its "building up," as Anni Albers put it, which could only be properly explicated by one of its founders, namely Theodore Dreier.[45] In *his* reply, Dreier was also self-nominating, though he was more generous toward Richards, encouraging her to write a book about her own "fresh teaching. You contributed the latter to the college with a full glow, I am ready to believe, as long as you could. But you participated only in the decline and death of the college."[46]

Somehow these responses must have only strengthened her conviction, because Richards went on to apply for grants, to no avail. Both the Ford and Guggenheim Foundations declined her grant applications of 1962 and 1963.[47] In the end, Richards gave up her Black Mountain book but not before being courted by three other researchers. The winner proved to be Martin Duberman, the pioneering scholar of gay cultural history, who published the first account of the college, *Black Mountain: An Exploration of Community*, to great acclaim in 1972. As Mary Emma Harris noted, "M.C. gave her papers and the work which she had done on a Black Mountain book to Duberman, not to me. Although I might have pressed him for the material, I did not. I cannot write M.C.'s book on Black Mountain."[48]

It seems, however, that Duberman could. And did. Admittedly, his book was a self-described "subjective history," one that dispensed with the formalities of neutrality and instead reframed the writing of history as an active, opinionated practice. This was received with some trepidation, of course. One reviewer called it "embarrassing, pretentious, the very epitome of bad taste . . . Duberman erroneously promotes self-intrusion as a new type of history."[49] This same writer did also offer a dose of irony-tinged praise: "With the help of Martin Duberman, Black Mountain has finally bared its soul."[50] For the most part, his book was praised for its "intensely personal" insights.[51] Another astutely noted, "For Duberman the book was an experience in getting *close*, as a historian to something, and he did indeed get close, not only to the something but to practically everybody seriously identified with it . . . Insiders may have more to complain about his account than I."[52]

This might refer to the copious telephone interviews, from which no known written notes or records exist, entirely dependent upon his interpretations. But more likely, it refers to Duberman's own entanglements. In his footnotes, he writes: "a contemporary of the Levis at BMC told me that I had become 'besotted' with M.C.—and proportionately unfair to her husband. I don't accept that judgment, but thought it best to include nonetheless."[53] As Duberman and Richards grew close (she affectionately referred to him as "Marty"), much of this "intensely personal material" was willingly bequeathed by Richards herself. Why would Richards not just give over her notes but, arguably, also bequeath her own intellectual labor and allow her book to be subsumed by Duberman? Was she also besotted? Was it an issue of not receiving grant money and various personal approvals, or, in the end, did she lack the confidence and commitment that either scholarship or memoir demands? If so, why Duberman? Why not Mary Emma Harris?

The answers to such questions can be traced through Richards's correspondence, where a pattern in her behavior clearly emerges. A heterosexual woman with a penchant for gay men, Richards was attached domestically to a series of gay or bisexual men, all products of the same 1950s circles: the pianist David Tudor, the composer John Cage, the novelist and playwright James Leo Herlihy, and the former Martha Graham dancer Paulus Behrenson.[54] Even in avant-gardist circles, gender norms often prevailed; platonic exchange between men and women was not always optional or even possible. The secondary social status of being a woman artist was magnified by the predicament of being a woman. With the exception of Tudor, who was her lover for some but not all, of their ten years together, Richards's movement within a world of differently oriented men held at least the promise of mutuality, freedom from the burden of female sexuality and a release from the marriage entitlement of coitus. In a diary entry, Richards called this "enforced sex in the most miserable spirit—most negative. I felt like a whore. Debased. It is hard to forgive—that is hard to get out of the body memory."[55]

Above all, this other, alternative world of men she sought to join held the potential for true collaboration and parity within artistic production. Richards could produce freely, without being confined to either the traditional roles of wife or mother. Like women, homosexual men also struggled against society's recriminations. But their strategy was separatism: to develop distinct urban communities to initiate a way of living that would eventually epitomize the term "lifestyle." In *Queer Beauty*, his wide-ranging treatise on sexuality and aesthetics, Whitney Davis argues that mid-twentieth-century queer identity was wholly invested in establishing a social, rather than a sexual, identity, or what he calls "new territories of . . . social relationality."[56]

While same-sexed, the paradox of Cage's circle is that it was highly dependent upon Richards aesthetically, intellectually, and domestically. Richards's experience as a woman artist at the center of a homosocial community connotes a fluid gender identity, one of the salient features of queerness. Working in the interstices between the self and the social, Richards's own aesthetic commitments enabled a space of collective alterity.

John Cage: Self-Styled Avant-Gardist

It was due to Richards's influence that the Black Mountain community was able to experience French absurdist theater and its radical doctrines: Richards translated Erik Satie and Jean Cocteau for collective theater performances, directed several joint student-faculty productions, and, most notably, began translating Antonin Artaud's *The Theater and Its Double* in 1951, work that continued throughout the 1952 summer and into the winter of 1953. Periodically, she would read and discuss sections aloud. In her memoir, the dancer Carolyn Brown recalled, "The weekly gatherings had been arranged so that `M.C. Richards could read aloud her in-progress translation of Artaud's *Theatre of Cruelty*. After dinner, she would sit on Merce's [Cunningham] bed, knees drawn up, back propped against the wall, and the rest of us would sprawl around her."[57]

At Black Mountain, during the same summer of 1952, Richards also participated in Cage's *Theater Piece #1*, a multimedia performance intended for multiple performers, widely designated the first-postwar "Happening" in modern art.[58] Both Mary Emma Harris and Branden Joseph credit Richards's translation as a catalyst for Cage's performance:

> However spontaneous its inception, Cage's *Theatre Piece No. I* was not conceived in a creative vacuum . . . . Antonin Artaud's *The Theater and Its Double* exerted a critical influence on the Happening, as well as on [Wes] Huss's theatrical curriculum and on the general discussions of the role of the arts in American culture. M.C. Richards first heard about the book when reading Jean-Louis Barrault's *Reflections on the Theater* (1951) . . .[59] (Harris)
>
> Because Rauschenberg was present at Black Mountain College in the summer of 1952 when M.C. Richards translated and discussed *The Theater and Its Double*, he had been exposed to Artaud much earlier and much more thoroughly than most of his peers in the world of postwar performance . . . . It was, in part, in order to put these aspects of

Artaud's work in to practice that Cage organized an untitled performance that summer which came to be known as *Theater Piece #1* or the *Black Mountain College Event.*[60] (Joseph)

But *Theater Piece* was not Richards's and Cage's first aesthetic alliance: in the summer of 1948, at Cage's request, Richards translated what became the first English-language production of Erik Satie's *The Ruse of the Medusa* (1913), performed August 14, 1948 (figure 4.10). Still, she has never been officially recognized as one of Cage's collaborators.

At Black Mountain, Richards's interdisciplinary pedagogy served as the bridge between the literature and the visual arts, which manifested itself as an intensive commitment to theater, particularly in the absence of a formal drama teacher in the years between 1945 and 1950. She was also a key participant in Black Mountain's Light Sound Movement Workshop, which produced short, multimedia theatrical works using projected imagery, dance, music, and oration between 1949 and 1951.[61] Such experimental juxtapositions of media, perspective, and space not only preceded Cage's *Theater Piece #1* (1952) but also involved overlapping participants, including Richards herself.

In the annals of art history, *Medusa* has been lauded for its memorable cast full of luminary artists: Buckminster Fuller played the lead, opposite Elaine de Kooning and the now-forgotten novelist and *Partisan Review* writer Isaac Rosenfeld. Cage played the piano, Merce Cunningham danced (in a monkey suit), and Willem and Elaine de Kooning collaborated on the set design. The play was directed by two of Richards's students at Black Mountain, Helen Livingston and Arthur Penn, the latter of whom would go on to a distinguished film career in Hollywood, director of both *The Miracle Worker* (1962) and *Bonnie and Clyde* (1967). In 1950, Richards translated Cocteau's *Knights of the Round Table* (1937) and translated and directed his surrealist comedy *Marriage on the Eiffel Tower* (1923), performed that June, also with a mixed student-faculty cast that included her husband, Bill Levi, and her stepdaughter, Estelle.[62]

Knights of the Round Table (*Les Chevaliers de la Table Ronde*) was a fitting parable for Black Mountain's own struggles. Based on the medieval myth of Camelot, King Arthur, Lady Guinevere, and Lancelot are complacent in the face of their own addictions, egos, and all-consuming obsessions: King Arthur is blind and drugged, Queen Guinevere is tempted by religious fanaticism, and Sir Lancelot is consumed with guilt over his affair with the queen. Camelot's gardens are barren and brown, and the protagonists wait for the Holy Grail that they believe will restore the kingdom and its inhabitants to its previous prosperity and happiness. Instead, violence and murder wreak havoc: King Arthur murders Lancelot and Guinevere drowns in the lake. Written during Cocteau's own struggle with opium

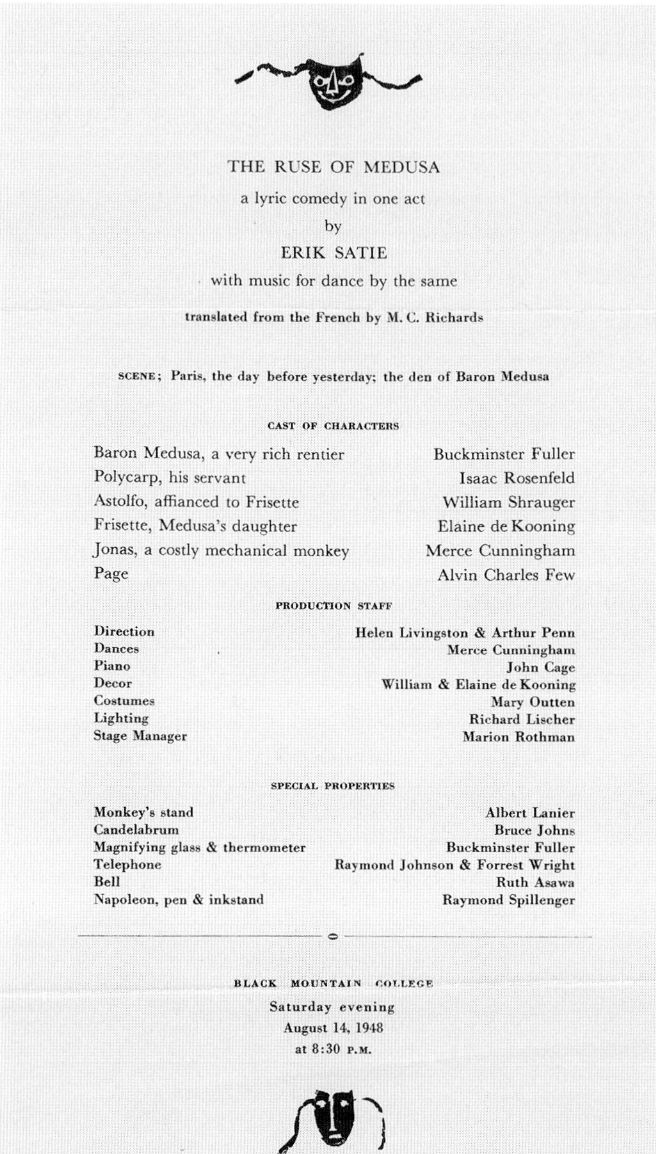

THE RUSE OF MEDUSA

a lyric comedy in one act

by

ERIK SATIE

with music for dance by the same

translated from the French by M. C. Richards

SCENE; Paris, the day before yesterday; the den of Baron Medusa

CAST OF CHARACTERS

Baron Medusa, a very rich rentier	Buckminster Fuller
Polycarp, his servant	Isaac Rosenfeld
Astolfo, affianced to Frisette	William Shrauger
Frisette, Medusa's daughter	Elaine de Kooning
Jonas, a costly mechanical monkey	Merce Cunningham
Page	Alvin Charles Few

PRODUCTION STAFF

Direction	Helen Livingston & Arthur Penn
Dances	Merce Cunningham
Piano	John Cage
Decor	William & Elaine de Kooning
Costumes	Mary Outten
Lighting	Richard Lischer
Stage Manager	Marion Rothman

SPECIAL PROPERTIES

Monkey's stand	Albert Lanier
Candelabrum	Bruce Johns
Magnifying glass & thermometer	Buckminster Fuller
Telephone	Raymond Johnson & Forrest Wright
Bell	Ruth Asawa
Napoleon, pen & inkstand	Raymond Spillenger

BLACK MOUNTAIN COLLEGE

Saturday evening

August 14, 1948

at 8:30 P.M.

Figure 4.10 Drama Program for *The Ruse of the Medusa*, performed August 14, 1948. Mimeograph on pink paper. Black Mountain College General Collections, box 28. Courtesy of the Western Regional Archives, State Archives of North Carolina.

addiction, the play is a surrealist's morality tale, emphasizing the false happiness that such an illusion provides. Each protagonist must work toward his or her own personal version of a truth fraught with psychic difficulty.

Within the Black Mountain community, however, *Knights* was a pronounced form of *remediation*, offering an immediate and immersive environment of participation that reinterpreted the play, using the central metaphor of salvation as an admonition against the self-centeredness and low morale that much of the campus experienced during the early 1950s. Cocteau's play exemplified the sorts of lessons that Richards herself later imparted in *Centering*, writing, "Even disaster is meaningful; but its meaningfulness does not save it from happening, nor us from suffering."[63]

In creating collaborative, multimedia stage productions, Richards's theatrical work was the gateway between an illegible text and the immediacy of Black Mountain's performance events. Her process-based pedagogy, steeped in rehearsals, workshops, readings, and tutorials, offered a participatory experience to students wholly new to performing avant-garde works, or what Antonin Artaud himself called "that sort of theatrical language foreign to every *spoken tongue*, a language in which an overwhelming stage experience seems to be communicated, in comparison with which our productions depending exclusively upon dialogue seem like so much stuttering."[64] It was this exact sort of foreignness that Richards sought to overcome in her teaching, stripping away fear, pretension, and uncertainty to get to an authentic self as a means of partaking fully in the world and in the making of a *live form*, which theater itself embodied.

However famed *Medusa*'s players, in its audience was another of Richards's students, James Leo Herlihy, who, like his roommate Arthur Penn, would also become well known in theater and film. Herlihy credited Richards as being "very simple and helpful" to him as a young writer.[65] His novels *All Fall Down* (1960) and *Midnight Cowboy* (1965) both became acclaimed feature films by the same names.[66] The latter (directed by John Schlesinger and released in 1969) is often considered the first mainstream queer film, characterized by an intense relationship between two male protagonists with complicated social and sexual identities. The film was the runaway winner at the 1970 Academy Awards, for Best Picture, Director, and Screenplay. Long after Black Mountain, Richards's relationship with Herlihy was among her most enduring and rewarding, perhaps her only queer relationship of mutual exchange, in which they shared and commented upon each others' manuscripts, including *Centering*, of which Herlihy wrote, "You've done a large and dangerous and rich thing: Putting yourself to paper like this. And it is good. It, the book, is its own demonstration of what you <u>say</u> in it."[67] During the 1970s, she lived for a time with him on his rural Pennsylvania farm.

While rehearsed, theater at Black Mountain was unlike traditional theater in that it was performed only once, for the entire community, in the dining hall after dinner, usually commencing at 8:30 pm. Satie (1866–1925) and Cocteau (1889–1963) were virtually unknown in America, and the performances of their work were the first-ever English-language productions, using Richards's scripts. Such one-off events foreground *Theater Piece*, which was also an "unknown" work.

While there is no photographic record of *Theatre Piece #1*, there are many firsthand accounts, none of which correlate entirely, but which at least reach consensus on the roster of performers: Cage himself, reciting a speech; Tudor playing the piano; Cunningham dancing; Robert Rauschenberg, whose *White Paintings* were suspended from the ceiling, playing records on an old gramophone; and Richards and Olson reciting poetry from the top of a ladder.[68] Duberman has done a noteworthy job of compiling and overlapping eyewitness accounts (from Francine du Plessix, Caroll Williams) some of which were recounted to him in interviews fifteen years after the event itself (David Weinrib, Katherine Litz, Merce Cunningham). Merce Cunningham's online archive describes the event as such:

> [Theatre Piece], 1952 arranged by John Cage, poetry by Charles Olson and MC Richards, film by Nicholas Cernovitch First Performed: Black Mountain, NC; 16(?) Aug 1952 Music: David Tudor Design: Robert Rauschenberg Dancers: Merce Cunningham[69]

In a 1981 interview, Cage confirms this list of participants.[70] Remaining in dispute is the exact date of the piece (possibly August 12, 16, or even early June), its length (varying in length from 45 minutes to two hours), and, strangest of all, the content and whether or not a script existed. In various descriptions, Cage either recited a speech on Zen Buddhism from a lectern at the back of the room, or atop a ladder; Tudor played the piano; Cunningham danced, or was chased by a small dog, or danced *while* being chased; Nick Cernovitch screened a film he had made; Robert Rauschenberg played (possibly Edith Piaf) records from the position beneath his *White Paintings*, which hung from a string; and Richards and Olson recited poetry from the top of a ladder. While legend has it that all of the events coincided, according to most of the participants, this was not the case, as they occurred in predesignated intervals. Duberman argues that by imposing time limits and a specific order upon his collaborators, Cage went to great lengths to impose "an intentional structure" upon a work of art that has long been celebrated for its experiments with randomness and chance.[71]

In the late 1980s, William Fetterman, a performance studies scholar, picked up where Duberman left off, continuing to excavate the event through firsthand interviews with participants Cunningham, Richards, and Tudor. In his resulting dissertation, Fetterman includes a drawing that Richards made during a 1989 interview with the author, in which she indicated one of Franz Kline's paintings was also hung from a string. Here, Richards recalls reciting her own poetry, contradicting other renditions, which had her reciting Edna St. Vincent Millay's poetry (David Weinrib) or something in French (Katherine Litz).[72] Five years before her death, Richards again reflected back on the work, stating: "We were all somewhat contaminated in those days by our studies of Zen. Unpredictability, indeterminacy and paradox—all those sorts of experiences were very congenial to me."[73]

The Ruse of the Medusa was the final event of a weeklong Satie festival that

Cage had initiated and performed, both as concerts and lectures, during which he berated Beethoven and upheld Satie's antirhythmic style as a standard of achievement, much to the consternation of Black Mountain's more traditionally inclined classical music enthusiasts, all of whom were German and rather pointedly referred to by Cage and his cohort as "the Germans."[74] These were the Alberses; the musicians Fritz Cohen and Erwin Bodky, who was teaching on Beethoven at the time; and Johanna Jalowetz, voice teacher and widow of the famed conductor Heinrich Jalowetz, a contemporary of Schoenberg. Not coincidentally, Arnold Schoenberg had been Cage's teacher in Los Angeles, so Cage's rejection of a classical genealogy had roots in his own formal training. However, Cage and his circle were not the only ones to poke fun. The college's resident carpenter and furniture maker Molly Gregory reported that the kitchen work crews complained of "the endless sauerkraut, which smelled."[75]

As an influential figurehead of the French avant-garde, Satie had a profound effect on twentieth-century music, both a writer and a composer of minimal compositions, many of which contained bizarre Dada-like instructions to the performer at a time, a purposeful rejection ofromanticism, which was the pervasive turn-of-the-century style. Satie is also remembered for "furniture music," composed as a background accompaniment to modern life. In the vast body of Satie literature, Satie figures as a Duchampian personality, a trickster whose legacy stems from his constant break with and rediscovery by successive generations of the avant-garde, which included a young Cocteau.[76] At Black Mountain, it was even rumored that Cage had visited Satie and acquired scores and scripts firsthand. But even Cage could not have been that precocious; he would have been only thirteen when Satie died. It would have been more likely that he had been to see Cocteau, who lived until 1963, though there is no documentation that the two ever met. Openly homosexual, Cocteau was an iconic figurehead to the generation of artistically inclined gay men who came of age pre-Stonewall. Unconstrained by discipline or allegiance to any one medium, Cocteau, as an artist and a personality, embodied liberation, prior to the existence of liberation politics.

While Harris and Joseph have pointed to Artaud's (and, by extension, Richards's) influence on Cage's *Theater Piece*, I am proposing an additional, Richards-inflected source: Satie and Cocteau's famed Paris collaboration, *Parade* (1917), known as the first modern, non-narrative ballet. As Richards was Black Mountain's reigning expert on modern French theater and Cocteau's oeuvre, she would have been well aware of *Parade*'s impact on the history of avant-garde French theater and its intersection with ballet history.

Throughout the sabbatical year she spent in France (1948–1949), she had voraciously attended film, theater and dance performances, including Cocteau's film *Les Parents Terrible* and play *L'Aigle à Deux Têtes* [The Double-Headed

Eagle].[77] As the only faculty member who had seen Cocteau as it debuted in Paris, she was the conduit for these experiences at Black Mountain. Likewise, she was also shaping Cage's own sensibilities as a conduit, translating her own experiences for him, which he in turn incorporated into his own writings. For instance, her interest in Russian ballet, found at the end of his essay "The Future of Music: Credo" (1957) Cage notes:

> M.C. Richards went to see the Bolshoi Ballet. She was delighted with the dancing. She said, "It's not what they do; it's the ardor with which they do it." I said, "Yes: composition, performance, and audition or observation are really different things. They have next to nothing to do with one another."[78]

Performed May 18, 1917, at the Théâtre du Chatelet in Paris, *Parade* was devised as a meeting ground of disparate avant-garde and popular sensibilities, incorporating dance-hall music, cubist costumes and stage-sets, and a metanarrative by Cocteau, essentially, theater about theater. The production was staged to benefit disabled World War I veterans, in attendance alongside Parisian society. Now legendary, the audience's negative response to the production was long attributed to its radical aims. However, in his groundbreaking social history of the period, art historian Kenneth Silver counters this assessment, arguing instead that the ensuing chaos and negativity of the crowd toward the performers stemmed from the fervor of French nationalism, a consequence of the creators' collective non-veteran status (none had served) and the dual foreign identities of both Picasso and the Ballets Russes itself.[79]

Parade was commissioned by the Russian ballet impresario Sergei Diaghliev, with stage sets and costumes designed by Pablo Picasso, music by Satie, choreography by Léonide Massine (the Ballets Russes soloist who took over the male roles after Nijinsky departed), and the libretto written by Cocteau. Richards spent her 1948–1949 sabbatical year in France and England. Coincidentally Cage also came to Paris, on a Guggenheim Fellowship in 1949. As she preceded him in Paris, she might very well have seen Picasso's stage curtain (figure 4.11). The work is enormous, weighting in at eighty pounds and impossible to miss at 34 by 55 feet, which was at that time permanently on view at the Musée d'Art Moderne, and she may have recommended the work to Cage. Indeed, her journals of the period map her museological forays throughout postwar France, including numerous repeated visits to the Louvre, the Musée d'Art Moderne, and the Impressionist Museum.[80]

In this way, Richards herself becomes the conduit between two transitory, Dada-suffused theatrical events, traversing the space of thirty-five years by offering to Cage the combined artistic legacy of his heroes Satie and Cocteau at their peak collaboration, in concert with Picasso and his outsized theatricality.

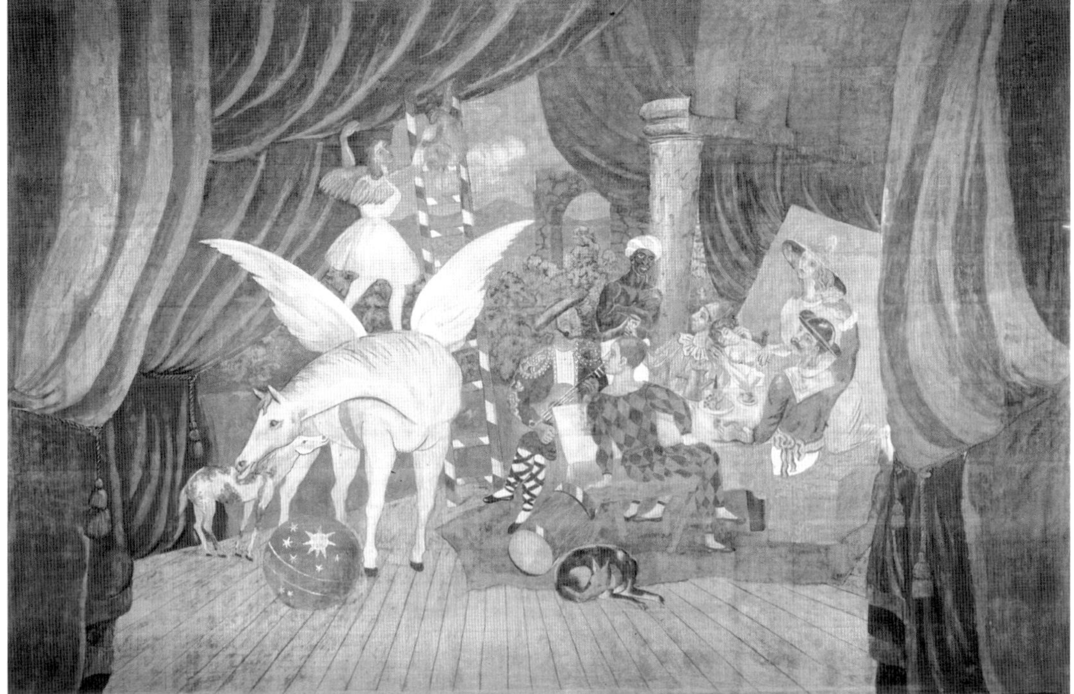

Figure 4.11 Rideau de scène du ballet "Parade." Overture curtain for the ballet, *Parade*, 1917. Pablo Picasso (Spanish, 1881–1973). Tempera on canvas, 10.5 x 16.4 meters, 40 kilograms. Musée National d'Art Moderne, Centre Georges Pompidou, Paris. © CNAC/MNAM/Dist. RMN-Grand Palais/ Art Resource, NY. Photo by Christian Bahier. © 2015 Estate of Pablo Picasso/ Artists Rights Society (ARS), New York.

This is suggested by the visual tropes Cage employed in *Theater Piece*, which directly correspond to Picasso's imagery. Picasso's theater curtain remains an iconic visual symbol of *Parade* and of the artist's immediate postcubist production. A narrative work, the curtain is a *tromp l'oeil* of a stage, midproduction: the players, including a troubadour and two harlequins, break for an impromptu meal behind a set of parted stage curtains. A slanting ladder divides the space, and on its other side is an even more fantastical grouping: a ballerina standing upon a winged mare who is busy grooming her foal. Perched high in the ladder is a monkey wearing a crown of laurels, the poet's symbol in the classical world.[81] The monkey was a recurrent image in both Satie's and Picasso's oeuvres; in *The Ruse of the Medusa*, Cunningham had danced in a monkey suit, and in her exhibition on *Parade*, curator Deborah Rothschild interpreted the monkey to be resolutely symbolic of the artist, making reference to Picasso's own self-caricatures, in which he repeatedly drew his own likeness as an ape.[82]

In *Theater Piece*, Cage literalizes the image of the poet on a ladder, stationing

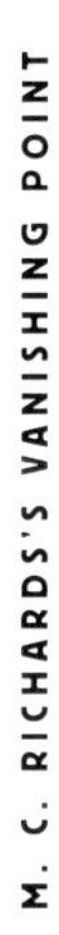

his two poets, Olson and Richards, atop ladders, where they performed their own work. Picasso's fanciful image is extracted and realized as an actual poetic recitation, rather than a metaphoric one. From high up upon her poet's perch, then, Richards herself was a *remediation*, a rupture in the perceptual immediacy of the event while embodying a symbolic representation of something in the past. Such "seeing through" is emblematic of the perspectival force of Picasso's imagery, as well as the transference from one event to the other, a visual imaginary haunting *Theater Piece*, insuring that, as Bolter and Grusin write, "the new medium remains dependent upon the older one in acknowledged or unacknowledged ways."[83] Thus, *Theater Piece* itself remediates, collapsing formal concerns, historical specificity, and live performance into a single entity that defies and competes with the form it appropriates.

In Cage's work, the poets atop the ladder were meant to be *heard* but not necessarily *listened to*, as they coincided with or preceded other spectacles. In this way, Richards and Olson become the epitome of Picasso's monkey artist—an overlooked soothsayer with a semiattentive audience of players. In Picasso's narrative, the players are serenaded by a troubadour seated among them. They are at the end of their meal, at a table outfitted with a plate of bread and fruit, a wooden-handled coffee pot, and a single demitasse cup and saucer. As a participatory gesture, Cage *also* utilized coffee cups in his *Theater Piece*. Each member of the audience received one. As he described in 1965:

> In each one of the seats was a cup, and it wasn't explained to the audience what to do with this cup—some used it as an ashtray—but the performance was concluded by a kind of ritual of pouring coffee into each cup.[84]

Perhaps these were even the same cups that Karnes and Weinrib had crafted earlier in the summer, further collapsing the media hierarchies that lingered at Black Mountain, for the cups themselves became bit players with a back story. What the audience members were "to do" with the cup was self-selecting and defamatory: to ash, or not to ash. Either decision had symbolic consequences; to do so was to reenact the Josef Albers's epithet "ashtray art"—the second abuse of Karnes and Weinrib's dining hall gift that same summer, sanctioned by the context of *Theater Piece*, signaling the triumph of performance over craft, the visual over the tactile, sheer spectacle over artistic labor.

But *to not ash*, that is, to have chosen the hot coffee, was, in actuality, Cage's triumph: the ultimate remediation of Picasso, expanding the curtain's narrative in a new, interpretive framework. Such a recalibration was, in its own time and space, also an action of equilibrium, in the sense that the participants and audience were one and the same; as in Picasso's painted audience, Cage's audience was also comprised of other artists and cultural sophisticates. At midcentury, Dada was still very much a part of living memory. There is even cultural resonance between Picasso and Rauschenberg: *Parade* was the first stage décor ever created by Picasso, who had never set foot in a theater, and Rauschenberg was also trying his hand at theatrical design for the first time.[85]

Black Mountain's initiation of an American model of multimedia performance offered Cage a platform for experimentation. As a translator, director, and player with a history in theater, Richards was at the forefront of this innovation, an uncredited collaborator. *Theater Piece #1* can thus be construed as Cage's own self-conscious rendition of mythmaking, an attempt to join his own personal canon of much-admired predecessors in the most unassuming of ways: the piece occurs as any other Black Mountain College performance, a singular event, without documentation but for its legendary audience, better than Cocteau's, the *right* audience, this time without need of sets, or costumes, or a theater curtain for their word-of-mouth telling repeated versions of it, over and over. Converted into an oral tradition, through retellings and, later, numerous formal interviews, the work itself slips into history, fully assimilated.

Gatehill Cooperative Community, Stony Point, New York, 1954

From Gatehill's origin, there were disparities regarding the philosophical premise of the cooperative. Richards had devised the formation of the community as early as 1949, writing to Paul and Vera Williams from her sabbatical in France: "From time to time, when we are feeling strong and courageous, we speak of coming together with you and with others who are interested to talk about starting another, a new place. I wonder if we could. A place to live and work together."[86]

For the first year of the cooperative's life, Cage, Richards, Tudor, Karnes, and Weinrib all shared a tiny cottage, with only a hand pump for water and an outhouse, while the other homes were being built piecemeal. The close proximity between Richards and Cage, and their intimate domestic arrangement at Gatehill, can itself be construed as influence. In her memoir, Carolyn Brown describes Stony Point's unbearably close quarters:

> For more than a year the five of them lived together in their cramped and awkwardly intimate proximity while Paul set to work designing their houses. They were the pioneers who began "The Community on The Land." The Williamses, who had children, rented a larger house down the road.[87]

Cunningham never officially lived at Stony Point, unable to reconcile country living with the urban audience required by a fledgling dance company. Cage himself once said that all his works composed before 1970 were written either for Tudor or with Tudor in mind.[88] In Cunningham's absence, Tudor quickly became Cage's muse, or, as one commentator put it, the "ideal interpreter" of Cage's own works; the two couples' lives became even further enmeshed.[89] In subsequent letters to Olson, Richards reports the cooperative arrangement: each person put in ten dollars for groceries and drink. But it seems that, as the only woman in her immediate circle (Karnes had just had a baby), she was responsible for most, if not all, of the laborious domestic tasks.[90] She describes "getting this little house ready to be lived in—painting and sanding and furnishing and taking off old wallpaper and patching the plaster, etc., etc."[91] She also details how she had learned how to make cream cheese from yogurt and how she had planted her first garden.

Richards's social and pedagogical commitments were as clear as Cage's equal and opposing lack of interest. As Karen Karnes revealed, "I think our only 'groupness' was to do with MC. Yeah. Maybe Paul [Williams], but not with John Cage."[92] However, it was the continuities between Black Mountain and Stony Point that strongly defined Richards's pedagogical project, driven by her conviction that experimentation required community. Richards carried the idea of the college with her as a metaphor of failure and redirection. Trying to entice Charles Olson to join, Richards outlined her plans for the "weekend Black Mountain":

> And I have an idea for a building for Paul [Williams] to make in the theater field which would be useable for theater, studios, whatever. A circular space in the center, with small rooms around the circumference, which would open on the central court and could be used for audience & dressing rooms when they weren't used for study or sleeping. Performers, teachers, artists could be invited for regular or irregular week-ends. People could eat down at the Blue Note a short distance away. Fees would be low: something like $10 a week-end. If we used the fees from, say, 25 visitors, we could pay for the building in a year or two. And I would like to have institutes, which I would hope you would be interested in planning if not arranging and being present at. I would want any part or the whole, whatever you would be interested in. It could be a real open center for inquiry and work. The possibilities are endless, it seems to me. And we have a real potential source of funds, if Paul is interested. I would be willing to do administrative diddle-daddle (hardly the word for it, I know) . . . and I would like to talk with you or hear from you again on that whole curriculum re-structuring that was once on our minds. From scratch.[93]

Though there was clear interest in building a theater for Cunningham, Carolyn Brown describes him as a reluctant participant in the community, one who "rarely visited John out there, and when he did, he complained of allergies to flora and fauna, mosses and molds, and dampness from the brook."[94] Neither did Olson ever elect to join, possibly never even visiting. Instead, after Black Mountain, he returned to his hometown, the seaside fishing community of Gloucester, Massachusetts, in July of 1957, only a few months ahead of Richards's barrage of earnest invitations.

In subsequent letters to Cage, Richards indulged his obsession with mushroom hunting, offering him informal courses on musical composition *and* botany.[95] But for Cage, the real draw of Stony Point seemed to be the freedom to live and work cheaply, among a community of serious artists, without the distraction of teaching. Artist-driven colonies and utopian communities had long flourished on the East Coast and in the Hudson River Valley in particular. The Arts and Crafts colony of Brydcliffe (1902–1929) had thrived in nearby Woodstock, New York. Even closer to home—also in Rockland County, New York—was Threefold Farm (established in 1926), the anthroposophist community where Janet Leach (née Darnell) had lived and taught. Stony Point ran the risk of being one more underfunded vision. Yet Williams emphasized that the community was *not* "an artist's colony," rather ". . . a place of full and immediate use . . . what's required is the use of the artist as tool, to give form to material (human life) and to take its form from material and the work at hand. Then, to carry this out, a workshop."[96]

While Richards's ideas never materialized formally at Stony Point, it is clear she was far and away ahead of her own peer group in terms of envisioning the flexibility of nontraditional learning, via weekend retreats, institutes, and short-term programs. Since the 1970s, such programs have grown exponentially throughout the United States, from low-residency MFA programs to university extension programs to myriad new age organizations like Esalen (Big Sur) and Omega Institute (also in the Hudson Valley), which advocates for the sort of whole living and learning Richards herself had begun to conceptualize.

For most of the twentieth century, professional artists relied upon residency programs like Yaddo (Saratoga Springs, NY) or the MacDowell Colony (Peterborough, NH) for bouts of quiet retreat. These were live-work environments that expressly prohibited classes or instruction of any sort. But Richards seemed unconcerned with offering opportunities to professional artists, who already exceled at self-discipline. Rather, she seemed to thrive on the educational and creative enrichment for self-proclaimed students of any age. "Weekend Black Mountain" would provide situational learning aimed at the sort of population with which she herself readily identified: the amateur.

At Stony Point, Richards also mentions creative work, writing the libretto to cellist and composer Seymour Barab's one-act opera *Chanticleer* (1955).[97] Though she lived with and among musicians, she was not invited to collaborate with them. Perhaps participation came with conditions, such as in November of 1957, when she made a turkey for Cunningham's entire company because everyone was too busy with rehearsals to stop for Thanksgiving.[98] Reciprocity was obviously not part of the social contract. However, she continued to perform unacknowledged intellectual labor:

> But anyway then yesterday Merce came out here to the country where I am and we all are and he brought me a present of $25—if you can imagine that, I hardly can, it's the first cash I've seen like that for so long—only it wasn't altogether a present; you see I ghost-writ an article for him for the 7 Deadly Arts publication, so he split the fee with me. I'd forgotten about it. Anyway, that's how I happen to have a fiver on me . . .[99]

The fiver Richards promptly sent off to Olson, to put toward the *Black Mountain Review* fund, now being run by Robert Creeley, whom Olson had installed as editor—a journal which, to Richards's great disappointment, never once published any of her poetry, though she had established it. She recounted the indignity to Olson two years later, in 1956:

> And it hurt my pride no little bit that he [Creeley] could find nothing of mine for the Review—My Review, which I and my students started. I think, if you will pardon me for saying so, that you should be careful about putting words into these boys' mouths . . . I don't like to be so drearily treated by such and [sic] underling. I am no schoolgirl.[100]

Neither schoolgirl nor housewife, but sometimes treated as both, Richards's role among her peer group is vexing: was the avant-garde as sexist as the rest of 1950s America, or was it more so? In his study of queer artistic networks in 1950s New York, Gavin Butt remarks upon "the generally disadvantageous condition of being a woman artist—gay or straight—in the masculine fifties art world."[101] Richards's circumstances affirm the era's now widely acknowledged sexism. But like so many other women of her generation, she too was implicated, looking to male artists for affirmation of her own creative work, the poetry she continued to write while at Black Mountain and, later on, her book *Centering*.

Earlier in time, Olson had approached Richards in a somewhat paternalistic manner, playing the part of the established, veteran poet in an attempt to instruct Richards about her own poetry. In 1951, she sent him a satchel of twelve poems for review, and he made deep, decisive, severe cuts, writing "NO! BAH!" in all caps on nearly everything. Of the twelve poems, only two passed muster. These were called “Organization” and “For Political Reasons,” both from 1947. Taken together, the two poems function as a critique of power and directly reference Richards’s own experiences at Black Mountain. Written as faux-manifestos, both are prose poems inflected with a dry humor, oblique narratives that describe how she came, ultimately, to view authority as ultimately authoritarian. “Organization” begins:

> Organization is not interesting, why.
> If I am the chairman and you are on my committee it is not very interesting.
> Or if you are the leader and we are your group and stay so who isn’t bored.

And an excerpt from “For Political Reasons”:

> I don’t think politics when a man intimidates, I think human;
> what he conceived politically grows face and hands,
> image of government is self-portraiture[102]

Given his own experience with the circumstances of administration, Olson must have appreciated Richards’s dry-eyed appraisal of the college. In response to these two poems, he praised her rather gruffly, encouraging her to forget her own education, and training, and write in a manner “not consciously poetic”:

> for the curious (becoz true) thing is, that, when you go by your own bent, when you are not consciously poetic (forget all previous patterns of such behaviour)—write either automatically (as you did that sonnet that day) or, as in the two best of the verse I took away, the two on committees, what one might call, with the head, you are quite effective and that is why you must, for sometime, forget poetics, and previous poets, and hew to these two paths.[103]

Richards’s poetic sensibility was more in concert with the visual and performing arts than Olson’s Black Mountain School. Her poetry tended toward ideas, rather than lyricism. As a poet, Richards was self-taught and distinctly out of place in the circle that developed around Olson: she was faculty and she was female. Trained as a scholar, Richards taught courses in Shakespeare and Greek tragedy, rather than Olson’s whiskey-fueled writing seminars that went on for eight or twelve hours at a stretch.[104] While Olson’s praise is genuine, it comes across as fairly disinterested; Olson was a man of his time, who did not necessarily believe that women could or *should* be artists, routinely directing his energies toward his all-male cohort, which included Robert Creeley, Robert Duncan, Fielding Dawson, and Michael Rumaker. Even James Leo Herlihy expressed disdain for

Olson's pedagogy, stating, "He had a coterie, and it was around him . . . Olson has an aesthetic that was like law, except that nobody would have ever said so. But, boy, they didn't break it."[105]

Yet during Richards's time at Black Mountain, and in the larger scope of the 1950s, there were very few *legitimated* women artists; very often, they were in secondary roles, wives, like Anni Albers, or helpers, like the woodworker Molly Gregory, who candidly recounted that Josef did not find her qualified enough to teach sculpture, though she had a strong teaching background.[106] But the real question remains whether he would have found *any* woman he believed qualified enough to teach sculpture and whether Olson would have found *any* poetess (as they were then contemptuously termed) worthy of his attention. Both Ann Eden Gibson and Michael Leja, in their revisionist histories of abstract expressionism, demonstrate the limited availability of art world roles for women beyond either the wife or the art patron/dealer.[107]

Indeed, there was another woman poet who arrived at Black Mountain just as Richards was leaving: Hilda Morley (1919–1998), who came in 1952 with her husband, the German avant-garde composer Stefan Wolpe (with whom Tudor had studied). Morley taught literature and Hebrew during the four years she was at Black Mountain, but she has also gone unacknowledged, a mere footnote in the literature, or "little more than a piece of luggage brought along by her husband," as scholar Brian Conniff noted.[108] In his careful reconsideration of Morley, he comments further on Olson's sexism:

> Even a brief consideration of Olson's career—with his proclivity for manifestos, his macho posturing, his "great man" approach to teaching, his tendency to construct militant counter-canons, and his often excessive attempt to influence his protégés—makes it seem just about inevitable that women's voices would be silenced in the construction of a Black Mountain canon . . . and the usually reluctant consideration of Levertov as a token woman in (or out of) the movement.[109]

Certainly Richards can be claimed here as well, a suppressed and overlooked voice influential only as a teacher—a conventionally acceptable female role—rather than as an artist. During this era, it was the few women she encountered who affirmed her artistic practice. Richards shared a pottery studio with Karnes until 1964 (figure 4.12), producing eclectic functional wares, such as *Dragonfly Tiles* (ca. 1960s), which could be either wall mounted or used as a decorative trivet (plate 6). Others in her orbit included Peggy Penn [wife of Arthur], who also made pottery; Anaïs Nin, who had been encouraging about her poetry years before, when she had briefly visited Black Mountain in 1948; and the English-born poet Denise Levertov, of whom Richards wrote, rather pointedly, to Olson: "She

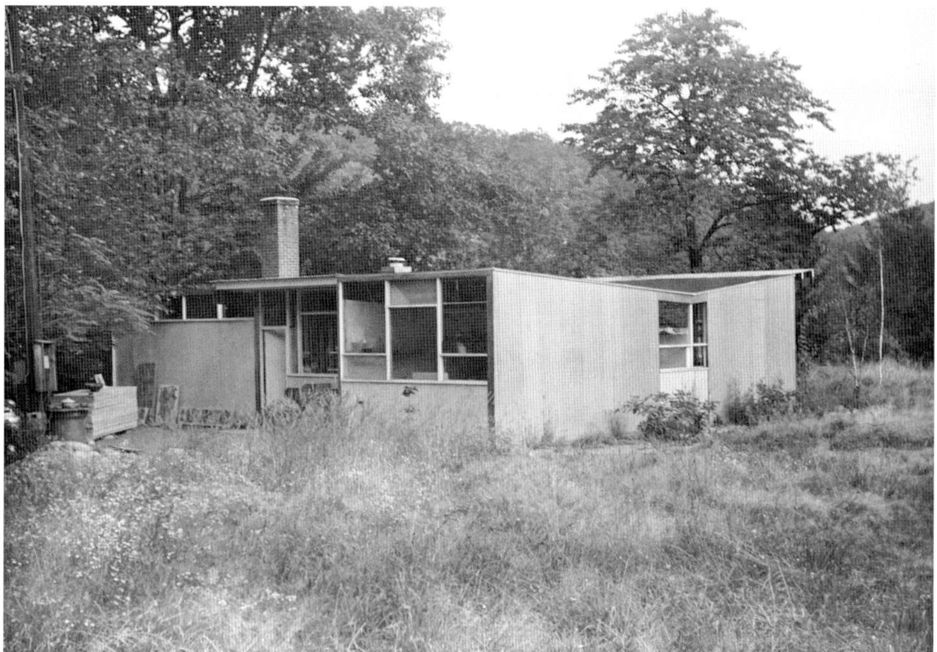

Figure 4.12 Pottery Studio at Stony Point, New York. Gatehill Cooperative Community, ca. 1960. Designed by Paul Williams. Collection of Mary Caroline Richards Papers, the Getty Research Institute, Los Angeles (960036 ADD2). Courtesy of the Estate of M. C. Richards.

talks to me about my poems. Nobody else does. She is very downright. I like that. She likes some of my work—thinks it should be published. Music to my ears."[110] Levertov also filled the role of the female saboteur. A longtime correspondent and close friend of the poet Robert Duncan, she sent Richards's poems to him without permission in 1958, writing the following:

> I enclose copies (wd. you return please?) of the first 2 poems by M C Richards that I really like. Others seemed in a curious way, that is in a way disagreeable to me, female poems, something menstrual or hysterical about them, not in their content but in the very language—something I couldn't possibly say to her because it was something no-one, I'm sure, could possibly smell about themselves.[111]

Levertov's objection to Richards's poems leads down a dark path: her characterization of female subjectivity as "menstrual or hysterical" is a toxic conflation of internalized misogyny, drawn out into a metaphor in which Richards is perceived as having an offensive (menstrual) odor or air about her, a poetic persona that was acrid and shrewish. Arguably, Duncan himself was highly influenced by Levertov's critique in that he returned also to the language of the odorous. In

1963, as the anonymous peer reviewer for Richards's *Centering*, he describes her prose style as containing "gassy losses of form."[112]

Born in 1916, Richards was in the same peer group as artists such as Hedda Sterne (b. 1910), Elizabeth Catlett (b. 1915), and Elaine de Kooning (b. 1918), one generation younger than Louise Nevelson (b. 1899) and one older than Levertov (b. 1923) and Joan Mitchell (b. 1925). Overall, prewar women artists were staunchly antifeminist, notorious for denying the art world's entrenched sexism. Largely male identified, their basic professional strategy was to form deep professional and personal alliances with male, rather than female, peers. This worked better for some than others but, as a pattern, seemed to simultaneously launch and inhibit them as young women (Catlett and the Spiral Group, O'Keeffe and the Steiglitz Circle), which led to the breakdown of their fundamental alliance, most often a marital or mistress relationship (Frankenthaler and Motherwell, Mitchell and Jean-Paul Riopelle), followed by a kind of endurance—a long period of isolation and ambition as she aged, culminating, ultimately, in triumph, outliving, and, ultimately, outproducing most of her peers. Thus, in the twilight of her life, she amassed an overlooked, but important, body of work, leading to her eventual reclamation (Dorothy Dehner, Pearle Fine, Lee Krasner, Sterne) and, at the very elite levels, the canonization of her oeuvre (Louise Bourgeois, Joan Mitchell, Alice Neel, O'Keeffe, Nevelson).

If the path of the prewar modernist woman artist is habitual enough to constitute its own dramatic genre, then Richards's also fits this pattern. Like so many other women before and after her, Richards also reached an impasse. As a trio, Cage, Cunningham, and Tudor performed all over Europe for six weeks in 1954. The tour is legendary for its influence on the German avant-garde composer Karlheinz Stockhausen, who was prompted to begin working with open-ended forms. As it was Tudor's first trip to Europe, it seems pointed that a more seasoned and bilingual Richards, who had already spent a long period in Europe, hadn't come along. In this very fruitful collaboration, Richards is the silent fourth, at home, tending the garden. As a principal dancer who did tour frequently with Cunningham's company, Carolyn Brown provides a glimmer of Richards's bitterness at her own situation, recalling: "M.C. Richards once said that she wished I would come back from one of our tours 'scathed.'"[113]

Richards's spitefulness is an indication that she felt herself to be damaged in some profound way by the obvious intimacies of the group and her own outsider status, no matter how tirelessly supportive she was. Unlike Brown, she was not in a long-term marriage and was clearly not enjoying the social privileges that marriage offered. In fact, Richards was twice-divorced (Levi was her second husband), an anathema to the rigid social conventions of the 1950s.[114] Nor could she depend upon the iron-clad heteronormativity of the abstract expressionist set, as the "art-wife." During the trio's frequent travels, many of her letters are addressed to "Dearest John and David," as though they were a distinct pair, with "kisses to Merce and the girls [Brown, et al] and Bob [Rauschenberg]," offered in closing.[115] If Tudor was Cage's muse, his subject position was, according to the tenets of modernism, feminine. This particular queering of the male artist/female muse dialectic—a longstanding source of feminist consternation—can be interpreted as differently oppressive, a clear source of Richards's personal and professional frustration, thwarting her attempts at collaboration. This was perhaps even more painful in that she had been welcomed early on, as a central intellectual figure at Black Mountain. Only three short years after *Theater Piece*, being a woman in an all-male peer group actively excluded her from meaningful artistic participation.

Writers as Readers

Written during the period from 1959 to 1964, *Centering* made its public debut as an invited lecture given on January 28, 1962, in Middletown, Connecticut, at the regional meeting of the Society for Connecticut Craftsmen and Friends, with the Wesleyan Potters as hosts. The group was looking for inspiration, and Richards provided it. She was, by all accounts, a dazzling lecturer, interspersing her poetry, always alternating between the registers of artist and intellectual. Perhaps she was still waiting for the *right* audience, as poetry forms a significant portion of *Centering*, encompassing two of six sections, one-third of the book.

Centering's commissioning editor was the philosopher Norman O. Brown, who taught at Wesleyan and with whom Richards had an intensive correspondence and possible affair during the early 1960s. At the onset, they exchanged manuscripts: he, his Phi Beta Kappa address at Columbia's commencement, she, her Artaud lecture, delivered at the Living Theatre, all while confiding in Tudor, ex-lover-turned-confidant, the extent of the liaison, writing:

> . . . By the way, N.O. Brown sent me a copy of his Phi Bete [sic] Oration at Columbia, and it is quite astonishing. All about how "mind is at the end of its tether," all about "second sight," the need to be "super-natural," the necessity to reenter the mysteries, etc. etc. "all his vocab." jeepers. How the god must come to birth within, the spiritual eye and the physical eye be one, etc. Goodness me. No wonder he's been pursuing me so zestfully, if you know what I mean, wanting to read my Artaud lecture, poems, etc. It's amazing, the letters I get from him; his crisis, and all. . . . Now of course the connections are clear: my interest, through you (?), in R[udolph] S[teiner], and my experience, however brief, with Jung's techniques for making contact with the unconscious, etc. My academic background, my work in poetry—it all gives NOB a turn. And he gives me

> a turn: he speaks to me about the deepest matters; asks me questions of the deepest "spiritual" kind. And that, of course, is very moving to me. He is so ambivalent and confused he doesn't know what to do with himself: is trying to gain some techniques of meditation, etc. I gave him my [Owen] Barfield books, and the Redemption of Thinking to read. Wish I had a copy of the Philosophy of Spiritual Activity to lend him, since it's out of print and he can't buy it for himself. It is all very odd and absorbing. Must be handled with care.[116]

Through Richards's description, we see Brown (or Nobby, as he was well known) as a permanent, if petulant, seeker, searching for direction and meaning, guided by the gentle hand of Richards, who understood and was intrigued by the delicacy of her new charge. Equally, Brown must have been smitten with Richards's free-wheeling bohemian lifestyle: his Ivy League Columbia, to her Living Theatre, is the epitome of New York's uptown/downtown dichotomy. Discovering that Brown's mother (with whom Richards also had a brief correspondence) was a member of the Theosophical Society (a nonsectarian branch of Christian philosophical thought), she directed him toward her own mystical inclinations, specifically, the English literary scholar Owen Barfield and the German theosophist and philosopher Rudolph Steiner, standards in her personal canon. Richards would come to write two books on Steiner's pedagogy, *The Public School and the Education of the Whole Person* and *Toward Wholeness: Rudolph Steiner Education in America*, both published in 1980.

Richards also had a hand in introducing Brown to Cage, as the two became quite good friends and mutual admirers, a friendship that deepened during their overlapping time at Wesleyan, while Richards, almost predictably, receded into the background. Brown appears to have ended their relationship very suddenly at the end of the summer of 1960. It is the only letter in which he mentions his wife:

> It feels good today to have been with you yesterday—the really true things come as surprises don't they? I got up this morning with well-being in my breathing and was surprised—And was surprised again when it expanded and included my wife as I sat at breakfast today telling her about you: and she caught it too and sent it back. Isn't that nice? So let's leave it right there. With one more plane lane?[117]

This abrupt retreat was no doubt shockingly hurtful and even callous. Only two weeks prior, he had written adoringly:

> Trying the meanwhile not to shrivel up with inadequacy. You are right: it is not the word it is myself I do not like. Trying not to shrivel: trying to bear the beams of love . . . For right now in my labyrinth you are the only clue: even if I don't see yet how to use it."[118]

Despite their fraught and severed connection, Richards's relations with Brown persisted into the 1960s, as did her relationship with Cage. The three were linked intellectually, their books all released as a unit, three of four authors (the fourth being Owen Barfield, whose presence can be credited back to Richards, having introduced it to Brown) on the Wesleyan University Press booklist known as "Interdisciplinary Explorations":

> Worlds Apart A Dialogue of the 1960's (Owen Barfield) @ $5
> Life Against Death: The Psychoanalytic Meaning of History
> (Norman O. Brown) @ $6.50
> Silence: Lectures and Writings (Cage) @ $10.00
> Centering: On Poetry, Pottery, and the Person (Richards) @ $8.00[119]

Brown had initially read, and rejected, Richards's manuscript, which she recounted to Olson:

> After a long, long silence, the Press suddenly came to life and it looks as if the poor thing will be published. It has been much hacked at. There is so little enthusiasm for it, I privately wonder why WUPress goes through with it. A strange experience. I lately learned that Robert Duncan has been hired to read it and advise on cutting the manuscript. The press has never given me the benefit of Duncan's estimate. I asked for his letter, and they sent it. It is the first real hard-boiled attention my ms has received. Too bad I didn't have it sooner. And of course Duncan doesn't like the book. It doesn't have the tension, the form, he requires of a "work." Its energy fluctuates, "gassy losses of form" as he expresses it, and all this I relentlessly insist upon. So, as I say, it's discouraging. The book goes forward, in its strange delirium, talking, talking, and who is listening. Nobby Brown also read it in ms for the press, it turns out, and didn't like it. John Cage read a chapter and found it boring. It tempts me to think there may be something in it after all. Something that does not confirm what people already think. Bores them, puts them to sleep.[120]

Clearly, Richards had not found the *right* audience. Amid the collective rejection of her work was the glimmer of something else—a form of refusal. First, her own refusal to be either a traditional potter, poet, or academic and write straightforwardly, rather than a text that had multiple registers, sections titled both simply ("Pedagogy") and dramatically ("Ordeal By Fire: Evolution of Person"). But moreover, a refusal of form as it stood: during the 1950s, Richards and her colleagues had successfully negated traditional disciplinary boundaries through their lectures, performances, and concerts. A decade later, such events were still limited by their inability to move beyond the sphere of their own tight-knit, self-sustaining circles, which, after Black Mountain's closure, easily re-formed and expanded in New York, revolving around spaces such as the Living Theatre, Pocket Theater, and Judson Memorial Church.

Such events, by a plethora of artists (George Maciunas, Robert Morris, Claes Oldenburg, Yoko Ono, Robert Watts, Robert Whitman), composers (Earle Brown, John Cage, Morton Feldman, Richard Maxfield, David Tudor, Christian Wolff, La Monte Young), and dancers (Trisha Brown, Merce Cunningham, Simone Forti, Yvonne Rainer) are, by now, canonical, forming the core downtown scene of 1960s New York.[121] But as Cage became the most significant progenitor of late 1950s-era performance practice or, as dance historian Sally Banes has argued, its "most influential father figure," Richards has been nearly excised from the history of the New York downtown scene, though she was an active presence, giving lectures at spaces such as The Living Theatre, attending meetings at The Art Club, and arranging ceramic-based Happenings at East Village galleries.[122]

In the following passage of *Centering*, Richards characterizes the art and literary worlds' lack of spiritual commitment and social participation:

> For the most part, in our times, among the sophisticated and godless, I find either a kind of sentimentality, with man standing tragic and brave, under the cold stars, as they say. Or I find, in the more hip circles, a sort of fey and delirious relish of what is usually called "dissociation." The person dissociates himself from himself, and floats in a kind of deliberate schizophrenia, enjoying "himself," as if he could drink himself in like a glass of water. A sort of inane hilarity breaks out in the gravest of circumstances. Or a remarkable coolness refrigerates natural warmth. Everything seems self-conscious and calculated—decorative, you might say, even the spontaneity. A third reaction to the loss of equilibrium in our culture—an attempt to right the human being, as it were, in a centered position where his dreams and his practices may speak together in a chorus—in the mystical intoxication of the so-called beat generation. These individuals look for ecstasy, for mystical orgy, for deity . . . They are God-seekers hysterical with desire and with fear of frigidity . . . The atheism of our times wears the look of a religious fraternity, as it cast about for some symbol of union which all men wear.[123]

Richards' barbed descriptions implicate many of her own colleagues, friends, and associates: sentimentality could easily be attributed to abstract expressionist painters, dissociation could be an accurate depiction of the self-indulgence and narcissism of the Factory scene that orbited Warhol, and, finally, the "inane hilarity" and "remarkable coolness" could easily be leveled at Cage. The Beats—the circle from which she was furthest removed—she castigates outright. In this passage, Richards seems intent on exposing the falseness of 1950s spirituality—

pointing out its limitations and conformities and registering her discontent at the moment when her immediate peer group was becoming increasingly legendary and historicized. Even Olson had his moments of crystalline self-mockery when he wrote to Richards in 1965, before accepting what would be his last teaching post, at the State University of New York, Buffalo, five years before his death: ". . . I am in only the sadist stages of seeking to become ensconced . . . Yeah! A fortification! How about that? Or to settle comfortably Or smugly??"[124] The irony of Olson's self-recrimination was surely not lost on Richards. Richards refused to rest on the laurels of Black Mountain, which she considered a failed enterprise, and similarly resisted participating fully in the traditional circuits of life and livelihood available to a middle-aged writer and potter—tenure, gallery sales, craft fairs, and summer programs—which amounted to a refusal of security, either financial or emotional, risking, instead, the possibility of permanent obscurity and eventual disappearance. Was she, in fact, a blatant moralist, condemning her peer group's lack of engagement with the world beyond libraries, museums, and concert halls? Or was she critiquing the limits of the avant-garde artist's own tolerance for *response*—a socially engaged reception beyond the conventional networks and, moreover, *responsibility* in the creative act, where the communal need is privileged over the individual effort or ego? As T. J. Clark famously observed, "the real history of the avant-garde is those who bypassed, ignored and rejected it; a history of secrecy and isolation . . . we need to search for the conditions of this distance."[125]

Centering, in effect, renounces the avant-garde establishment. Richards's usage of *live form* is that she sought to take form live, unmediated by the artist, amounting to a communal impulse she calls "wholeness," where soul, body, and mind are conjoined and in concert with the vitality of community, a wholeness that is morally incompatible with either the heroic (or merely ironic) aims of the avant-garde. For Richards, an artist is compelled to form as a willfully *ethical* imperative. As she wrote in an early draft of *Centering*:

> There is some kind of life-form at work which makes a baby turn into an adult, physically, in the form of its species, but a unique one; any seed seems to be endowed with this life-form that it knows what to become, and this knowing is not conscious. It is a natural intelligence staggering in its consummations. And there is also a moral form which seems to evolve, in man as a species and in the individual, through time. It is a form which one might speak of as conscience.[126]

While this passage was eventually excised, in his review of *Centering*, Daniel Rhodes praised Richards for this very virtue, writing:

> M.C. Richards does not shrink from a consideration of moral value; she tackles it fearlessly and passionately. There is more than of a little of Oriental philosophy expressed in the ideas of the book. In certain ways it parallels the sense of Tao and of Zen. For example, there is the stress on the primacy of experience, the need to give of one's self and to loosen the attachments which inhibit new connections, and there is the discernment of wholeness, or the possibility of wholeness, achieved through growth. As in Zen, there is the insight that actions, forming of some sort, is not a result, but a part of wisdom and health.[127]

Given the Zen-like qualities Rhodes identified, it is more than a little ironic that the American Zen master Cage himself felt "bored" by *Centering*. In a rather revealing letter from this time, Richards unequivocally states:

> I have not been interested in John Cage's work, nor he in mine for several years. I gave him, at his request, my chapter on Pedagogy from my book to read while I was writing it; he returned it saying that since he was not interested in the ideas expressed, he found it too boring to read. We have known each other a long long time and are in some way very dear to each other, but we have had no real contact for some time . . . Merce Cunningham, on the other hand, I still see. He lives around the corner, and real sharing flows between us, upon rare and treasured occasions.[128]

Or perhaps Cage's purported "boredom" was actually a defensive response to Richards's work and personal commitments. Cage's interests—Zen, chance, mushrooms, cooking, atonal music, simplicity, rigor, discipline—these did not change. But perhaps over time, he hardened in them, his immense productivity hardening into a kind of myopic arrogance and, finally, detachment—his failure to be interested, or even feign interest, in the creative production of an old, dear friend such as Richards. Instead, Cage became deified, like Satie, by the next generation's eager disciples, the boy-artists of Fluxus, performance, and early electronica—Claes Oldenburg, Al Hansen, Alvin Lucier, Allan Kaprow, Dick Higgins, Jackson Mac Low—all young men who continued to cite Cage as a critical influence, having drunk from the fountain of his wisdom during a single semester at the New School in 1958.

So Cage was not only an artist-genius but also a genius-pedagogue as well, exceeding Richards in reputation and worth in her chosen profession, perhaps the one thing—the only thing—in which she could claim autonomy separate from her Stony Point circle. So *Cage*, rather than *Richards*, is not only the most important avant-garde artist of the 1950s but its most important teacher as well. Like Brown and Marcuse and McLuhan and even Olson, Cage had found the *right* audience.

His lineage, of course, is one of patrimony. To find the *right* audience means to have produced from a position of power, as a figurehead. Cage's legacy has not been left to chance but rather, at every chance, glorified: each essay analyzed, each composition re-performed. Since Cage's death in 1992, there has been an enormous and prodigious scholarly and artistic output devoted to him. Cage chanced history at the *right* interval and won.

Cage was not immune to feelings of vulnerability and attack, as Christian Wolff, a composer who was part of Cage's circle, suggests:

> And there may have been other cases of anti-Cage sentiment then. The late sixties, early seventies, that sort of mid-Vietnam antiwar period was very turbulent, and even counter-cultural heroes like Cage were being attacked from their own camps, so to speak. So those people who had become "politicized" suddenly felt that Cage was not doing enough or was not committed or whatever. I think he felt on the defensive and kind of fragile to a certain extent during that period.[129]

But if chance dictated Cage's oeuvre, Richards herself was chancier, embodying real risk, with real consequences in her self-displacement and upheaval, leading to an eventual self-exile from New York. So maybe it was just by chance that Cage wrote, "Got caught in an elevator between 2 floors in Brussels and thought of you in that darkness. I am so optimistic [sic] I could not take it seriously . . . D.T. [David Tudor] will explicate. Love, J."[130] Or maybe it was in jest, a few hasty lines dashed off on a postcard sent from Europe, but the darkness (that Cage could not take seriously) was real.

Richards is distinctly different than the other women artists of her generation in that she did *not* persevere: she did not thrive privately, pursuing her own artistic practice, to eventual, if belated, recognition. Instead, she veered sharply off-course. Released from the insularity of her domestic entanglements, Richards had given up New York entirely by the end of the 1960s. She continued making pots, traveling, and teaching by way of itinerant workshops, efforts which led her to a community-centered, decidedly non-object pottery practice. The summer of 1968 finds her organizing a two-week "Festival of Kiln Building & Firing and Pottery Making," at Paulus Berensohn's farm in Uniondale, Pennsylvania. As she wrote in the flier:

> This is not a school or any kind of organization. It is just a group of people getting together in a certain spirit. This is the way it looks now:
> a festival of shared opportunities and shared costs
> community limited to 30 persons/no restrictions on age or background

Figure 4.13 M. C. Richards in her Stony Point pottery studio, 1956. Collection of Mary Caroline Richards Papers, the Getty Research Institute, Los Angeles (960036 ADD2). Photo by Valenti Chasin. Courtesy of the Estate of M. C. Richards.

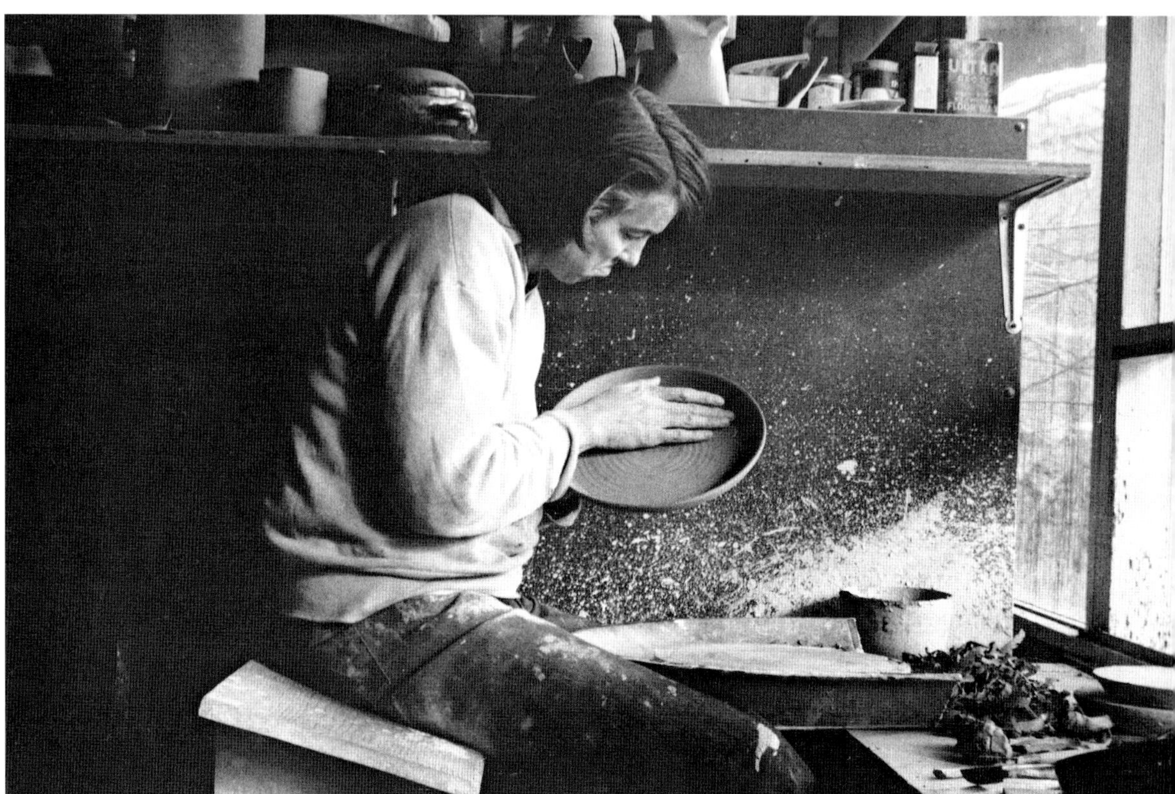

open workshops/collaborative learning.
Emphasis may be on hand-building since there are only 2 potter's wheels in the studio (unless we can get more)
$100 minimum covers:
CAMPSITE (field, water, toilet)
TUITION
MATERIALS (no firing charges)[131]

Such a festival not only extended the space for the 1950s Happening but also epitomized the collective yearnings of the 1960s. Staged in a rural field, with picnic-style communal meals, togetherness became the primary encounter, a "be-in" motivated by the crafting of handmade objects, unmediated even by the wheel. Such an event enacted Richards's model of wholeness "to take time, making the contact deep and personal."

This was *not* "non-object" pottery in the avant-garde sense, with the intent to rid the craftsman of the spiritual burden of producing objects. This *was* "non-object" pottery whose intent was to foster commitment and spiritual nourishment. Or was it? Couldn't Richards have employed the lessons of the avant-garde and brought them to bear within the scope of craft and radical community building? Or were these dual fates ultimately incompatible? For the last fifteen years of her life, she lived in a handicraft- and agriculturally based community, Camp Hill Village, a gathering of committed theosophists in rural Pennsylvania—one of eighty worldwide—where mentally handicapped people lived, farmed, and produced crafts alongside their teachers, all of whom were volunteers, including Richards.

For Richards, wholeness could not be achieved without a fully developed interior life, honed through the *live form* of the creative act (figure 4.13). This interiority radiated outward as an energy that fueled a classroom, or the mentally handicapped people at Camphill Village, who, as Richards believed, were gifted in their ability to experience the creative process and the work of art as simultaneous, rather than two distinct activities. And wasn't this simultaneity what New York's avant-garde, at its core, hoped to achieve? Fueled by repetition, irony, chance, and, most crucially, absence of content—Richards's peers privileged formal experiments in sensorium over the synchronicity of human experience.

Through the studied humility and pathos she had found in the role of the potter, M. C. Richards's own self-formation was achieved. Conceiving of process as a metaphor for self-actualization, she renounced the formalism in which she'd been steeped when she wrote, "It is not pots we are forming, but ourselves."[132] In such a phrase rings an intonation that coincides with the earliest inklings of gay liberation—a sphere she so clearly inhabited, circulating almost exclusively in a world of gay men. Eventually, even Cage conceded to having learned from Richards, writing in 1968: "The subject she teaches isn't listed in the catalogues. Sooner or later we know we're studying with her."[133] This set of remarks was eventually used as a back cover plug for the paperback version of *Centering*, allowing Cage to have, essentially, the last word.

Yet ultimately within Camphill Village, the bohemian identity she had allegedly forsaken was actually the backbone of her practice: the avant-garde potter, poet, and translator, translating one kind of creative existence for people that were instinctively creative and translating instinct for the creative audience that prioritized intention. It was in the difficulty of rendering these translations, in the gap of this silent effort, that M. C. Richards has vanished, consumed by her own ecstatic forms.

WOMEN KITCHEN POTTERS

Susan Peterson, "The Julia Child of Ceramics"

CHAPTER FIVE

Three years prior to the advent of public television and in the same year as Marshall McLuhan's influential text *Understanding Media* was published, the potter Susan Peterson (1925–2009) brought clay to an audience beyond the classroom.

Peterson created and starred in a fifty-four-episode television series titled *Wheels, Kilns, and Clay* (1964–65). Per McLuhan's definition, hot mediums reduced the possibility for participation, while cool ones increased it through widespread transmission. All hand-enabled, ancient technologies, then, such as fashioning stone tools, shelter building, and pottery, fell under the rubric of an all-encompassing "mechanical time," which left the whole world hot to the touch for two millennia.[1] By translating McLuhan's "hot" medium of clay for the widely disseminated "cool" medium of television, Peterson herself could be construed as an early adapter, an artist with an acute understanding of how an ancient technology like ceramics could be quickly outmoded unless reconfigured for an audience unfamiliar with the specialized techniques and processes of ceramics, that is, an audience of amateurs.

In her important revisionist history of artistic labor in the postwar era, *Machine in the Studio: Constructing the Postwar American Artist* (1996), Caroline Jones weds a history of "post-studio production" to the masculine bravado of the 1960s and, specifically, the continuing influence of Judd, Smithson, Frank Stella, and Andy Warhol. These artists transformed the artist's site of production from the studio, an historical oasis of isolated object-based creation, into a theoretical and critical sphere both nomadic and unfixed, a situation of "de-architecturing," to borrow Smithson's own term—altering the private, symbolic space of the studio and envisioning it as a series of procedures orchestrated by the artist and executed by others.[2]

But Susan Peterson *also* extended the space of production for a medium previously condemned to obsolescence: by the 1960s, ceramics was an outmoded technology, searching for a wider audience beyond its own professionals. Amateurs had always sustained the field, as the Black Mountain Pottery Seminar and its locale, rural North Carolina, demonstrated. A television audience was a McLuhan-esque extension of the live pottery demonstration, already primed for the ceramics process through the popular cooking show format, pioneered by Julia Child (figure 5.1). While Peterson is credited as the first educator to introduce an Alfred-style, traditional ceramics curriculum to the West Coast, far more is at stake in her multimedia ceramic-based practice.[3] She was, as Robert Smithson wrote of Donald Judd, "a whole artist engaged in a multiplicity of techniques."[4]

An accomplished potter, Peterson's own work is rooted in domestically scaled functional wares, such as *Bottle* (undated) (plate 7). Textured with a shower of black speckles that rain down upon the surface of a smooth and shiny red glaze, this small vase is an object to hold rather than behold. On four sides of the bottle's exterior are indentations that invite touch, areas of "hand feel," in which the experience of the vessel's symmetry in the round is highly considered. Such a tempting tactility contrasts with the vessel's very narrow neck, which would hold little other than a single bud.

As a professor, Peterson established two urban ceramics departments (at the now-defunct Chouinard Art Institute in Los Angeles and at Hunter College, New York). In between, she taught at the University of Southern California (USC) from 1955 until 1971. Beyond art in the university, she also spearheaded three diverse, nondegree programs: a summer ceramics and Native American program affiliated with USC, called Idyllwild Arts, in Riverside County, California; the Clayworks Studio Workshop in New York; and the Joe L. Evins Appalachian Center for Crafts in Smithfield, Tennessee. In 1972, she began writing the curriculum-version of her television series, *Wheels, Kilns, and Clay* and, eventually, monographs and surveys of the ceramics field, ten in all, including *Shoji Hamada: Life and Work* (1974), *The Craft and Art of Clay* (1992), *Working with Clay: An Introduction* (1998), *Contemporary Ceramics* (2000), *Jun Kaneko* (2003), and, most importantly, a number of volumes and exhibition catalogs on Native American women potters, including *The Legacy of María Martínez* (1977), *María Martínez: Five Generations of Potters* (exh. cat., 1978), *Lucy M. Lewis: American Indian Potter* (1984), and *Pottery by American Indian Women: The Legacy of Generations* (exh. cat., 1997).

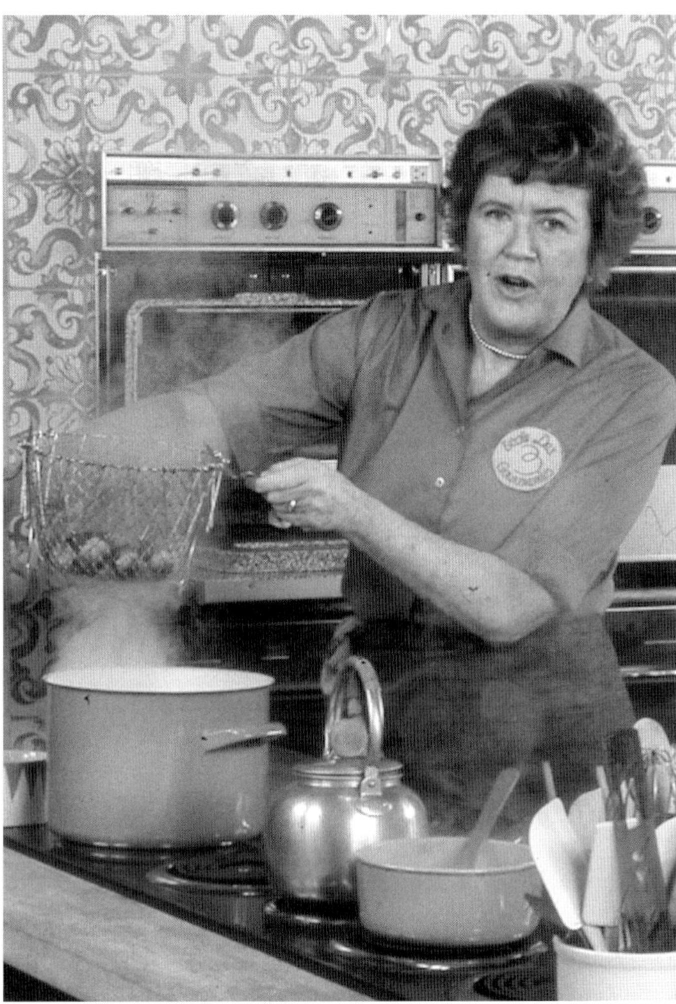

Figure 5.1 Still image of Julia Child cooking brussels sprouts from *The French Chef*, 1970. Courtesy of the WGBH Media Library and Archives and the Julia Child Foundation.

In this chapter, I will argue that Peterson, like the better known avant-gardists of the 1960s, can *also* be construed as an artist engaged in "post-studio production," albeit in a medium—clay—that has never been allocated such conditions of operation. Furthermore, the particularity of Peterson's form—process-based educational television that takes place in a staged, kitchen-like setting—destabilizes both the origins of the "post-studio" 1960s, which denigrated hand labor in favor of an industrial aesthetic, *and* predates the symbolic and gendered associations of the kitchen in the decade that followed, the 1970s, when the kitchen became the iconic site of women's universal oppression. Peterson's kitchen is a protofeminist space that anticipates early feminist video art by Martha Rosler and, in particular, stages a specific set of *values about form*, contiguous to an avant-garde moment—the 1960s—that had already severed its connection to craft (plate 8).

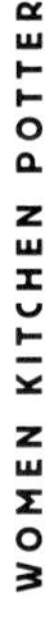

The Haptic Viewer

Since the post-2009 growth of digital television and subscriber-based cable channels, the half-hour "how-to" program has become a rather ubiquitous phenomenon, delivered via highly specialized cable channels, many with a do-it-yourself ethos, such as the Food Network, Home and Garden Television (HGTV), and The Learning Channel (TLC). Rather than indulgent housewife and dating dramas, this is the other world of reality television: perfection and accomplishment in the domestic sphere. Cooking and home repair shows have become palatable background noise for busy Americans whose aspirations for creative, self-actualized living can be indulged in thirty-minute fantasy blocks, from Martha Stewart's creation of a Joseph Cornell-like souvenir display box for lasting family memories to Rachael Ray's popular cooking show titled—what else?—*30 Minute Meals*.

But of all such manifestations, it is the cooking show that best embodies the theatrics of the canned demonstration—the raw pie goes into the oven and, seconds later, a perfectly baked one emerges. Since clay "bakes" at temperatures well beyond those of pies, cakes, or casseroles, Peterson came armed with a variety of ceramics processes to display to her viewers—raw clay; the formed but unfired leather-hard pot; bisqued wares that had undergone one firing but were unglazed; and, finally, fully fired ceramic works that had been trimmed, bisqued, decorated, *and* glazed, such as *Platter* (figure 5.2), a serving dish in which the wheel-thrown texture of concentric circles is contrasted with a decoration of bold beige stripes against a black background.

Like cooking shows, Peterson also arrived with a plethora of tools, many of which were taken from the domestic sphere and imported for artistic use, such as the meat cleaver or the wooden spoon, used to bat down clay. As she recounted in 2004:

> I would practice in my own studio about 18 hours of time with a tape recorder listening back to whatever I was saying, trying to explain in the best possible way whatever it was. Of course I'd seen Julia Child. She was already on television. I had seen her . . . (but) because I was moving everywhere they had to light the whole set. So it was so hot that I had to have three or four pieces of everything that I was putting together or trimming, or whatever, under the table so while I was doing a slide or something else I could pull up the next piece because they would dry so fast under these lights. So I had a lot to learn about that kind of thing.[5]

Perhaps there was a learning curve in performance for television, but *Wheels, Kilns, and Clay* had a long gestation, rooted in Peterson's years of classroom

Figure 5.2 *Platter*, undated. Susan Peterson (American, 1925–2009). Stoneware, 1.25 (depth) x 14.25 (diameter) inches x 44.5 (circumference) inches. Nora Eccles Harrison Museum, Utah State University, Logan, Utah, Gift of the Artist. Photo by Andrew McAllister. Courtesy of Jill Peterson Hoddick, Jan Peterson, and Taag Peterson/the Estate of Susan Peterson.

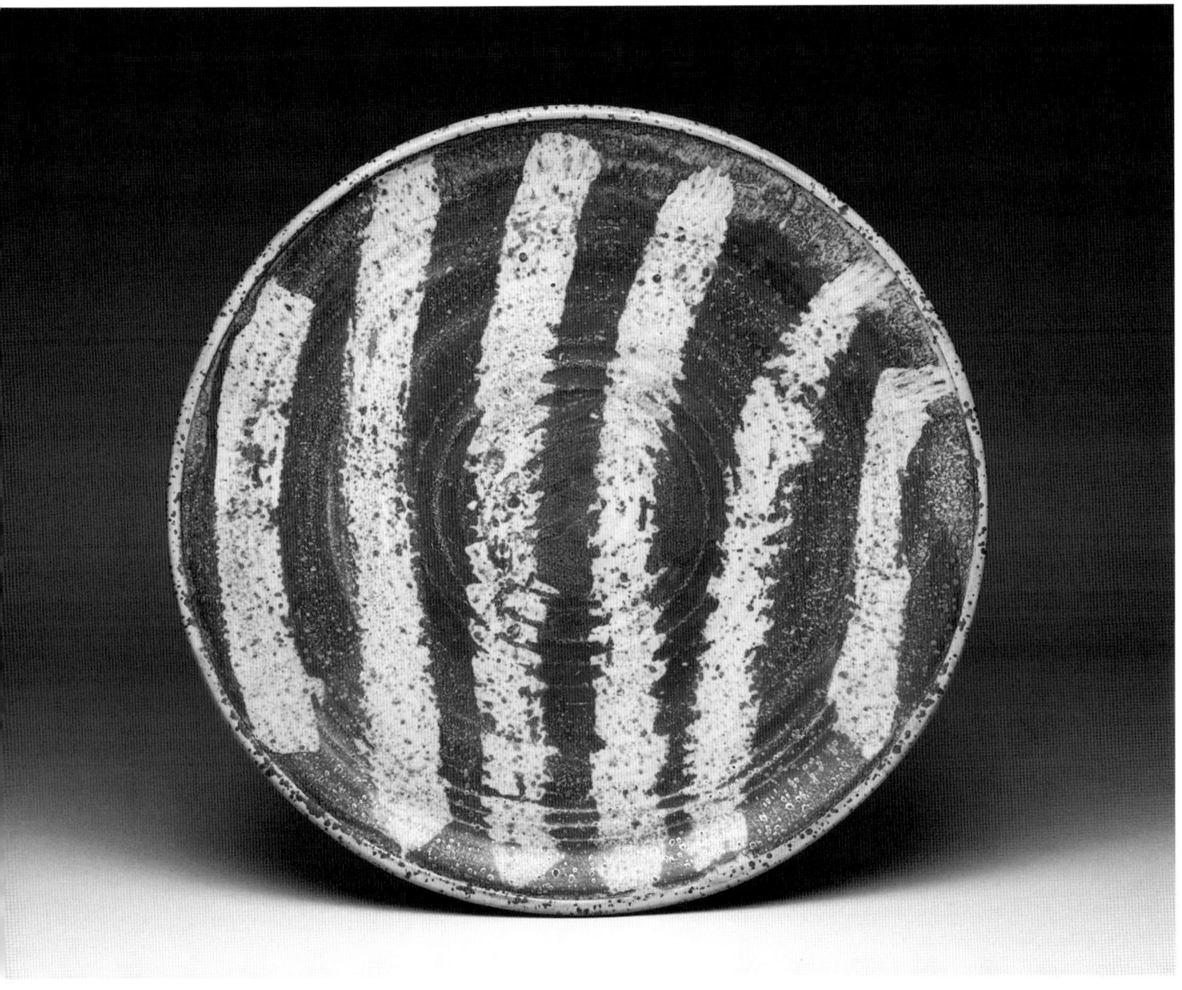

experience and her Amish-Mennonite religious upbringing, in which communal bread making was a gendered ritual practiced in conjunction with Communion.

Peterson began her artistic career as a painter, studying art as an undergraduate at Mills College with its luminary faculty: drawing with Claire Falkenstein; photography with Imogen Cunningham; printmaking with Cunningham's husband, Roi Partridge; and pottery with F. Carlton Ball (figure 5.3). She graduated from Mills in 1946 and went on to earn her MFA in ceramics at Alfred in 1950 before settling with her husband, a ceramic engineer, and the first of three children in

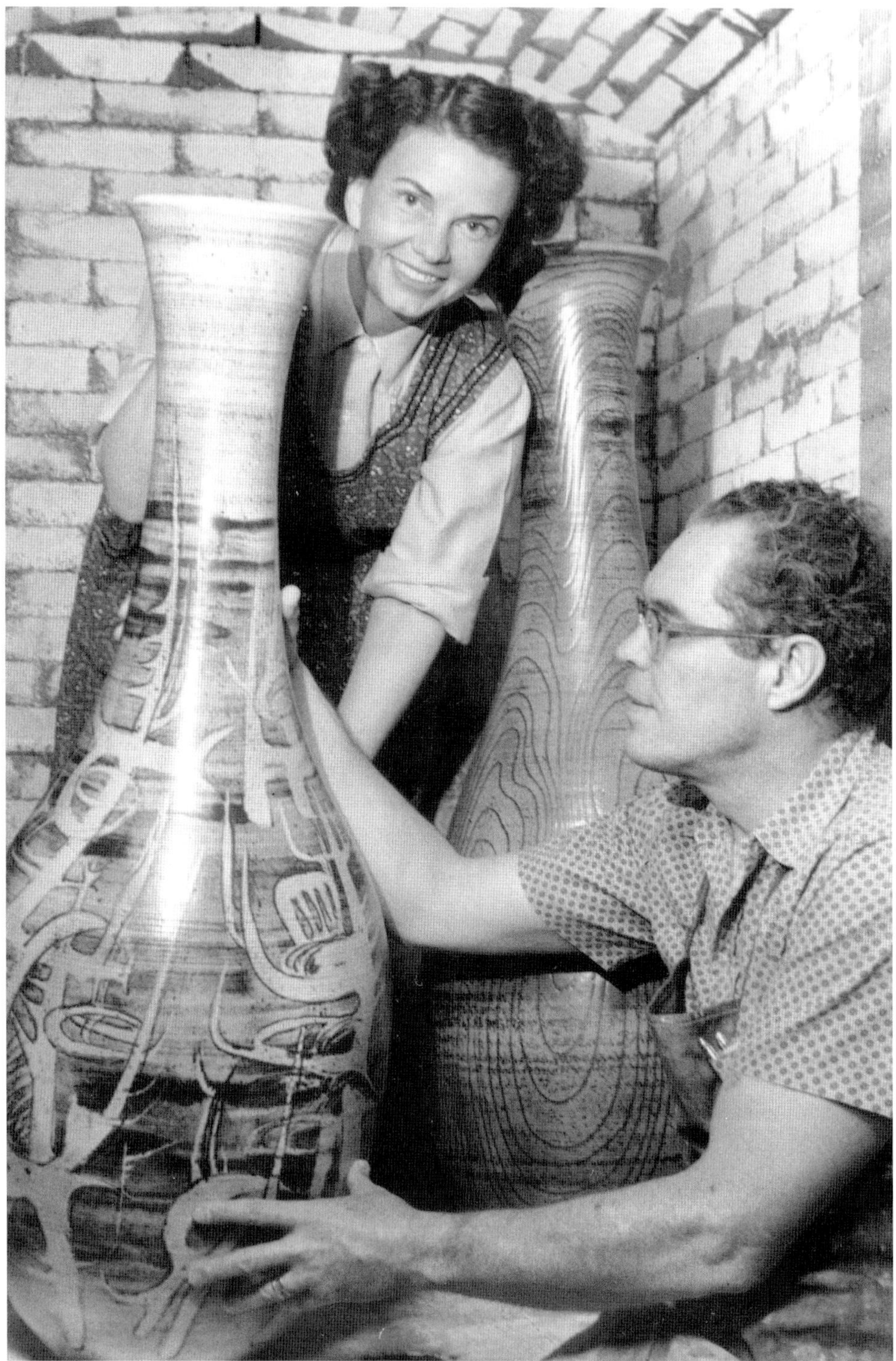

Figure 5.3 Carlton Ball and Susan Peterson, posing inside a kiln, 1957. Susan Harnly Peterson Archives, Ceramics Research Center, ASU Art Museum, Arizona State University, Tempe. Courtesy Jill Peterson Hoddick, Jan Peterson, and Taag Peterson/the Estate of Susan Peterson.

Los Angeles in 1952. As she reflected back, in 1984: "When we were young, *ceramics* as we know it today was also young. Most of us began as painters or had other occupations and found clay in some sort of epiphany.[6]" Peterson's epiphany became her ability to utilize the ceramic process as a locus for experiencing and building community beyond the confines of master-student lineages, as a structuring device for nonhierarchical and participatory experiences. Through her lifelong commitment to nontraditional education, Peterson was instrumental in promoting the nonhierarchical nature of studio pottery, which she felt was exemplified by Hamada, with whom she first came into contact in 1952. Late in life, she took a completely different viewpoint, turning her back on *mingei* or what she called "faux folk art" in favor of Native American pottery made by women. As she characterized this earlier moment:

> I was still young when I brought Leach, Shoji Hamada, the most famous Japanese potter, and Soetsu Yanagi, the well known exponent of Zen Buddhism and a writer on folk art, to Chouinard Art Institute in Los Angeles where I was teaching in 1952. Youth has its advantages but experience isn't one of them. Neither I nor my colleagues knew what folk art was except perhaps early American weather vanes.[7]

Chouinard and the West Coast Pottery Seminar

While the Black Mountain Pottery Seminar expanded the opportunities for emerging potters to gain exposure to *mingei* principles, it wasn't until the end of their tour that the Leach group found its true audience: in Los Angeles, where there was already a Japanese presence.

In December of 1952, Peterson hosted Leach, Hamada, and Yanagi at the Chouinard Art Institute, Los Angeles's oldest private art school. This was the culminating event of their ten weeks, a workshop similar to the Black Mountain Seminar but longer by one week: three weeks, instead of two, and by invitation only, with nearly thirty professional participants who came from around the country, rather than an open call aimed at tuition-paying students.

During the summer of 1952, Peterson was hired by Nelbert Chouinard to establish what would become Chouinard's renowned ceramics program. The formation of two concurrent ceramics programs the same summer of 1952 is notable and underscores the improved academic status of craft within the immediate postwar period. Peterson was a colleague of Karen Karnes and David Weinrib and also a recent alumnus of Alfred. She too was a novice teacher, in the first full semester of her first professional teaching post.

But Chouinard had a much longer lifespan than Black Mountain, running from 1921 until 1972.[8] Founded by Nelbert Chouinard, a Pratt-trained artist and educator, it holds the distinction of being the West Coast's first woman-run art school. Chouinard had been a student of the legendary pedagogue Arthur Wesley Dow and was also a veteran of the settlement house movement. Both patron and director, her establishing mission was deeply invested in the fate of the student-veteran: the focus of its pedagogy and instruction was consciously directed toward its male students, with an emphasis on imparting specific skill sets geared toward attaining civilian employment in commercial design or advertising. Susan Peterson described her own role as part of Chouinard's desire to have something "more vocational" on hand.[9]

While Black Mountain's war era curriculum was short lived, Chouinard's was long term, designed to appeal to veterans beyond World War II. Still to come were the hundreds more who had served in the Korean War (1950–1953). In 1958, one of the few years for which hard data exists, Chouinard's total student population consisted of 456 men and 282 women. Of these, 210 men were "vets, rehabs, or national defense loan students."[10] As Chouinard wrote in her 1947–48 course catalog introduction:

> Progress does not permit aimless experimentation. While the talents and ambitions of individuals vary greatly, the same intellectual and spiritual development, plus a knowledge of art fundamentals, is necessary to the successful portrait painter and the applied artist alike. Commercial courses in Chouinard Art Institute are built on an art education which implies an understanding of drawing, composition, anatomy, perspective, color, design etc. The commercial courses are closely tuned to the demands of the business world and are directed and taught by men of standing in their chosen fields. Chouinard invites investigation of the results they have achieved.[11]

It would seem, then, while Black Mountain not only encouraged "aimless experimentation," its notoriety as a radical institution depended upon it. Chouinard, on the other hand, rejected experimentation outright as an impediment to the "progress" of an artist and his outcomes and successes. Though Chouinard was a four-year school, ceramics was the only two-year major.[12]

The Chouinard seminar was markedly different from the Black Mountain seminar, in that it was jointly organized by three potters without the help of their respective institutions: Richard Petterson (no relation, previously introduced at the end of chapter 2) of Scripps College, Laura Andreson of UCLA, and Peterson. The California potters themselves shouldered the responsibility of feeding and housing the Leach group. As Vivika Heino described it:

> Susan had the facilities on Saturdays or during Christmas at Chouinard's. Jane Heald said she and Nancy Hitch [local potters based in Venice] could house them. They slept

at one house and they ate at another house and they brought them into Chouinard. I've forgotten what I was to do, publicity or something.[13]

Based upon Heino's account, we can ascertain that the Chouinard seminar was very different than the Black Mountain one. First of all, the level of personal investment was greater: a gathering of like-minded peers, operating independently, absent a profit motive or any compensation whatsoever, during the holiday recess. The Chouinard seminar was held at the school by happenstance—because the facilities could be used discreetly over the academic break.

Rather than Black Mountain's Pottery Seminar-as-profit, meant to bring publicity and revenue to an ailing college, Chouinard's, in contrast, was truly an underground venture: both highly informal and extremely specialized, limited to an elite audience cultivated through peer networks throughout Southern California, unaffiliated with Chouinard but for its physical dependencies (space and equipment). As such, there are no formal documents to be found in Chouinard's archive regarding the seminar: no records of enrollment, or correspondence, or even any advertising. Yet, as Vivika Heino recalled, "Susan had the place there, and Lennox Tierney was in the class and Ed Traynor and Al King came. Just about everybody went to that class who were potters you would know."[14] Of these attendees, Albert Henry King was a British potter who had emigrated and helped to found Art Center in Pasadena; Ed Traynor taught at Pasadena City College and had studied for several summers with Marguerite Wildenhain; and Tierney was a former student of Laura Andreson at UCLA, who went on to curate as an East Asian specialist, serving as arts commissioner during the Allied Occupation of Japan (1945–1952) and later as the founding director of the Pacific Asia Museum in Pasadena in 1971.[15]

The Chouinard seminar is distinct from its Black Mountain counterpart in that it happened entirely by word of mouth. Like Cage's *Theater Piece #1* (1952) earlier in the year, it too had found the right audience, an informal community of Los Angeles potters who managed to grow a mythic reputation for the Leach attaché without money ever changing hands. A far cry from the outrageous financial demands Leach had made at Black Mountain, we can presume that Leach, Hamada, and Yanagi considered themselves to be among peers of a much higher caliber in California. In this sense, the Chouinard seminar seems closer in sensibility to a musical jam session than a pedagogical event.

Almost like a rock group, the Leach trio emerged from their cocoon to make just one scheduled public appearance during the course of their three-week sojourn. This might have been where the money was: a joint lecture and demonstration at the Los Angeles County Museum of Art (LACMA), organized on their behalf by their American counterparts. The event drew unprecedented numbers,

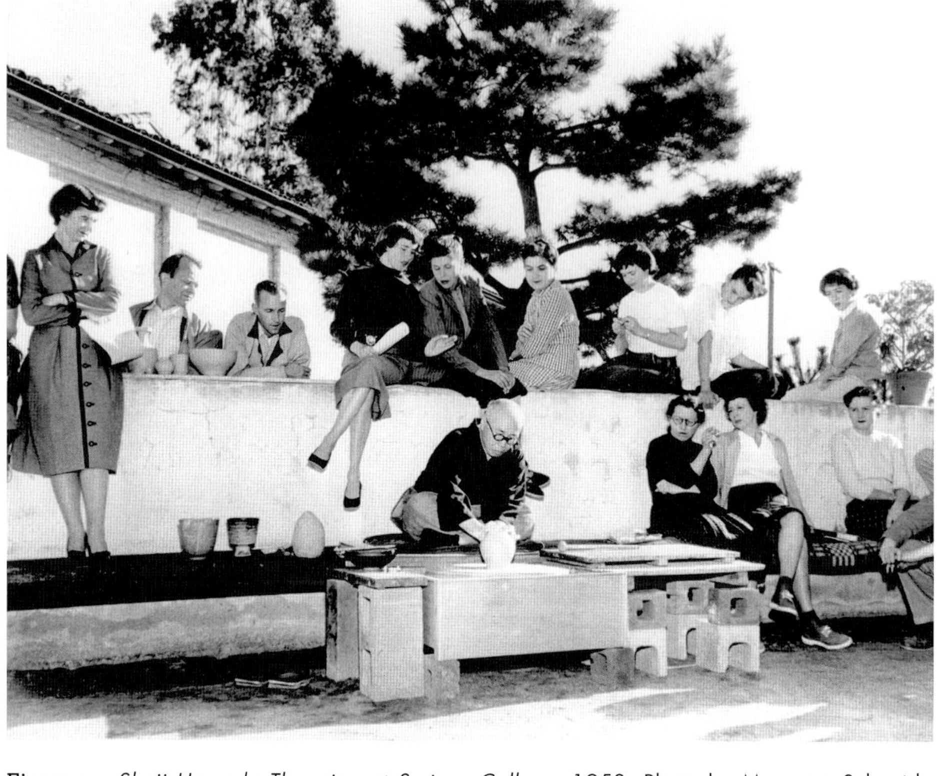

Figure 5.4 *Shoji Hamada Throwing at Scripps College*, 1953. Photo by Margaret Schnaidt. Courtesy of Archives Scripps College, Claremont, CA.

a massive crowd of over two thousand, as Richard Petterson described it, several years later, in 1956:

> Indicating the cumulative interest, [in craft] one rainy evening a hall at the Los Angeles County Museum accommodating 1200 was filled to overflowing and 1000 people turned away from a program presenting the famous potters, Bernard Leach and Shoji Hamada.[16]

It is no wonder, then, that Leach and Hamada's reputations were solidified in California, reaching mythic proportions through the live demonstration of Hamada performing in exquisite silence, punctuated by Leach's commentary. Though he spoke (and wrote) English fluently, he performed his silent, Zen-like persona (figure 5.4).[17]

This was the same sort of silence Cage traded in: an American-type Zen that disrupted the sensory expectations an audience had in relation to a particular medium. One does not need to hear Cage's *Silent Piece* to experience it; it is a visual piece predicated on the self-reflexivity of an audience.[18] Likewise, to handle a Hamada pot—the haptic experience of the object—was not necessary, but watching him was, since it was his particular performance of form that elevated his practice. In both examples, medium specificity is *denied in favor of an alterity* that registers as "revelatory" for its primary audience, the experience of which was, in turn, disseminated widely through oral or written explanation. During 1952, then, Hamada, like Cage, had found the *right* audience: through his American demonstrations, pottery became an experiential, rather than a productive, entity, yet the performance and its object were wholly inseparable, a *live form* held in perfect Zen equilibrium.

Wheels, Kilns, and Clay

Peterson broke with Hamada's silence. Her own wheel throwing eschewed the solitude and self-containment of Zen spirituality and regarded the throwing demonstration as an explicit pedagogical event. This was, in fact, closer in spirit to Wildenhain's communal setting, predicated on collective skill building, and yet like Richards, Peterson was firmly committed to the amateur and the experiential qualities that performance offered. Peterson's ceramic practice occurred in an entirely new and enlarged framework: television.

Broadcast throughout 1964 and 1965, *Wheels, Kilns, and Clay* spanned fifty-four distinct episodes and screened in half-hour segments once a week. The show was produced in Los Angeles by KNXT-TV Channel 2, the CBS-owned and operated station. As Peterson explained its initiation:

> There wasn't really a public television at that time, not as there is today and CBS was feeling that they needed an educational—some educational kind of program, so they started with a Shakespeare lecture by one of the SC [University of Southern California] faculty, and that was very successful, and then CBS decided they wanted a more visual program, so the head of the theater and film department recommended me. They came, we talked . . . and I thought this was going to be a terrific opportunity.[19]

In the early 1950s, television was still fairly new and unregulated; in 1951, CBS had split its radio and television production divisions. In 1952, as an alternative to commercial broadcasting, the Federal Communications Commission (FCC) reserved numerous channels for noncommercial operation.[20] As economist Martin Mayer wrote in 1972, "Though no fewer than 242 channels had been set aside for educational stations in 1952, by the end of 1959 only forty-four were on the air, and most of these were broadcasting only a few hours a day."[21] It was not until 1967 that public television, or the Corporation for Public Broadcasting, also known as PBS, was born. PBS was founded by congressional mandate, an umbrella organization for member stations with interconnected facilities and screening rights, funded through a combination of federal funds, Ford Foundation grants, and program-based subsidies underwritten by major corporations, such as Mobil and Xerox.[22]

Educational television was lauded for its democratic potential: the ability to reach a vast array of households regardless of socioeconomic status, through the apparatus of the television. But as David Joselit has pointed out, even commercial television had an educational imperative, influencing, if not standardizing, the consumerist middle class on everything from mealtime etiquette to familial roles and behavior, ascribing a uniformity of values between programming and advertising, a symbiotic relationship between television content and the products it endorsed. Further, since television narratives were (and arguably still are, today) premised on harmonic resolution, there is a ruthless efficiency at work when addressing domestic and social ills through, as Joselit writes, the "romance between people and things."[23]

In *Wheels, Kilns, and Clay*, however, this romance not only persists, it is suffused with an unbridled urgency to communicate an alternative set of values. As Art Seidenbaum of the *Los Angeles Times* wrote in his review of the show, "'Wheels, Kilns, and Clay' is a funny example of the way TV serves the public when cereal-makers and car dealers aren't looking."[24] Peterson translated the pottery demonstration into a medium that was performative: the didactic cooking show. In twenty-eight-minute episodes, Peterson performed ceramic processes while narrating the history and culture of pottery. Referred to as "The Julia Child of Ceramics," Peterson, through her television presence, synchronized the link between people and the things from which they were most certainly estranged: hand-based skills and knowledge, while using her own body as the conduit of production (figure 5.5).[25]

Each episode goes something like this: the program's title sequence is superimposed over Peterson's hands compressing clay on a slowly rotating potter's wheel, overlaid with tinny classical music and a male, off-camera announcer who introduces Peterson: "And now, your host, Professor Susan Peterson." This odd dialectic, host-professor, is reinforced throughout the series, as the camera pulls back to a long shot, which establishes the scene: Peterson next to her wheel, introducing the topic in the form of a lesson. The full-body view is intentional, as each episode features her in a different, color-coordinated apron-skirt set combination, often with matching flats. Entirely unsuitable for making pottery, her outfits are the apotheosis of fifties-era housewife femininity, with her bare legs offered as a focal point beneath an open table construction, performing the motion of the kickwheel.

Wheels, Kilns, and Clay offered a combination of technique and practical

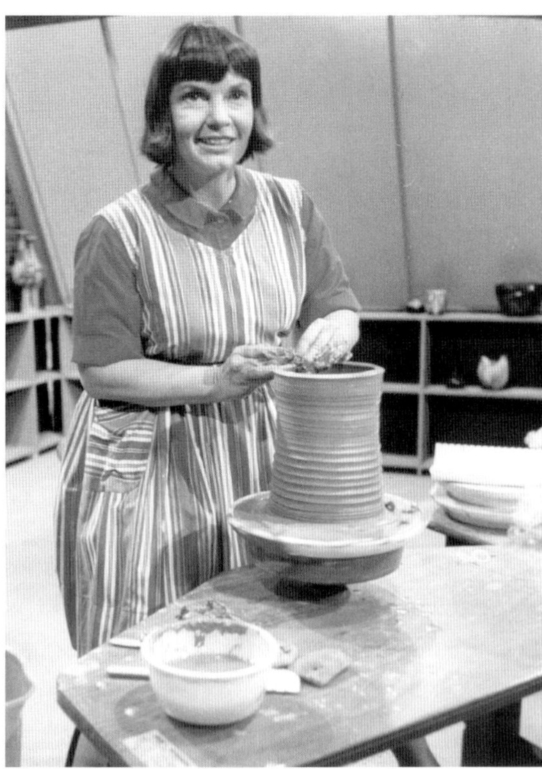

Figure 5.5 Throwing a cylinder, *Wheels, Kilns, Clay*, 1964. Susan Harnly Peterson Archives, Ceramics Research Center, ASU Art Museum, Arizona State University, Tempe. Courtesy of Jill Peterson Hoddick, Jan Peterson, and Taag Peterson/ the Estate of Susan Peterson

application. The programs can be subdivided into roughly six basic categories: materials, kiln technologies, ceramic histories, forms, glaze chemistry and techniques, and, finally, contemporary artists and issues. Each episode is divided between the ceramics demonstration and a historical treatment of an individual theme: its ancient heritage, primary sources, and, finally, modern usages. The programs were based upon her own classroom instruction, which drew heavily on conveying technique through the lens of history. As John Mason recalled of his time as Peterson's student at Chouinard:

> She was very well informed, both about the history of ceramics and about technical information, which I found very attractive. Here was somebody that you could talk to and discover answers for your questions. So, beginning with the night class, and then later when I looked around and saw all the activity there, I said, Susan, you need a TA [technical assistant] around here. She immediately responded to the idea.[26]

In episode one, "Overview of Ceramics," Peterson wears a dark pink pinafore over a pale pink blouse and skirt, standing in a television studio that looks and feels like a makeshift kitchen, complete with shelving, a slab of plywood that serves as a countertop, several plastic buckets filled with water, and an array of kitchen tools

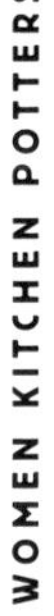

Figure 5.6 *Vessel*, undated. Susan Peterson (American, 1925–2009). Stoneware, 10.16 x16 x 50.25 inches. Nora Eccles Harrison Museum, Utah State University, Logan, Utah, gift of the artist. Photo by Andrew McAllister. Courtesy of Jill Peterson Hoddick, Jan Peterson, and Taag Peterson/the Estate of Susan Peterson.

and instruments, such as sponges, a chamois, and wooden mallets. Introducing ceramics as an ancient art form with fully contemporary applications, Peterson models a series of sample vessels, as well as still photos or single film clips of low tech, present-day ceramics production in Kathmandu, Cairo, and Oaxaca. The photo from Egypt features child laborers, which Peterson describes only as ". . . these children put clay in the mold, making thousands of bricks by hand."[27]

Despite her absence of commentary on this issue, Peterson conveys the laborious nature of clay and the way ancient and modern cultures alike mitigate its sheer physicality through the advent of technologies such as molding and casting or, for example, machines that process raw clay, such as mixers and pug mills. Throughout her series, Peterson does not regard clay solely as an artistic pursuit but treats it instead as an evolving set of technologies that offered a dependable livelihood to countless people throughout human history. According to Mason, one of the directions in which she encouraged him was industrial design, ostensibly because it would have provided a stable and steady income.[28]

It is not well known that Peterson trained two famed ceramic sculptors, John Mason and Ken Price. According to Price's wife, Happy, "Kenny never forgot the importance Susan had on his life at that time. Since his student days they didn't cross paths often but he has [sic] always been grateful to her."[29] While both Mason and Ken Price were indebted to Peterson for their technical training, as artists, they publicly allied themselves with Voulkos: Mason as a studio mate on Glendale Boulevard and Price as a colleague who showed alongside painters at the Ferus Gallery. Such a distinction points to the vast disparities between the 1960s-era reception of functional pottery versus ceramic sculpture. Sculpture didn't insist upon the stodginess of a clay-rich technical vocabulary; it allied itself with the minimalist vocabulary already in circulation. To the untrained eye, non-functional sculpture was eye-grabbing and seemingly more innovative, full of discontinuous forms.

Peterson experimented with brushy and expressionist glaze chemistries, such as in an undated *Vessel* form (figure 5.6), but she did not drift toward sculpture. She was, by contrast, process rich, reveling in technical instruction as a means of resituating clay as a vastly humanistic medium. While her male students

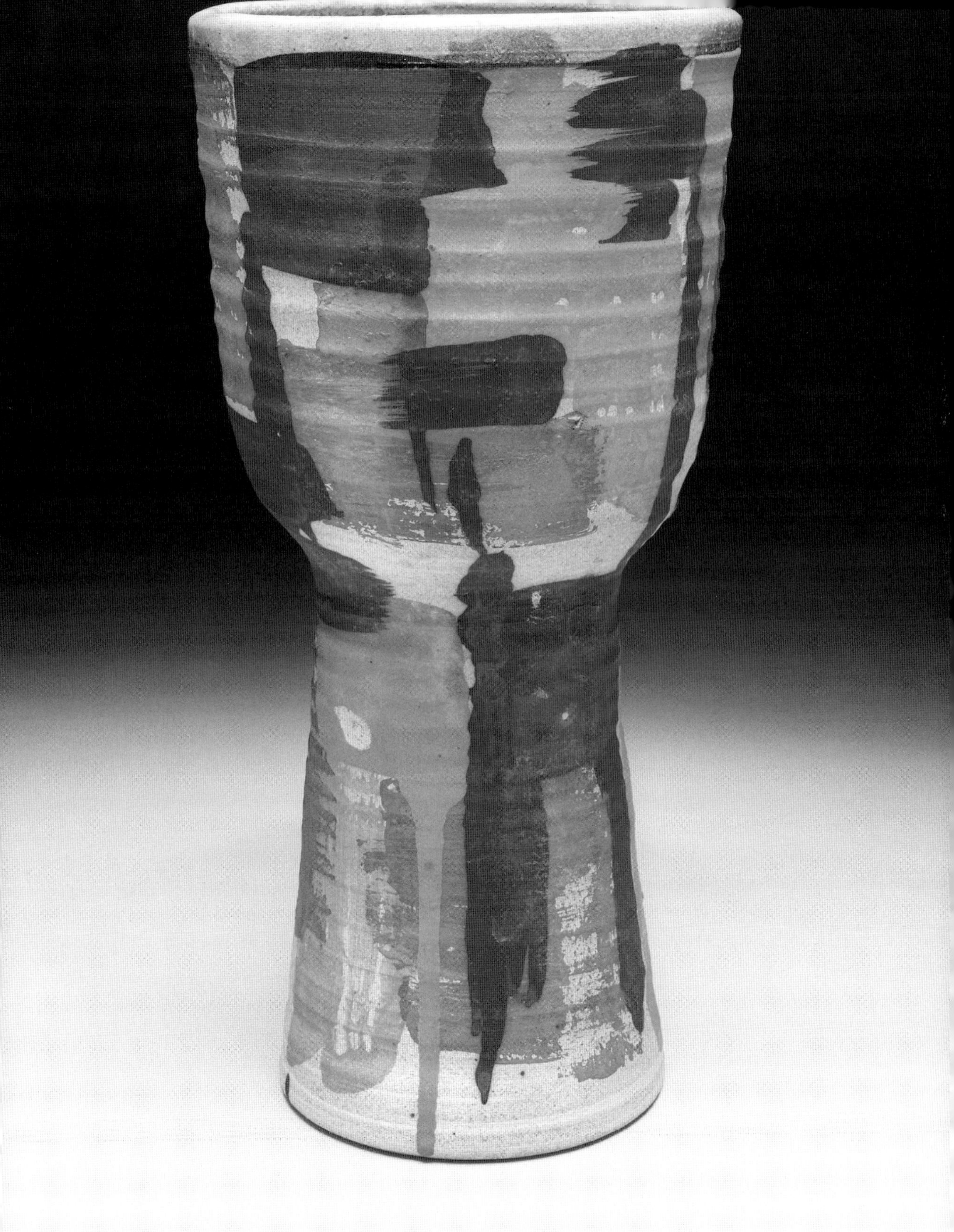

and peers grappled with their despised collective status as the artisans of the art world, Peterson engaged with communities far beyond the scope of urban contemporary artistic practice.

Women Kitchen Potters

Raw clay is a dense and heavy material, in need of priming, or what is known as *wedging*—that is, purging the clay of air bubbles through a push-pull kneading process that demands a great deal of upper-body force. Then there are the fire motor skills required of throwing and shaping clay. The primary method by which Peterson, and indeed all ceramists, teach students is by example. Simple methods of coiling, pinching, and slab building are the most basic techniques for vessel construction, which is why they are so readily introduced to children at the elementary level. Students who graduate to wheel throwing utilize the standard gestures of kneading, pulling, pushing, and compressing in order to master the difficult and much-practiced technique of centering the lump of spinning clay and then pulling up on the wheel, growing the raw clay into a vessel form.

Cooking, and in particular baking, depends upon a similar range of techniques: rolling, rising, kneading, stretching. Clay and dough have remarkable formal similarities not only in consistency but also in their respective compositions (dependency upon compound chemical properties) and pliability (based on the proportion of ingredients). As Mason recalled, "Susan was a person that did give assignments, and they could be to formulate, from scratch, some glazes for a specific temperature and test them . . . . And you say, oh well, I have to go through this stage, and you learn certain techniques with this amount of clay and learn different techniques with greater amounts of clay."[30]

The principles of forming a clay body are the same regardless of whether the result is earthenware, stoneware, or porcelain; a single clay from the natural world (i.e., straight from the earth) will seldom have all of the characteristics upon which the potter depends. Thus, both studio potters and professional bakers purchase raw materials (feldspar, alumina, silica, ash; flour, yeast, sugar, eggs) or premixed clays sold commercially to create composites that form their own signature recipes (i.e., the Barefoot Contessa Ina Garten's triple chocolate brownie mix), in order to control for shrinkage, plasticity, hardness, and texture. Like dough, too much dryness can hamper formation, while too much moisture can create brittleness and warping in the kiln.

In both fields, at some level, standards, mixes, recipes, and formulas are prohibitive: working with fine clays or pastry dough is a skill of both intuition and timing, or, as Peterson described it, "an empathy with the consistency and condition of the clay."[31] In both instances, overworking ruins the elasticity, and

Figure 5.7 *Untitled*, ca. 1975. Lucie Rie (British, 1902–1995). Glazed stoneware, 10 1/4 x 4 x 4 7/8 inches. Collection of the Arizona State University Art Museum, Ceramics Research Center, Tempe. Photo by Craig Smith. Courtesy of the Estate of Lucie Rie.

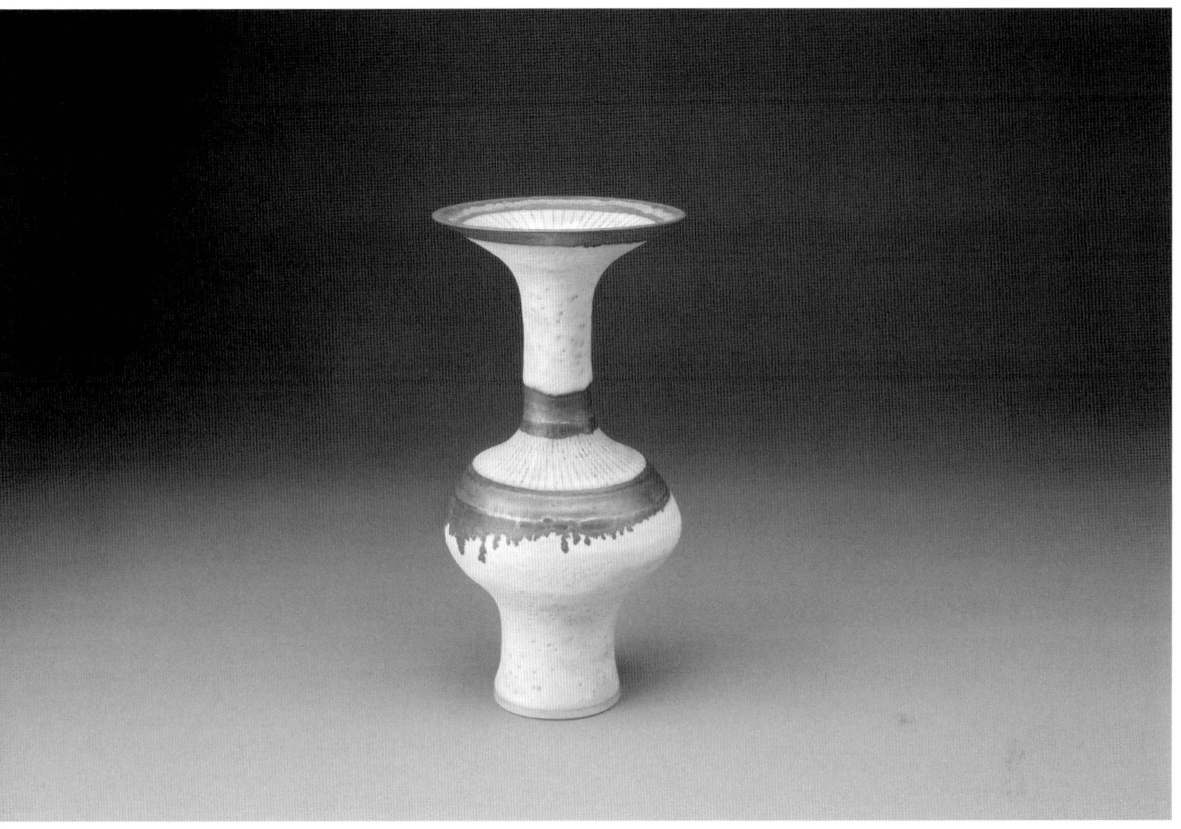

temperature and measure are crucial to success. Just like throwing on the wheel, anyone can learn to bake, but not everyone can bake well: skill building is reinforced through failure, learning through touch (the haptic) rather than vision, a frustrating negation of conventional wisdom and traditional education. As Peterson wrote, "Learning to feel is one of the big issues in clayworking . . . the pleasure of throwing will be enhanced as control and skill improve."[32]

Like cooking, clay is rooted in the dialectic of communal knowledge. Janet Leach (née Darnell) proposed the term "women kitchen potters" to describe the cultural traditions in which limitations of space and materials were embraced by women potters in order to produce functional domestic wares throughout the twentieth century: the stunning numbers of coffee pots, urns, lemonade pitchers, breakfast trays, tea sets, and cereal bowls. Specifically, Leach was describing her friend and colleague Lucie Rie (b. Vienna, 1902–1990) (figure 5.7).

As an urban potter based in London, Rie purposefully embraced her limitations of space and materials and used them to her advantage in her whimsical, thin-walled pots:

> The Western world is full of women "kitchen" potters attempting to work in urban condition. Lucie's solution to this problem could be held up as an example to other professional female potters, were it not for the fact that she is a very private person and prefers not to have detailed biographical descriptions written of her life-style. Her house and the room where she works are the essence of simplicity. The masculine concept of digging clay, chopping wood, etc, etc, as essential parts of the making process has been answered by her. She proves it is not necessary to dig, chop or use heavy machinery. To me, she is important as a feminine "no-shovel" potter. I have observed, many times, that the studio workshop requirements of the craftswoman are often different from those of the craftsman. Lucie is the epitome of this observation. The male potter usually likes the feeling of "going to work," whereas Lucie has integrated her studio and her home in a feminine manner.[33]

Leach's astute observations chart a linked and varied history of the "post-studio" moment in Peterson's pottery practice. During the 1970s, the kitchen became a feminist symbol and strategic alternative to the virtually all-male land-based experiments of 1960s practitioners like Michael Heizer, Judd, and Smithson.

Moreover, we can repurpose the term "women kitchen potters" as a particularly significant way of describing the central role that the domestic sphere played in dictating functional pottery made by women, both in traditional and modern societies. The British anthropologist Moira Vincentelli, whose two studies on women potters center on preindustrial, non-Western pottery traditions in Africa and Asia, observes, "The pottery that women make is easily taken for granted—humble cooking pots and simple water jars are everyday household equipment, not works of art."[34] Purposely reproachful, her meaning is clear: domestic-use pottery is a gendered craft with an established social role, equating women not only with the kitchen but also with its domestic labor: making cooking, eating, and drinking vessels (production) and then laboring again (and again and again) to fill them (reproduction). This cycle, then, is a metaphor naturalized over time into an archetype, permanently linking the molding of clay, the vessel as receptacle, and the female biological imperative of birth/nurture/nature.

Peterson had a heightened level of experience in her own upbringing regarding the coexistence of communal kitchen-based rituals and gender performance. Her Amish-Mennonite community of origin had also adhered to traditional gender roles, in which women performed as "wives, mothers, and foodgivers."[35] Peterson was born in 1925 in McPherson, Kansas, and raised there and in Grand Island,

Nebraska, among a deeply religious extended family of Dunkard Brethren Amish Mennonites:

> I grew up with a lot of long beards and black hats and little bonnets and the accouterments that went with those kind of people, and I had even my great-grandmother, who was over 100 years old, I had grandparents on both sides and great-grandparents on both sides, and a lot of relatives, because Mennonites had a lot of children generally.[36]

As modernization rapidly transformed Kansas's agrarian culture, Mennonite communities, and women in particular, underwent significant changes with regard to self-identification, religious dress, and work outside the home, with varying degrees of tolerance.[37]

Peterson's immediate family exemplified this disjunction: an aspiring painter, her mother ultimately assumed her traditional role of wife and mother, while her father was a high school principal who earned his doctorate in education under John Dewey and Alfred North Whitehead at Stanford University during the early 1930s. Through her father, Peterson was exposed to Dewey's progressive pedagogy at an early age.

Despite her parents' liberalisms, Peterson was barred from accepting the full scholarship she had been offered to study art at Carnegie Mellon in Pittsburgh. Instead, in keeping with Mennonite cultural values, she attended a women's seminary for two years, before transferring to Mills College in Oakland, California—far from home—to pursue an art career, perhaps against her father's wishes. As her own mother had abandoned a similar path, Peterson's father commented upon this choice in his unpublished memoir, found among his daughter's papers. As he wrote with deep pride before he died in 1988:

> Not long ago I asked Iva [Peterson's mother] whether she has ever regretted giving up what might have been a successful art career. She replied that rather than feeling regret she felt privileged to have been at home with her children during their very important formative years. At times it required real sacrifices but she always placed the children and her family ahead of any personal ambition.[38]

While Mennonite women occupy a subordinate role within their intensely patriarchal communities, anthropologist Margaret C. Reynolds charts their religious significance within an important weekly ceremonial rite of preparing Communion Bread collectively, an individual kitchen task transposed to a public forum, known as the bread making ritual, which takes place each Sunday morning at church, or what is known as the "experience meeting."[39]

Positioned between sessions of reading Scripture, group prayer, and solo testimony, women's ritual bread making is a departure from the male-dominated religious forum in that it is both experiential and female-directed, conducted entirely in silence but performed communally. One sister is assigned the task of preparing the utensils and combining the raw ingredients of flour, butter, and raw milk. Enough dough is mixed to serve the entire community, and after a ritual handwashing, women are called upon to assemble and perform before young, unmarried girls, children, and the whole of the male community, consisting of husbands, who, in their ultimate authoritarian roles, also double as church deacons, ministers, and elders.

For nearly twenty or so minutes, balls of dough are continually passed and kneaded, using a system of nonverbal cueing that indicates the pattern and order in which women should transfer their batch. A uniquely hands-on, sensory experience, the religious setting is transformed into a sustenance-bearing kitchen, filled with the sounds and smells of women working the dough in order to feed the community spiritually. In the end, each woman has handled all the dough for equal amounts of time, until the signal comes to combine their share, enacting unity. Silent and spiritually whole, the women themselves embody the holiness of the ritual, which is ultimately a communal affirmation. According to Reynolds: ". . . bread functions metaphorically as a test of faith for the participants. Overcoming obstacles in preparing the bread for Communion is a gendered metaphor for life's trials and a testing of sisters' faith."[40]

In this way, bread making becomes the epitome of a collective material that prefigures Peterson's own introduction to clay. Later in her career, Peterson was wholly cognizant of the medium's gendered constructions in relation to her American Indian subjects, using the universality of female domesticity as a strategy for transcending the obstacle of cultural difference. In her books on the potters María Martínez (Pueblo) and Lucy Lewis (Acoma), respectively, she uses the first-person plural to emphasize the inherently female bonds that form and endure in the kitchen setting:

> We were sitting in the kitchen, waiting for the bread in the outside oven to bake . . . Barbara [Gonzales, daughter-in-law and collaborator of Martínez] is mindful that the qualities of discipline and patience are legacies of Maria's art and of her diligence. Traditional techniques of Indian pottery making are demanding. Once a piece has been started, its construction must be finished in the proper time; once polishing is begun, it must be completed. Barbara knows that these qualities have other associations in daily life. "Clay is like dough; it remembers," she says. "It is special. It has character! If you don't knead it properly, it may not rise." She laughs. Likewise, your masterpiece could

explode in many pieces in the firing. We learn that a pot doesn't forget where it was ignored. It is pretty much the same with people . . .'[41] (1977)

The kitchen is off to the left, large and sunny. When Lucy [Lewis] is here for lengths of time she sits at the big table, near the wood stove, making pots. Emma and Dolores [her daughters] mix clay for their mother, on hands and knees on the linoleum floor or with their feet. There is space for all three women to work in clay at this table—and for Andrew's [her son's] pots, when he is here—with room left over for everyone to eat.[42] (1984)

Through her monographs on indigenous women potters, Peterson celebrated the traditions upheld by "women kitchen potters" with whom she fired alongside on New Mexico's reservations, taking part in the purity and spirituality associated with their domestic habits and rituals. But within *Wheels, Kilns, and Clay*, Peterson *also* performed the role of the "woman kitchen potter." In performing the visible labor of artistic practice, she produced a participatory series of videos, in a medium—ceramics—that conceived of itself as object only.

Audience and Feedback

Broadcast live, three mornings a week, at 6:30 am, to greater Los Angeles as well as eleven other Western states, Peterson's early morning viewership—an odd assortment of early risers, retirees, and medical professionals (Peterson received numerous enthusiastic letters from ER doctors and nurses)—would likely never make anything. Such amateurism is today often known as "enrichment learning," resulting in formalized educational endeavors often sponsored locally through public universities or community colleges that offer no-credit extension programs full of "how-to" seminars and workshops on a plethora of topics, ranging from motorcycle mechanics to landscape design.

In the episodes I am classifying as "materials," Peterson devotes a program each to the three formal clay bodies, earthenware, stoneware, and porcelain (episodes 5, 6, and 7, respectively). She also includes some late innovative twists, including a rethinking of glass as a ceramics process ("Glass, Another Phase of Ceramics," episode 43); a discussion of the role of ceramic molds in contemporary toy production ("Plaster Model and Mold Making," episode 45); and finally, relying upon the expertise she had acquired from her husband, a ceramics engineer, the high-tech industrial uses of ceramics at the height of the Space Age ("Electronics, Jets, 'Special' High Temperature Ceramics," episode 50).

On the other side of the spectrum was an audience of sophisticated viewers:

contemporary artists, educators, and, most crucially, a generation of student activists, whose fervent support of live and local television, or what came to be known as cable access, was an extension (McLuhanesque or otherwise) of their belief in democracy and the media's potential for social change. As one avid student-viewer wrote in 1967:

> We won't get college credit for it, but we expect recognition for the fact that three times a week, the alarm goes off to send us out into the morning darkness to school . . . Our instructor comes via TV, and the classroom is the living room of Florence Crosby where the three of us sip coffee and learn how to make pots . . . It is amazing what can be learned on TV before channels become clogged with news, commercials, suspense and galloping horses. All the bright people are talking at dawn about profound subjects. As we pried open our eyes for the pottery class this week, we picked up the tail end of a sociology lesson. Our whole culture and social structure may be undergoing significant changes while most of the nation sleeps.[43]

Clearly, this was a different sort of "tuning in," one that signified the makings of an alternate countercultural movement that had so much promise, as craft labor history alone was a subject area rich with potential, but never quite went the direction of radical politics. The 1960s and 1970s saw an explosion in experimental craft practice, in particular textiles, throughout Western and Eastern Europe, the United States, Canada, and Australia, but by the late 1960s, craft became in many ways politically soft, hippie-revivalist, rooted in an ethnographic revival of preindustrial craft processes (i.e., Indian caftans, macramé plant hangers, and handmade coffeehouse ceramic mugs). Finally, its discourse became increasingly, destructively insular. Its most important critics, like Rose Slivka, became obsessed with what Elissa Auther calls "the struggle for legitimacy," jockeying for equal status among the sculptural practices situated in the contemporary art world.[44] Within *Wheels, Kilns, and Clay*, Peterson makes few distinctions between making functional versus sculptural ceramics, showcasing both forms of production through her segments on contemporary artists. This is unsurprising, as the ceramics community itself, as discussed in chapter 1, did not adhere to such strict categories. The binary itself had always come from outsiders.

Contemporary production is the lifeblood of Peterson's series, another layer of *live form*, as each episode highlights an active instance of the particular episode's theme in action. Sometimes this means guest artists performing on-site, as with two Los Angeles-based ceramists in "Dora de Larios Builds a Slab Sculpture" (episode 16) and "Cliff Stewart Builds Wheel-Thrown Figures" (episode 17). Others highlight ceramists renowned for their eschewing of tradition altogether, such as Ken Price's use of autobody paint as a kind of "tough-guy" glaze, seen on

"Variety of Glaze Effects, Introduction to Glaze Melts, 1900 to 2300 °F" (episode 36). "Artist Potters Make a Living Today" (episode 49) features the microcosm of Los Angeles, including ceramic work by her predecessors Andreson, Ball, and Lukens, her best-known students, Price and Mason, and colleagues such as Voulkos. Pragmatic advice about business acumen and recordkeeping abounds in "How to Teach Ceramics and Set Up a Studio" (episode 51) and "The Future of Ceramics" (episode 53).

Because of the show's relative popularity, it seems that Peterson had high hopes for the series, fully expecting it to "take off" in some way, or at least develop further, into an expanded television series or to offer additional media opportunities:

> . . . of course I thought that I'd be hired by somebody quickly to have my own program in some way. As a matter of fact, this was before women were announcers or on television at all newscasting, and my time at the station often coincided with the male newscaster, whose name I can't remember but he was very famous on the West Coast, and he used to tell me-because he would watch my show and then he would go on doing his show-and he said, oh, you're just great; I'm going to talk to CBS; we'll have you on. Well, it didn't ever happen. I had to go back to the classroom, back to my studio, but I loved doing it.[45]

After *Wheels, Kilns, and Clay* ended its broadcast, Peterson authored a companion sourcebook with the same title, and reworked the series as a twenty-six-week course that could be offered televisually. The course was offered on KNXT-TV during the 1968/69 academic year, through USC's College of Continuing Education, and rebroadcast in 1970 and again in 1972. In this sense, Peterson prefigured the distance learning courses that currently proliferate at the university level. This is in tandem with Richards's own experimentation with outreach courses: both artists anticipate contemporary conceptions of alternative formats of learning by at least twenty-five years. Her course was listed as Fine Arts TV 408, and it could be taken for either one credit (listed as "watch TV course and pass exam") or two ("watch TV course, pass exam, and attend two 'hands-on' seminars"), augmented by the additional "hands-on" requirement, which consisted of a two-session attempt to learn the art of wheel-thrown pottery. This seems to complicate the role of the student-viewer, transforming him/her from an amateur spectator into a guided participant. As Peterson herself wrote in the course's study guide:

> *Anyone can learn to throw.* Everyone can feel the accomplishment which always results when something is made from nothing. We can all achieve the sense of oneness, of being "one" with the soft plastic clay, of its work under one's hands. Not everyone will make his clay into a real art form, or be able to say the things an artist can communicate in

> this medium, but all who work on a potter's wheel will know the tremendous expression of self which comes through in this technique and the personal degree of satisfaction which can make throwing on the potter's wheel an end in itself.[46]

But was throwing "an end in itself"? Was it enough to just throw, but not throw well? What was the good of "achievement"? When encouraged, did such achievement actually translate to a prolific form of amateurization, in effect, a well-trained hobbyist? Interestingly, in the first draft of the study guide, this paragraph was completed by the following thought:

> M.C. Richards, an unusual American potter-poet, wrote a book called "Centering" about the process of centering a ball of clay which she likened to centering oneself.[47]

But the final form of the *Wheels, Kilns, and Clay* study guide excises Richards and with it perhaps the dangerous idea that it was enough to just gain pleasure from throwing, an endorsement of the antiexcellence that Richards encouraged through her own ceramics practice, which laid the groundwork for her social engagement at Camphill Village.

While emphasizing the "empathy" of the hand in the clay itself, Peterson was also forging a haptic, community-driven practice, encouraging students to become active participants, or, as Richards had put it, "alive and awake to the best of his abilities." Yet *Wheels, Kilns, and Clay* operates on another register: introducing ceramics to a lay audience, and asking them to rethink their own household objects and architecture, without any expectation of making but rather as a twenty-eight-minute indulgence in the tactility of form, or as one viewer put it, "this is strictly a laymen's interest, and since it won't be long 'til I'm a little old lady, it might be pleasant to pinch pots in my retirement."[48]

One of the unexpected joys of the series is Peterson's curatorial effort, which results in a wide-ranging cache of pottery, both utilitarian and museum-quality. An outgoing personality, and well-known in Los Angeles's ceramics community, Peterson secured a dazzling array of temporary loans from her wide network of personal friends (many of whom were other artists), museum staff, and collectors, to loan not only their own pieces but also prized permanent collection works. These she brought onto the show in order to provide seemingly endless examples of faience, majolica, Greek amphoras ("those marvelous pointed jars from which large quantities of wine was drunk"), Japanese tea bowls, and Native American wares, secured from private and public collections throughout the city, including the Los Angeles County Museum of Art.[49]

However, when dealing with history before the twentieth century, Peterson tends toward an overarching chronological narrative that is rooted in a quasilinear ancient material history. Complete with charts, graphs, and explanations that often begin with the generic phrase "for thousands of years," Peterson cobbles together a media-specific narrative of Greco-Roman, Minoan, Egyptian, Indian, pre-Columbian, and Ottoman ceramics, without regard for other forms of human expression. This creation approach reaches back into the muddle of prehistory to establish the necessity of vessel, brick, and tile formation to the most basic human endeavors of building, dwelling, and feeding. Such a strategy not only assumes the timelessness and naturalism of ceramics while introducing its technologies as a progressive chronological force but also belies Peterson's own anxiety, relying upon the medium's once-glorious past as a way of silently refuting its waning influence in the contemporary marketplace.

The Factory

Ostensibly, the wider world of avant-garde sculpture had overlapping formal interests in relation to clay, particularly when it came to surfaces, finishes, and the problem of "decoration," all aesthetic concerns that Peterson addresses throughout the myriad episodes on glaze techniques (episodes 24, 25, 27, 36–41).[50] An example of this is Peterson's extended discussion of modularity in "The Potter's Wheel, Ancient Man to Modern Day" (episode 19), in which she discusses the usage of the potter's wheel in the factory setting, showcasing manufactured ceramic units which can be stacked and fitted together at the opening or lip, used domestically by masons or for commercial purposes by builders and construction firms.

Peterson was certainly no stranger to factories, on or off camera. In several episodes, she showcased her own travels abroad, featuring a slew of 16 millimeter footage she had shot herself while on sabbatical in 1962, documenting stoneware and porcelain factories in Sweden and Finland and communal village potteries in Japan, India, and Egypt. "Folk Potting Around the World" (episode 48), for example, utilizes these films, but so does "Kilns for Firing" (episode 10), showcasing a variety of kiln constructions, in particular primitive firing pits lined with discarded license plates to retain heat (New Mexico). The second half of "The Potter's Wheel" (episode 19) features film shot on the factory floor in Sweden, featuring women production potters engaged in architectural commissions, pouring liquid clay slip in molds made for casting, resulting in walls of gleaming tile and ceramic sculptures intended for a public fountain. The episode concludes with footage shot at the home/studio of Hertha Bengtson, one of Sweden's leading potters. But the "modern day" portion of episode 19 also dispenses with the wheel completely, focusing on mold-made pottery not just in Sweden but by local Southern California firms like Hoffman Tile (machine-pressed stoneware tile),

Interspace (larger modular stoneware tile), and Architectural Ceramics, which, in Peterson's description, "has been doing units which fit together to make walls or room dividers, can be used either vertically or horizontally, are sand cast, and textured, but are completely modular and uniform."

Such ceramic work is entirely comparable to the objectivity and modularity that had gained currency among the emerging sculpture avant-garde in New York: Eva Hesse, Donald Judd, and Sol Lewitt. Due in part to the hegemony of abstract expressionism, many artists working in the 1960s turned to sculpture, feeling that painting was no longer a meaningful or viable art form. Alongside *Wheels, Kilns, and Clay*, 1965 is also the year Judd published his landmark essay "Specific Objects," in which he advocated for a new kind of sculptural production that made extensive use of industrial materials and processes to achieve a much sought-after neutrality, in which the sculptural form resolutely embodied its own formal properties (color, shape, and surface). It was also the year he began constructing his "stack" pieces, industrially fabricated boxes made of galvanized-iron sheet metal.[51] By and large, this meant employing new materials that had not yet been used to make "art," those from industry (Larry Bell, Dan Flavin, Robert Morris) or found objects (John Chamberlain):

> Materials vary greatly and are simply materials—formica, aluminum, cold-rolled steel, Plexiglas, red and common brass, and so forth. They are specific. If they are used directly they are more specific. Also, they are usually aggressive. There is an objectivity to the obdurate identity of a material. Also, of course, the qualities of the materials—hard mass, soft mass, thickness of 1/32, 1/16, 1/8 inch, pliability, slickness, translucency, dullness—have unobjective uses . . . The form of a work and its materials are closely related.[52]

Were it not for the list of industrially produced materials, Judd could have been describing the kind of ceramics production that Peterson taught: the primacy of simple, often repeated forms and an attentiveness to the aesthetic properties of surface and density. Both advocated for construction *as a kind of content* unto itself, where the formal properties of the materials are self-determining and aesthetic choices such as texture or color are made purposefully. Peterson's history making, then, is a history of "specific objects" a lá Donald Judd. My misreading of Judd's taxonomy is purposeful: though he valued ceramic objects domestically, Judd did not intend his classification of "the new work" to include ceramics.

Yet Judd was actually an avid collector of pottery, something that gets mentioned over and over again in his writings. In his essay "Marfa, Texas," he admits that one of the reasons he was motivated to leave New York for the arid emptiness of West Texas was that "Also the new and old pottery I had bought in the Southwest was spalling in the humidity of New York, and the cacti I had collected were dying."[53] The importance of Native American pottery in Judd's life is underscored by its prominent display in the artist's own bedroom in Marfa, even after his death. Peter Schjeldahl writes of the preciousness of Judd's own preserved sleeping quarters:

> At night, a specially invited artist (not me: I stayed in the Riata Inn Motel) may dream on an islanded mattress in one or another building that permits no distinction between art space and space for living. Navajo blankets, pottery, and turquoise jewelry, in Judd's own-as-he-left-it bedroom, mark the places' closest brush with gemütlichkeit [coziness]. The art-designer's rectilinear, anti-comfortable furniture is ubiquitous.[54]

Judd's bedroom is thus transformed into the private, symbolic space of the artist's studio, wedding his artistic production to his domestic life—exactly the sort of separation that Peterson sought in her own production. Schjeldahl's offhand observation is the ultimate reversal and points to the erosion of the way 1960s minimal sculpture has been traditionally received. As a ceramist performing the "woman kitchen potter," then, Peterson's work is an incursion, upending the traditional narrative of big sculpture.

But Peterson's notion of empathy is diametrically opposed to the dominant methodologies of the 1960s, as Peterson's sculptural counterparts Donald Judd, Robert Smithson, and even her 1970s-era colleague at Hunter College, Robert Morris, were *not* form givers. That is, they did not perform *live form* as a means of pedagogical instruction. Rather, their artistic practices were predicated upon a rhetoric that rendered highly skilled labor obsolete, or as Smithson wrote:

> Donald Judd has set up a "company" that extends the technique of abstract art into unheard-of places. He many go to Long Island City and have the Bernstein Brothers, Timsmiths put "Pittsburgh" seams into some (Bethcon) iron boxes, or he might go to Allied Plastics in Lower Manhattan and have cut-to-size some Rohm-Haas "glowing" pink plexiglass. Judd is always on the lookout for new finishes, like Lavax Wrinkle Finish, which a company pamphlet says, "combines beauty and durability" . . . Judd is busy extending art into new mediums. This new approach to technique has nothing to do with sentimental notions about "labor." There is no subjective craftsmanship.[55]

Peterson's gendered, "sentimental" labor was also busy extending itself into new mediums. The 1960s-era male artist's studio and her feminine, craft kitchen had more in common than at first glance: both had dramatically shifted roles in the tumult of the postwar era, from being a center of production to a center of consumption. The 1950s art market, with its distasteful and voracious appetite for abstract expressionist painting, had propelled artists to find alternate locales and means of production or, at their most extreme, like Judd and Michael Heizer, to flee New York entirely. In the American postwar kitchen, a glut of processed food and TV dinners resulted in a generation that lacked the "subjective craftsmanship" of cooking skills. Add this to the proclivity for marketing campaigns advocating convenience food, such as those by Poppy Cannon, *House Beautiful*'s food editor who, in her bestselling book *The Can Opener Cookbook* (1951), wrote, "It's easy to cook like a gourmet, though you are a beginner."[56]

But rather than commissioning "cut-to-size" pieces, or having iron boxes shaped and seamed like Chamberlain, Judd, or Heizer, Peterson *did the work herself*. Her *live form* ceramics demonstrations were not just educational; they were also strategic, entirely predicated upon her public persona of femininity. Plying her trade in a made-for-TV kitchen, Peterson might have been a "no-shovel" potter, but, like the men, she also "went to work" as an exercise in camp, playing the hyperfeminine role of the smiling "host-professor" festooned in an apron.

In this sense, Peterson was *most* like her predecessor, Marguerite Wildenhain, who, we might remember, could not effectively play "hostess" at the Black Mountain Pottery Seminar and who was publicly lampooned for her authoritativeness (and by extension her masculinity) by Ken Price and John Mason, not coincidentally two of Peterson's earliest and most successful students.[57] Or perhaps Peterson really *was* the "Julia Child of ceramics," since Child herself—perhaps because of her unusual height and dramatic baritone vocal range—was also taunted for her "masculinity," most famously by Dan Aykroyd in drag in a 1978 episode of *Saturday Night Live*, in which he parodied her PBS show *The French Chef*.[58]

Television as Video: Live and Local

While Peterson transmitted the *live form* of the pottery experience through the filter of the didactic television show, *Wheels, Kilns, and Clay* is distinct from film or video art in that it was actually broadcast live, as the technology was early, without the potential for editing. As Peterson recalled:

> They didn't edit television at that time. It was a live program going out live. It would be put on film, the tape-the two-inch tape-so that CBS could rerun it, and they did. But it went out live and never with any editing at all.[59]

The two-inch video format to which Peterson is referring was two-inch quadruplex, the earliest video format created for broadcast television.[60] It is important to note that two-inch tapes were commercial products intended for use by television networks rather than by individuals (unlike the later invention of the Sony Portapak in 1965), for whom the cost of tape would have been prohibitive. As such, the expense led to the common practice, throughout the 1960s, of reusing tape, effectively erasing a great deal of early noncommercial programming, more often than not community documentary or educational endeavors, like that of Peterson's show.[61]

Thus, network television was a redelivery system until the late 1960s, when one-inch and three-quarter-inch tape—which took up less reel space—became standardized technology, used both commercially *and* experimentally by artists. Media historian Deirdre Boyle argues that due to the technological constraints and often-utopian beliefs invested in television as a populist endeavor, the categories of "art" and "activism" were not differentiated, nor were those of "video" and "television," until the early 1970s, when the Portapak went into wide circulation.[62] After that moment, video art was made as a transgressive act, meant to circulate in a space *beyond* television.

While Peterson's television broadcast was not video, her half-hour show *circulated* as video, since "bicycling tapes" was a common practice, dependent upon a combination of personal networks and the postal service as the primary method of transmission and dissemination during the 1960s. But Peterson's work is distinct in that she was not ". . . using the tools of television production to make art," as David Ross, one of the pioneering curators of the medium, classified the category of video art. As he appraised the situation in 1974:

> The situation that existed before the introduction of the relatively inexpensive consumer-grade videotape production equipment was roughly analogous to that of a society that possessed a tightly controlled radio industry and no telephone service. Few people had any ideas that they could *make* TV as well as watch it.

Peterson chose to "make TV," but with a pedagogical intent that distinguished her work from that of her video art peers—though "peers" is actually a misnomer, since Peterson was not directly connected to, nor affiliated with, the video vanguard. Instead she came out of Los Angeles's ceramics scene, a parallel avant-garde, whose immediate discourse was *also* preoccupied with a distinct set of technical concerns and challenges. While Peterson's series parallels Nam June Paik's mid-1960s formal avant-garde television experiments with color television and magnetic interference, I am not suggesting she was a video art innovator.

Peterson's television series is significant, however, in that it predates the earliest forays in conceptual—and televised—video art, situating her as a process-based predecessor to California's illustrious history of experimentation by such disparate artists as John Baldessari, Chris Burden, Bruce Nauman, and Martha

Rosler. Burden also experimented with television as a medium, albeit much later, purchasing thirty-second spots in between longer commercials that jump-cut suddenly to his experimental body art, such as *Through the Night Softly* (1973), when the artist crawled naked through broken glass. As in Burden's piece, video art of the 1970s—the first full decade of its existence—came to be associated with a difficult viewing experience that is antithetical to Peterson's fundamental didacticism: the purposeful disjuncture between image and sound, nonlinearity, and the use of static, volume spikes, and out-of-focus effects were all conventions of early experimental video as a medium distinct from either film or television.

The early 1970s can be characterized by the primacy of the artist's body as subject material and the use of the camera as a staging device or a tool of surveillance, exploring themes such as perception, the limits of language, social satire, and artistic labor. But Peterson's process was none of these: largely informational, her ceramics demonstrations were neither social parodies nor critiques. Rather, they were instructional broadcasts made dynamic by the subtle, even unconscious, gender constructs being communicated: Peterson's femininity and her kitchen are premised upon artifice. But in the tumult of the late 1960s—the Civil Rights movement, the formation of Students for a Democratic Society (SDS), the Vietnam War and its ensuing student protests, all of which served as both inspiration and fuel for the burgeoning feminist movement of the early 1970s—a sea change took place: women artists also donned aprons and set themselves to the "task" of domestically inclined, labor-intensive artwork, this time embracing the kitchen as a site of resistance, rather than as the most readily available sphere of production.

Feminism and the Oppressive Kitchen

It was not until 1970s feminism that the woman artist's studio, with its domestically inclined nomenclature, found a "home" of sorts, borrowed from modernism; Virginia Woolf's treatise *A Room of One's Own* (1929), one of the hallmarks of modernist literature, was lauded as a protofeminist manifesto. In a succinct essay format, originally delivered as a lecture, Woolf advises women of their need for personal independence, both financially and spatially, in order to create works of art. As Miriam Schapiro noted in a 1989 interview:

> We were very proud of Virginia, you know. She really is a heroine to creative feminists, in the sense that her life is so instructive to us. Her devotion to her art. The fact that she was so original. Also the tragedy of her life. You know, these things belong to us. Like Sylvia Plath's life, or like Frieda Kahlo's life. It's always somehow inspiring to think of

someone who had great suffering and yet great achievement at the same time. We are inspired by their redemption. And all of that made for a genuine kind of myth for women who were at that time being so conscious of the fact that they were women, and that they had a history. We needed to have role models. Everything came together at the same time: Art and consciousness, myth and reality. You asked me before, and I don't think we went into it, why did I think that California was a particularly fertile ground for all this to happen? I don't think it could have ever happened in New York. New York is hell-bent on separating one person from another because of the competition. And in California there was some chance for us to get together. And it was as simple as that.[63]

While Woolf addressed the predicament of the woman writer, the concept was felt acutely among visual artists, whose "room of one's own" equivalence was the artist's studio. But this was exacerbated by the predicament brought on by the 1960s: male artists such as Judd and Smithson had already discarded the studio as a legitimate sphere of production a decade before women artists had even *gained* the social privilege of inhabiting their own studios.

Among early feminist artists, then, the "women kitchen potter" existed in full force. The ardent feminist critic Lucy Lippard also took note of the phenomenon, augmented to include a variety of different media:

In the winter of 1970, I went to a great many women's studios and my preconceptions were jolted daily. I thought serious artists had to have big, professional-looking spaces. I found women in corners of men's studios, in bedrooms and children's rooms, even in kitchens, working away.[64]

Indeed, the prevalence of using the kitchen as a studio space quickly gave way to its usage as a targeted site of tyranny throughout feminist performance art. The feminist 1970s was obsessed with reproducing versions of the oppressive kitchen, as a space to explore domesticity and the essential labors of femininity but also a site of the psychological exploration of motherhood, marriage, and maternal relationships.

Furthermore, feminist artwork about the kitchen is split between actual labor and labor's approximation. Actual labor included Mierle Laderman Ukeles's Maintenance Art performances (1973–1979), in which the artist most famously scrubbed the museum's steps at Hartford's Wadsworth Athenaeum, or a later, less-known incarnation of this type of artwork, performed collectively by Mother Art (1977–1979: Jan Cook, Gloria Hadjuk, Christie Kruse, Suzanne Siegel, and Laura Silagi). Other artists of the era approximated feminine role-playing in the public sphere, such as the collective The Waitresses (1977–1985: Jerry Allyn, Anne Gauldin, Chutney Gunderson, Anne Mavor, Patti Nicklaus, Jamie Wildman,

Figure 5.8 *Red Stripe Kitchen*, from the series *Bringing the War Home: House Beautiful*, 1967–1972. Martha Rosler (American, b. 1943). Photomontage. Courtesy of the artist.

and Denise Yarfitz), which performed waitressing actions publicly, highlighting issues of rampant sexism and low wages, and acted as activists publicly campaigning for the passage of the Equal Rights Amendment (ERA). In 1972, within the exhibition space of Womanhouse, Vicki Hodgetts, a student in Judy Chicago's Feminist Art Program, reconfigured the kitchen as an installation exploring the nature/nurture dialectic, in which sculpted fried eggs, made to look like breasts, were pasted up all over the walls, while the dining room table was set with innumerable plates of food.[65]

A decade after *Wheels, Kilns, and Clay* had first aired, the format of the cooking show was also embraced by the artist Martha Rosler (American, b. 1943), who perfected its parody in a series of videos made over a period of three years at an early moment in her career: *A Budding Gourmet* (1974), *Semiotics of the Kitchen* (1975), and *The East is Red, The West is Bending* (1977). Born and bred in Brooklyn, Rosler moved to San Diego in 1971 with her then-husband and small child to be among the first students to attend the newly formed MFA program at the University of California, San Diego, where she joined a thriving community of Marxist artists, filmmakers, photographers, and intellectuals that included other displaced New Yorkers such as Paul Brach and Miriam Schapiro, Eleanor and David Antin, and Allan Kaprow, as well as Babette Mangolte and Allan Sekula. Additionally, the philosopher Herbert Marcuse was a critical and iconic presence on campus; his lectures were widely attended by a range of students in the arts and humanities. San Diego's proximity to numerous military bases and defense operations made the politics around the Vietnam War a particularly potent discourse.[66]

Rosler's own art work from this period responded to these conditions, in particular, her photomontage series *Bringing the War Home: House Beautiful* (1967–1972), in which she collaged found war images from Vietnam, as seen in *Life Magazine*, and inserted them into the tranquil suburban settings reproduced in the shelter magazine *House Beautiful*. The luridness of *Red Stripe Kitchen* (figure 5.8), for example, comes from the perspectival focal point located in the back of the kitchen, in which American marines are seen crouching, looking for an invisible enemy. The decorative red stripe is reproduced in the whole of the kitchen décor, from the cherry-hued modernist ceramic dinnerware, to the fire-engine red interior of the sink, to the enameled cookware sitting on the

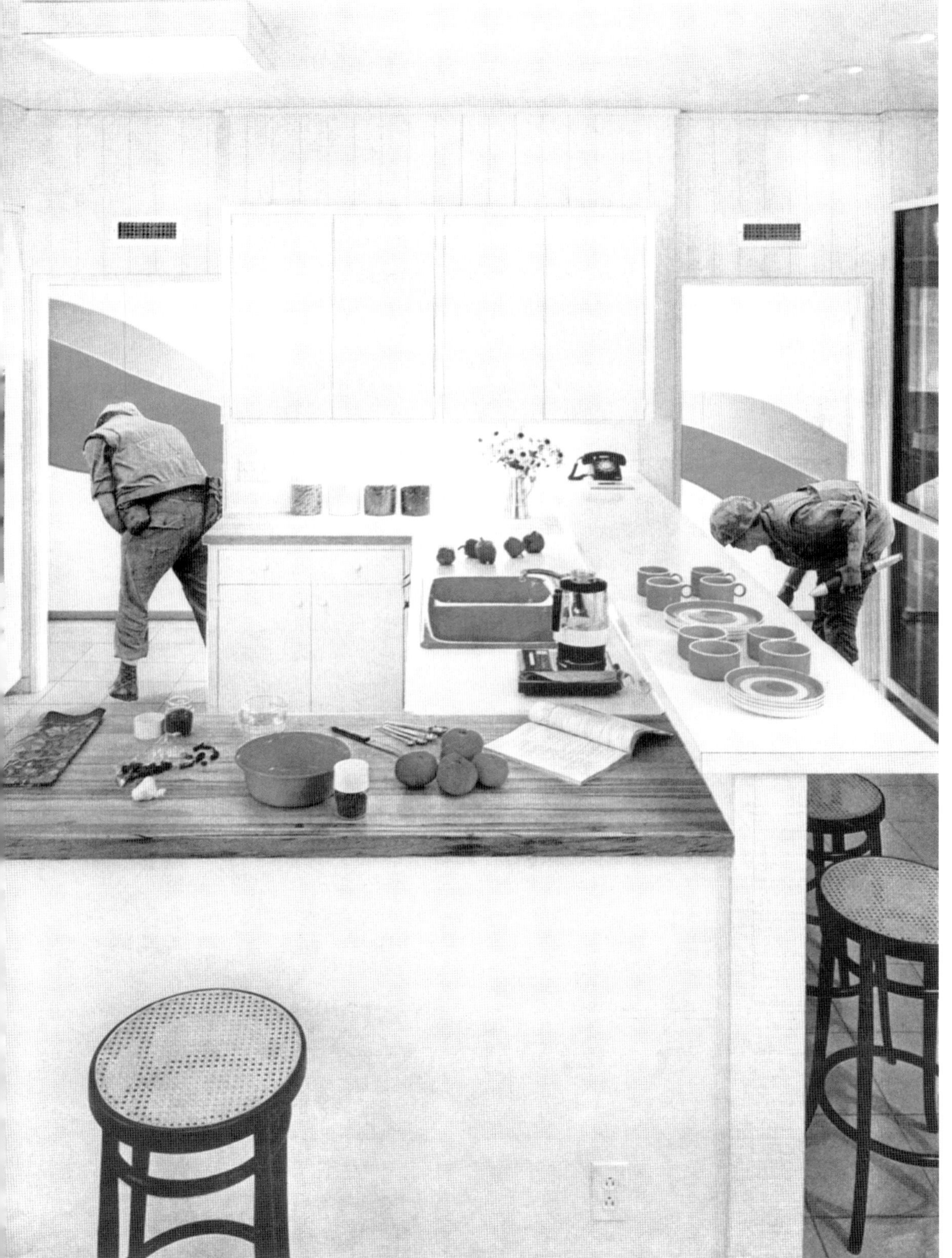

front counter, surrounded by an open cookbook, a trio of apples, and cooking utensils. The ceramics in particular are implicated both by color and their formal properties, awash in a slick red glaze that highlights all the trappings of consumption. What should be an impossibly cheery kitchen is irrevocably tainted, compulsively bloodied by all that red, an immobile, matte blood stifled in both its response and its self-containment, implying the kind of violence far from home. Ironically enough, in the early 1970s, the modernist industrial designer Russel Wright, who had designed the famed Fiestaware line of the 1930s, actually made several ultimately unsuccessful attempts to set up a handicraft factory to manufacture his goods in Vietnam at the close of the war.[67] *Red Stripe Kitchen* is not only antiwar; it is also antidesign, a campaign against senseless luxury during a time of national crisis.

With a strong penchant for activist and accessible works of art, Rosler's career has been written about extensively in relation to feminist and socially engaged artistic practices, and she has made it a point of her practice to work outside the confines of the traditional art world, beyond museum and gallery spaces.[68] But it is virtually unknown that Rosler had a relationship to ceramics: she studied pottery and had an early interest in the medium, cultivated at Eramus Hall High School in Brooklyn and later as a summer camp instructor in Connecticut (again, a rural, nontraditional site). As a high school student, she had a non-functional sculptural ceramic work included in a regional competition at Lever House, her first publicly displayed artwork. As Rosler states:

> I definitely felt a lingering affection for pottery, especially the glazes producing magical color, but I decided against it in favor of painting expressly for the arduous nature of the material practice that came with pottery—purchasing clay, forming it, grinding it, storing it, firing. And then there is the question of function.[69]

Rosler may have left behind pottery as a medium, but the kitchen figures prominently in her *oeuvre*.[70] According to Lisa Bloom, it also loomed large in her upbringing, considered by her émigré parents to be "the only secure domain for Jewish women."[71]

To that end, *A Budding Gourmet* (1974) (figure 5.9) is a nearly eighteen-minute black-and-white video in which Rosler, in character, is seated at a plain table, playing a nameless bourgeois housewife, known only to the audience as the wife of "Len," whom she mentions frequently, along with their circuit of suburban dinner parties. Bracketed by the handwritten, cursive sign "I wish to become a gourmet" while classical music plays throughout, Rosler's character embarks upon a self-transformation, and as the video unfolds she plaintively reveals her desire to join a world at once removed and tantalizingly close, represented

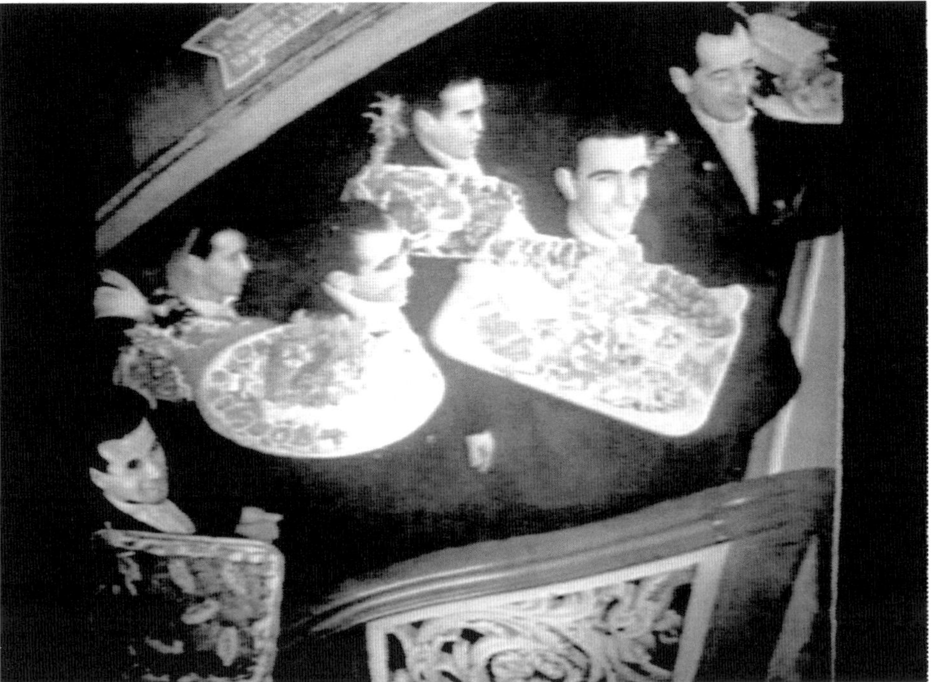

Figure 5.9 Still from *A Budding Gourmet*, 1974. Martha Rosler (American, b. 1943). Video, black and white and sound, 17:43 minutes. Courtesy of the artist.

through pictures in mainstream magazines and newspapers: the elite world of the very wealthy, featuring servants and butlers who cater to every need and offer luxury food to match, carrying trays of prawns and champagne, perfectly molded tarts and cakes, and whole fish, decorated with julienned vegetables. The images come from the world of professional food styling, yet they are sometimes blurry or grainy and purposefully shown in black and white, denying the audience the wealth of color that come with such laborious food preparation. Rosler also prominently features the service, reinforcing the concept of luxury through decorative ceramic plates, copper kettles, and baroque silver platters.

Throughout the video, Rosler's character keeps up a steady patter, telling stories and reciting recipes, playing the part of a well-bred woman cultivating her gourmet sensibilities as well as expanding on her imperial views of the world. She describes cooking as ". . . broadening my horizons, thankful that in America we can take the best of all times and all places and make them our own." Food, then, becomes an entertainment, a hobby, and a social gathering in which to show off new cooking skills (Brazilian, French, and Indian cooking!) as a way to impress or compete with so-called friends. Interspersed throughout are occasional maps of Africa and Southeast Asia, along with images of malnutrition

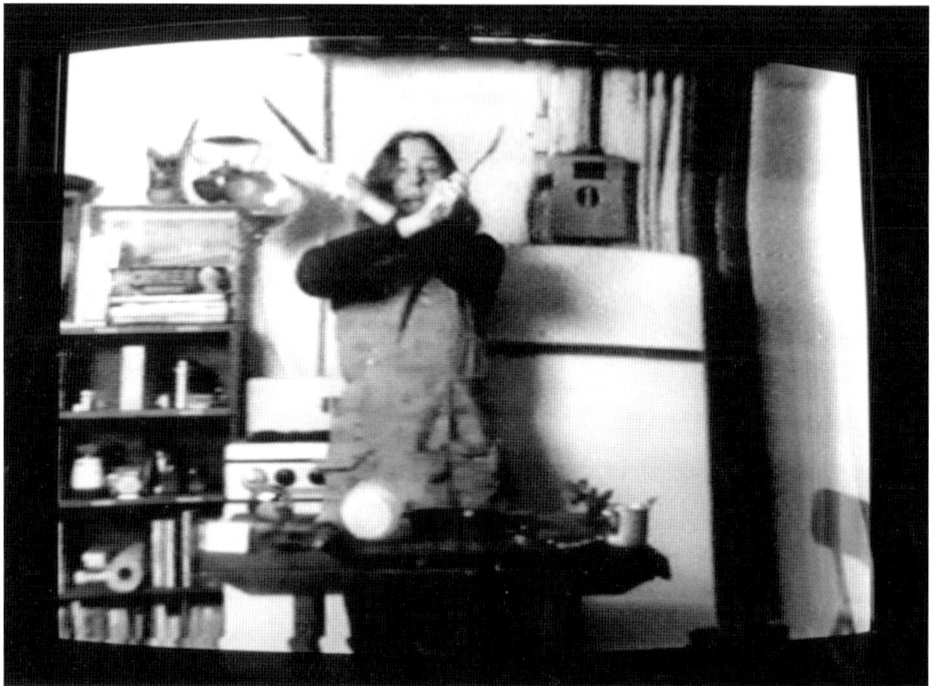

Figure 5.10 Still from *Semiotics of the Kitchen*, 1975. Martha Rosler (American, b. 1943). Video, black and white and sound, 6:09 min. Courtesy of the artist.

and hunger—the distended stomachs and saucer-like eyes of African children. Yet, becoming a gourmet, both as cook and connoisseur, is a redefinition of the consumer experience, transforming it into the kind of self-education project that Peterson herself might have championed, were the character not so naïve and self-contained. If Peterson identifies the kitchen as a site of use-value, Rosler's *A Budding Gourmet* negates the inquiry, questioning the ways in which women, through seemingly harmless exercises in taste, are overtly conditioned (or condemned) to traditional notions of femininity.

A Budding Gourmet was an outgrowth of Rosler's MFA thesis exhibition, held the previous year. Titled *A Gourmet Experience* (1973), the installation was even more low-tech than the video, a slide show of gourmet food displays also interspersed with images of hunger and malnutrition, but placed before a formal table setting. The slide show was paired with an audio in which the artist recited recipes from cookbooks, meant to critique the problematic relationship between producers and consumers, as well as first-world/third-world dependencies.[72]

An even shorter video, Rosler's six-minute parody *Semiotics of the Kitchen* (1975) (figure 5.10), has become an iconic video work. A humorous take on feminist rage, Rosler stands before the viewer in a kitchen, with an array of

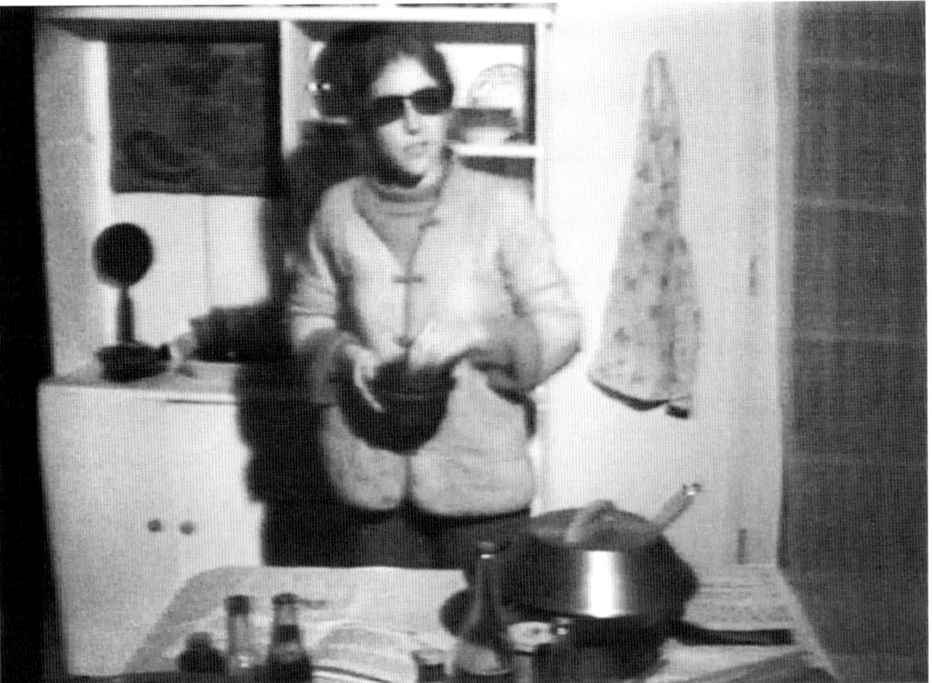

Figure 5.11 Still from *The East is Red, The West is Bending*, 1977. Martha Rosler (American, b. 1943). Video, color and sound, 19:57 minutes. Courtesy of the artist.

kitchen instruments set before her, proceeding to demonstrate their usage along with the requisite kitchen skills required of all modern women. Adorned with the theatrics of anger, the video comments upon the failure of language to properly convey the pent-up anger and frustration of the Everywoman. With the neutrality of each carefully enunciated word—Apron, Bowl, Chop—each is made pregnant with meaning, as the artist announces the word, then demonstrates its specificity with an exaggerated gesture, such as miming tossing hot soup over her shoulder, flicking the ladle harshly. The last four letters of the alphabet, W, X, Y, Z, are a parody of Zorro-like sword skills slashing through the air. Rosler's domestic paradigm instigates a previously unseen bravura, a distinct combination of parody and gravitas, weighting the situation and making it, ultimately, watchable, a diffusion of tension through humor, such as the final moment in *Semiotics*: the video seems to be over, until it is not. The artist steps out of her meat-cleaving angry housewife role to shrug at the camera, communicating directly to her audience a signal of being in cahoots, letting the audience in on the joke.

Of the three videos, it is *The East is Red, The West is Bending* (1977) (figure 5.11) that comes closest to replicating the format of the television cooking show. Nearly twenty minutes in length, this video performance also takes place in a small, urban kitchen but is shot in color. The artist stands with her back to the sink and stove before a draped table, upon which a West Bend electric wok is the centerpiece, ostensibly for a cooking demonstration. Wearing sunglasses and her hair in an "oriental" style bun, Rosler proceeds to read from both the West Bend's manual and an English-language cookbook about Chinese cooking. As in *Semiotics*, a wide assortment of cooking utensils and ingredients are at her disposal. She showcases a bottle of sake, Chinese beer, and performs silly stunts like putting the wok's lid on her head, a mimicry of the classic wide-brimmed peasant hat, or lifting a foot to highlight the flat black "one-size fits all" shoes sold widely in urban American Chinatowns.

Lists are a prominent feature of the video, such as reading aloud from a list of ingredients utilized in Oriental cooking (soy sauce, tamari, sugar, sake, cornstarch, onions, scallions, fresh whole ginger root, star anise, bamboo shoots, dried seaweed). Rosler uses her litany of items and aesthetic requirements—such as dinnerware with a "pleasing" lotus pattern—to critique the inherent Orientalist tendencies of the West and its endless fascination with Chinese cuisine, but also to call attention to the pervasive binary thinking that feminizes the East: praised for its "natural" and "primitive" affinities by the masculinist West, a society more "advanced" for its modern technological directives.

Rosler reads from the section of the cookbook that features food preparation, and the manner in which food is cut before cooking. As in *Semiotics*, she demonstrates uniform diagonal chopping on a carrot that then scatters all over the floor and table. The wok as an all-purpose vessel features prominently in her rhetoric, but she never actually cooks in it, becoming more and more heated in reading from the instruction booklet, which leads up to the finale, a recitation that is a conglomeration of found texts and a final breakdown between East-West relations:

> Remember, Oriental tastes are refined, but they are essentially primitive. They don't have our technology. Primitive thinking has led the Chinese toward communism! But the more advanced Japanese are now turning to the West. We have improved on the clever idea of the wok and moved it out of stagnation. We have even added a flat bottom to make it more like a real pot. . . . the primitive wok is cheap and hard to care-for, needing to be wiped and dried. The West Bend Wok represents the marriage of American know-how with the honest authentic simplicity of the mysterious East.[73]

Rosler's dry delivery uses the iconic cooking vessel to channel a profound rage about American imperialism, its "improvements" upon Eastern sensibilities and traditions, and an assumption about hierarchies of technologies. Her video plays on multiple levels: first, as an irony, in that it was a Japanese company,

Sony, that invented the technology that enabled American artists to begin working with Portapak video. Secondly, Rosler's commentary parodies McLuhan's "hot" and "cold" categorization, showcasing the ways in which so-called primitive, "authentic" tools—the wok—come to have cultural signification, circulating in more nuanced ways than straightforwardly teleological notions of progress.

In this case, McLuhan's progression becomes the American "improvement," the foolhardy entirely corporate electric wok, with the flat bottom, which contributes to a denial of the vessel's original form, or as Rosler describes her own performance, "A woman in a red-and-blue Chinese coat, demonstrating a wok in a dining room and trying to speak with absurd voice of the corporation."[74] Unwittingly, however, *The East is Red* also inserts itself into a larger national dialogue. Chinese cooking became a stand-in for East/West relations in tandem within a pressing political debate, as 1977 is also the year US President Richard Nixon triumphed in reestablishing diplomatic ties with communist China.

But *The East is Red* is the video that most closely parallels Peterson's practice, though entirely through satire: Asian cooking vessels, the Chinese cookbook, the fetishization of Asia, and the Orientalization of American pottery. Oddly enough, though, it was ceramics that sought to preserve such forms, salvaging them from "progress," which would have irrevocably altered their production and usage.

Leach and Hamada—and by extension, Peterson—worked to reinvigorate traditional forms, but they too were caught up in the binaries of East and West. Without knowing it, *The East is Red* empathizes with the formalist inclinations of the ceramics avant-garde while simultaneously pulling the plug on Leach/Hamada hero worship roughly in the twilight of their lives (Hamada died in 1978, Leach in 1979). Peterson eventually followed suit, though not for another twenty years. In 1999, however, Peterson publicly denounced Hamada as a "pseudo folk artist" and upheld María Martínez and Lucy Lewis as "forging new paths from communal traditions."[75]

Both Peterson and Rosler appropriated the same genre, the cooking show, using opposite strategies. Peterson's intent was sincere: performing a dynamic medium that had not only the same attributes of minimalist sculpture but also surpassed it as a profoundly more humanistic medium, threaded through history as a useful artifact, domestic object, and building material. That she chose to perform in a nonthreatening housewife's apron to get her message across is part of the campy aftereffect of the series: through the ironic role of the woman kitchen potter, she highlights the irony of the medium's secondary stature. Through her cooking show parodies, Rosler takes the genre a step further. Through the camp ploy of bold, obvious impersonation, she creates a repertoire of caricatures and likenesses that critique the earnestness of Peterson's own pedagogy. Video is the antithesis of narrative film. Rather than building a true theatrical character, fleshed out over time, Rosler's videos are closer to what her mentor Eleanor Antin spoke of in her own video works: "I denied my self-image."[76]

All three performances are a denial of self in favor of a performed personification of domestic labor and, with it, a biting social critique, downgrading the kitchen as a site that feeds, rather than transcends, oppression. As Rosler has observed:

> The same way performance could not be judged by theatrical standards, video could not be judged by the standards of cinema. Most importantly, it was easily duplicated and transmissible by mail, and therefore, it could evade commodity fetishes of the moment and simply be a work.[77]

Such "standards" could be applied to ceramics as well: Vincentelli maintains that one of the key patterns of women producers has been their continuous preference for ". . . low technology in all its various forms."[78] The simultaneity of "low" technologies is, at heart, a struggle over representation. With little or no commercial value, video quickly became a medium prized for its purity, shown and disseminated through peer networks and at alternative spaces and festival venues. Peterson's ceramic practice *also* had no commercial value, as was she producing a participatory series of videos in a medium that only valued object-based production.

But as a predecessor to either process-based video or participatory feminist art, Peterson's videos engaged the viewer as an equal participant. Since her audiences tended to be amateurs, Peterson's pedagogical instincts insulated her from the harsh realities that her practice, over time, had become outmoded, in effect supplanted by feminist artists such as Rosler. Long after *Wheels, Kilns, and Clay*, Peterson continued to fashion her own practice as a community-based endeavor, fostering not-for-profit organizations even as she continued to teach full time, eventually leaving USC for Hunter College in 1971. At the dawning of the feminist art movement in Los Angeles, Peterson swapped coasts, oblivious to the changes happening around her.

However, she may not have had the luxury of remaining in California, as USC's then-provost had moved to terminate the ceramics department and disband its faculty. The precise reasons for this are unclear, but perhaps functional pottery had outlived its use-value as either an occupational therapy or as a utopian value. In 1971, Peterson resigned in the midst of this heated debate. She retained her summer position at Idyllwild and moved to New York in the fall of 1972.

The West Coast

As Miriam Schapiro vehemently described it, New York was "hell-bent" on fostering unnecessary competition and friction between women artists. In theory, Woolf's manifesto rallied on behalf of the lone female creative, but in practice, it became a nascent battle cry to aid in feminist institution building—formal and informal—throughout Southern California. This includes both its networks *and* its spaces, often with pedagogy as a hub: just as ceramics built its profession presence during the 1950s, many feminist ventures, such as consciousness-raising groups and early all-women cooperative galleries (Womanspace, Grandview), also had origins in various art school or professional networks, such as the artist June Wayne's "Joan of Art" seminars, in which she instructed young women, through a "high priestess" persona, on matters of professionalization in their artistic practice. Actual physical spaces like university art galleries also had a pedagogical mission with women at the helm. Through their curatorial efforts, their profiles became markedly feminist, such as the gallery at University of California, San Diego (UCSD), programmed by art historian Moira Roth, and the California State University, Fullerton, gallery, programmed by Dextra Frankel. Other early curators worked itinerantly, such as Josine Ianco-Starrels, who organized the region's first all-women exhibition, titled "25 California Women of Art," in 1968 at a private commercial gallery.[79]

But by far the most comprehensive and well-known women's space in the region was the Woman's Building, a public center for women's culture, established in Los Angeles in 1973 by the artists Judy Chicago and Sheila de Bretteville and the art historian Arlene Raven. Occupying a site known as the Grandview Building, the Woman's Building opened November 28, 1973, in a building that had a previous illustrious history as the original site of the Chouinard Art Institute (1921–1972). Both psychically and physically, Chouinard—where Peterson had taught—is an important but forgotten predecessor to the burgeoning Woman's Building and its separatist art school, which came to be known as the Feminist Studio Workshop.

As established earlier in the chapter, Chouinard itself can be interpreted as having a single-sexed pedagogy, oriented toward men, training a generation of Los Angeles's most prized artists throughout the late 1950s and early 1960s. Chouinard had schooled a diverse population of both commercial and fine artists, including Harry Diamond, Walt Disney, Joe Goode, Robert Irwin, and Ed Ruscha. Its fashion design program, headed by Eva Roberts, had produced Bob Mackie and Rudi Gernireichs. At one time or another its distinguished faculty had included Robert Irwin, Millard Sheets, and Emerson Woelffer.

This period, as Judy Chicago described it in her popular autobiography, was an era of extreme sexism that ultimately created an intense need for feminist advocacy in the art world, leading to her establishment of the Feminist Art Program at CalArts.[80] Such insular artistic training and subsequent networks emphasize the circular relations between the three schools (Chouinard, Cal Arts, and the Feminist Studio Workshop), though these overlaps had an historical precedent as well: earlier in Chouinard's history, a group of its faculty had broken away to found Art Center in Pasadena in 1930. As Chicago writes:

> Chouinard [Art Institute] was well known but Nelbert Chouinard was NEVER mentioned—another indication of the many ways in which women's achievements were ignored by the L.A. art world (as well as the larger world) . . . The reputation of the school was part of the pleasure of taking it over for the Women's Building.[81]

Part of the legacy of 1970s feminist art has been its expansion of the scope of artistic practice to include the domestic sphere. The idea of "home," particularly the search for shelter, can be seen as a larger metaphor for community building. As a result, the artwork from this period was often collective and performative, in which artists used domesticity as a broad thematic upon which to project their abundant and often mordant desires and grievances. In her essay "House Work and Art Work," Helen Molesworth observes, "To position this work as negotiating the terrain of public and private is to establish the links to, as opposed to its separation from, other postwar art practices."[82]

The same case can be made for ceramics. By making process-based ceramic videos, Susan Peterson had captured the labor-intensive process of her craft by using the *language* of contemporary art. Had feminist video makers such as Susan Mogul or Suzanne Lacy (who also did a video parody of Julia Child, titled *Learn Where the Meat Comes From*, in 1977) at the Woman's Building seen her television programs, surely they would have delighted in Peterson's gender performance. They might have even offered her an exhibition, since recouping the prior generation of protofeminist women was a crucial feature of the Building's pedagogy.[83] Unfortunately, these introductions and overlaps were never made. In New York, Peterson took on the City University of New York's intensive bureaucracy to install kilns and initiate a degreed ceramics program—New York City's first—while also pursuing her own independent writing projects, such as those on Hamada and Martínez, throughout the 1970s.

Yet the stakes of Peterson's *Wheels, Kilns, and Clay* cannot be underestimated: while short-lived, her experimental television show fully destabilizes the constrictive genealogies of ceramics, sculpture, and video art, pioneering an intensely original interdisciplinary and protofeminist model before such ideas had scarcely

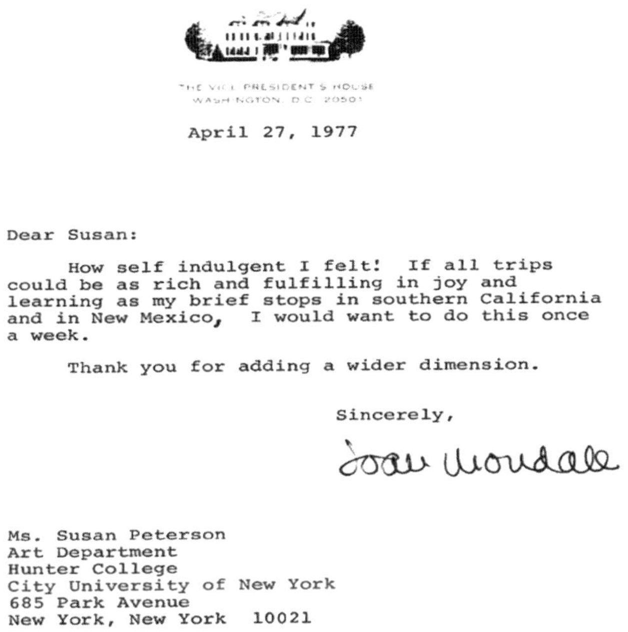

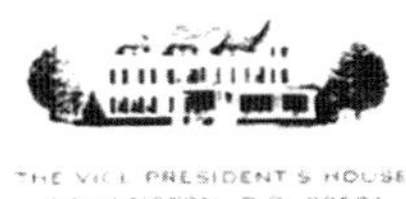

THE VICE PRESIDENT'S HOUSE
WASHINGTON, D.C. 20501

April 27, 1977

Dear Susan:

How self indulgent I felt! If all trips could be as rich and fulfilling in joy and learning as my brief stops in southern California and in New Mexico, I would want to do this once a week.

Thank you for adding a wider dimension.

Sincerely,

Joan Mondale

Ms. Susan Peterson
Art Department
Hunter College
City University of New York
685 Park Avenue
New York, New York 10021

Figure 5.12 Letter from Joan Mondale to Susan Peterson, April 27, 1977. Susan Harnly Peterson Archives, Ceramics Research Center, ASU Art Museum, Arizona State University, Tempe. Courtesy of Jill Peterson Hoddick, Jan Peterson, and Taag Peterson/the Estate of Susan Peterson.

materialized. As a *live form*, her ceramic performances engendered a sophisticated and radical pedagogy fully enmeshed in social and aesthetic innovation.

In 1978, a year after Rosler's *The East is Red*, Peterson enlisted a group of longtime friends such as Rose Slivka and the artist June Wayne to be on her advisory board and founded Clayworks on the Lower East Side of Manhattan. Like Margie Hughto's *New Works in Clay* series at Syracuse University, Clayworks paired contemporary artists with their ceramic artist counterparts as a collaborative gesture. Clayworks lasted only two brief years, folding in 1980, when universal cuts were made to American arts programming when the National Endowment for the Arts budget was slashed.

Over time, however, Peterson's community-building endeavors crystallized into a self-replicating system that continually reinforced the secondary status of craft while seeking to elevate it. The inherently minor associations of women, pottery, and the kitchen are epitomized in a 1978 *New York Times* piece on Joan Mondale—the wife of then-Vice President Walter Mondale. A self-appointed "advocate for the arts" on behalf of the Carter administration, Mondale met Peterson in California and became a powerful friend, selecting Clayworks for an orchestrated photo-op alongside the painter James Rosenquist.

Peterson was likely pleased with the attention surrounding Clayworks. But Mondale, as its supporter, was the ultimate amateur: a well-connected, well-meaning Yankee, whose own benign interest in American crafts in no way challenged the cultural authority of the Southern president Jimmy Carter, and First Lady Roslynn Carter. Mondale's Clayworks appearance was a way of epitomizing community-based ceramics as New York's own version of a popular art form:

> Joan Mondale, wife of the Vice President, sat in dungarees and sneakers at a potter's wheel yesterday morning pressing her thumbs into a lump of clay. She rolled, sliced, patted, punched, squeezed and pounded the clay into a "stovepipe hat." Then she stood, flashed a smile, lifted her creation as if it were some freshly baked torte and waited for the applause.[84]

In 1978, at the height of a sophisticated feminist art discourse, then, Peterson reverted to an outmoded language of domesticity, replicating the smiling "host-professor persona" she had sought to cultivate in the early 1960s, passing the torch to Mrs. Mondale, heir-apparent, who gladly played Peterson's *own* role, the "Julia Child of Ceramics," testing the limits of empathy for an afternoon, while the cameras rolled (figure 5.12).

LIVE FORM
DISSEMINATED

EPILOGUE

Ceramics as a discipline is so undertheorized in twentieth century American art history as to have been nearly forgotten. As this book has argued, it is an inherently elastic medium that turned outward from an object-only orientation toward an embrace of community engagement, personal enrichment, and social participation during the postwar era. The collectivity in artistic practice that flourished during the 1970s could not have occurred without the pedagogical networks put into motion by craftswomen of the 1950s. Their relentless experimentation has gone largely unnoticed because they produced not from inside, but beside and alongside, other modernist practices.

However, if we reconceive of the 1960s as a "gap decade"—that is, a decade of masculinist production and market consumption—we can reconfigure the gendered lineages that began in the 1940s and 1950s and traversed the 1960s to reemerge in the 1970s, when feminist concerns came to the forefront of artistic discourse. With the exception of Richards's inclusion in Edelson's poster, there was no reclamation effort on behalf of craftswomen by a younger generation. Art world feminism bypassed craft as a pedagogical framework, instead selectively appropriating its skill and materiality into explicitly feminist narratives, largely in the form of textile-based work.

But, as this book has argued, women have performed the unacknowledged labor of craft, that is, not just the making of objects, but crafting entire programs and curricula, sometimes at the expense of professional recognition, but in service to their own autonomy. Rather than canonization, what many achieved was self-sufficiency: a living wage, a modest artistic career, and the respect of one's peers. Often this came in the form of a teaching position rather than an exhibition record. Yet Wildenhain, Richards, and Peterson eschewed this well-trodden path, choosing instead uncomfortable situations that culminated in personal and pedagogical growth, one of the key components of radical education.

The three careers examined here are case studies that stand in for the permeable and tenuous networks, connections, and influences of Black Mountain College as it outwardly expanded as a legacy and came to subsume the multivalent strains of radical pedagogy that had developed in tandem to, and in spite of, modernism. If Pond Farm and Gatehill Cooperative at Stony Point were the most place based, other kinds of teaching practices were itinerant, ephemeral, or disseminated in other ways altogether, such as the television series *Wheels, Kilns, and Clay*. As a *live form*, ceramics embodies the vibrancy of performance, teaching, and object making through communal practice and collective skill building. These three craftswomen dedicated their artistic labor to creating situations in which young men and women could learn, grow, and thrive, free of the obstacles of the past, offering access to an education absent the rigid ideologies that could potentially suck the lifeblood out of one's artwork, such as the hegemony of Clement Greenberg's all-over painting or Donald Judd's specific objects.

As teachers, they emboldened other artists and citizens alike, insisting on an ontology of form beyond the masterwork, the traditional orientation of ceramics. Above all, they exposed ceramics itself—a discipline mired in technical and individual achievement—to experience: the pot was only as good as the process. *Live form* was a genuinely transformative pedagogy: gestural, meditative, and playful, it was a powerful social tool that strengthened and repaired the bonds, as Richards put it, "toward self-definition and toward community."[1] In time, the process replaced the object, and it was the communal endeavor of throwing ceramics that propelled a collective awareness of the social body and its relation to clay.

Moreover, it was only through this pedagogy of *live form* that ceramics as a discipline was able to move forward, joining a theoretical current in the late 1960s of process art, which championed creative activity in whatever form was generated. For a medium embittered by its own erasure from the contemporary field, this was a path forward, one that offered future possibilities of inclusion. *Live Form* points the way toward difference in modernism, the illusion of medium hierarchies, and the distinction between the incomplete object—the thrown vessel form—and its vibrancy as a heuristic tool when molded onto particular communities and situations. The shape and pliability of the form itself yielded a wealth of educational formats that didn't stop with Wildenhain, Richards, or Peterson.

Their stories offer a shared history of exceptional but previously uncharted accomplishment: collectively, they pioneered a hands-on teaching style through summer workshops, offering therapeutic practices for returning war veterans, college students, the elderly, the mentally handicapped, hobbyists, and seekers.

Working at some remove from the art world, they create lasting pedagogical structures outside of the academy where they could innovate and educate, parlaying the best of its aesthetic discourses into more meaningful, community-based engagements, before such ideas had any traction at all.

M. C. Richards's life and work comes closest to contemporary notions of social practice. For Richards, ceramics was a vessel that held the capacity for renewed interpersonal contact and social commitments such as empathy, trust, and the dissolution of competition. Richards's philosophies of living-as-form anticipated the human potential movement of the late 1960s. In choosing to live and work among the mentally handicapped population of Camphill Village rather than among her avant-garde peer group in New York, she committed to an authentic form of making, creating a communal workshop that anchored itself to a situation of need, truly integrating art with social improvement and accepting the ensuing responsibilities.

Live Form anticipates the post-1990s participatory forms of art making that have become collectively known as social practice. Recent generations of contemporary artists and architects have unknowingly revived much of the craft-inflected instruction I have examined in depth. The ceramics practices I have examined anticipate the communal, participatory pedagogies of several current ventures afoot in the United States, such as Mildred's Lane, a rural artist-run residency program in Beach Lake, Pennsylvania, cofounded by J. Morgan Puett and Mark Dion, in which Puett fosters "workstyles," a method of making she describes as "an experiential making-doing-thinking process with our fellows, friends and visitors. It is rigorous and creative domesticating as a highly intuitive aestheticism of all things at all times, in every aspect of life."[2]

There are also bright echoes in Rural Studio, a design education program in which Auburn University undergraduates design sustainable and eco-friendly housing and local buildings for poor and rural African American residents of Hale County, Alabama. Feminist Felt is another such project, spearheaded by the Chicago-based artist Melissa Potter, who brought feltmaking to the Republic of Georgia in the former Soviet Union, in effect reskilling a marginalized community of women to engage with a traditional form of art making.

Another project is Poor Farm, a summer school and long-term exhibition space for noncommercial artists' projects initiated by Michelle Grabner and Brad Killiam on a historic government-run property for the indigent in Little Wolf, Wisconsin. The most urban of this genre is Theaster Gates, an artist with a background in ceramics who established the Rebuild Foundation, based in Chicago, in which he works with community-based artists and activists to restore arts and cultural programming in blighted urban neighborhoods. *Wheels, Kilns, Clay* and Peterson's commitment to Native American communities precede the work in which Gates has engaged in his own urban situation. For Peterson, the spectacle of ceramics became an enrichment activity aimed at a broad public. Her most important work was as a mediator, translating the dynamic history of ceramics into populist forms: written histories, television shows, and throwing demonstrations, all live forms that extended the reach and experiential possibilities of craft production.

But we are now in a time and place where bodies of technical, skilled knowledge are increasingly out of favor. Functional pottery as it was taught and disseminated by Wildenhain, Peterson, and, to an extent, even Richards, has been eschewed. Programs in ceramics, metals, glass, jewelry, fiber, and woodworking are under duress in academic settings: these are the places to learn hard skills, particularly at the undergraduate level, and yet they are closing with steady regularity.

This is one of the things craft is really good at: offering a sense of community. I might be less skilled than the person next to me, but the sense of competition and envy diminishes when the tenor of the room is focused toward learning a skill collectively. There is an unparalleled sense of satisfaction when one makes something tangible in the real world. The art department is increasingly one of the last places on college campuses of directed, hands-on learning. Craft has been a bastion of nonhierarchy in that students and faculty together forge new relationships to materials, through a specificity of technique, skill, and practice, working in a horizontal structure, rather than a top-down, or vertical, model of instructor-dominated thought and implementation.

Yet a situation like Pond Farm was, in fact, vertically structured. For Wildenhain, lack of form was a social pathology, a sickness of American society. She imprinted the disappeared ethos of high modernist principles onto two generations of American students, and yet ceramics for Wildenhain was an inchoate form of grief, in which the form negated the trauma of lived experience, making and unmaking clay, building up forms and grinding them back down, as a metaphor for her own circumstance and experience of exile, loss, and the difficulty in rebuilding, in a new place and time, an unintended life. The hierarchical structure of her teaching instilled in students a renewed capacity—and, arguably, deep love—of the process of throwing on the wheel, of developing the dexterity and muscle memory to form clay intuitively, without the interference of the conscious mind and its hesitancies.

But Pond Farm also offers proof that hardship breeds creativity. At both Black Mountain and Pond Farm, European émigrés were grateful for the opportunity to live and work in safety and freedom. It did not matter what the campus did not have or how little it paid. They adapted to the new and strange environment, as well as to a student body of happy-go-lucky American youth who were virtually untouched by the sorrow and death of a war fought on European soil. American students could not possibly have understood the hardships their teachers had endured. But what came out of such endurance was the ability to make this new and unexpected situation a time of productive optimism. Black Mountain and Pond Farm both were experiments in cross-cultural and intergenerational interactions, turning lack into a kind of intellectual surplus and gain.

Ultimately, the communal experiences found in craft's midcentury orbit offer not just purely historical evidence but also the nuance and texture of narrative: for its women artists, making and living were intimates, connected through the embodiment of form as a generative and sometimes therapeutic process, healing and quieting the mind from aliments as diverse as wartime trauma, sexism, and the psychosis of modernity itself.

NOTES

Introduction

1. See Claire Bishop, *Artificial Hells: Participatory Art and the Politics of Spectatorship* (London: Verso, 2012); and Boris Groys, *Art Power* (Cambridge: MIT Press, 2008).

2. See Nato Thompson, ed., *Living as Form: Socially Engaged Art From 1991 to 2011* (Cambridge: MIT Press, 2011).

3. Eve Kosofsky Sedgwick, *Touching Feeling: Affect, Pedagogy, Performativity* (Durham: Duke University Press, 2003), 21.

4. Ibid., 8.

5. Anne Middleton Wagner, *Three Artists (Three Women): Modernism and the Art of Hesse, Krasner, and O'Keeffe* (Berkeley: University of California Press, 1996), 188.

6. M. C. Richards, "Introduction to the Second Edition," in *Centering: On Pottery, Poetry, and the Person* (Middletown, CT: Wesleyan University Press, 1989), xvii.

7. See F. Carlton Ball, *Decorating Pottery with Clay, Slip and Glaze* (Columbus, OH: Professional Publications, 1967); Tony Birks, *The Art of the Modern Potter* (London: Country Life, 1967); Michael Cardew, *Pioneer Pottery* (London: Harlow Longman, 1969); Michael Casson, *Craftsman's Art* (London: Victoria and Albert Museum, 1973); Clary Illian, *A Potter's Workbook* (Iowa City: University of Iowa Press, 1999); Susan Peterson, *Working with Clay: An Introduction* (London: Laurence King Publishing, Ltd., 1998); Phillip Rawson, *Ceramics* (Oxford: Oxford University Press, 1971); Daniel Rhodes, *Clay and Glazes for the Potter* (New York: Chilton Company, 1957); M. C. Richards, *Centering: On Pottery, Poetry, and the Person* (Middletown, CT: Wesleyan University Press, 1964); and Marguerite Wildenhain, *Pottery: Form and Expression* (New York: Reinhold Ltd., 1959).

8. Bernard Leach, *A Potter's Book* (London: Faber and Faber, Ltd., 1940).

9. Ibid., 214.

10. Briony Fer, *Eva Hesse: Studiowork* (New Haven: Yale University Press, 2009), 15.

11. Wagner, *Three Artists*, 11.

12. This was preceded by a short-lived venture in the United States: Dr. Herbert Hall had established the Marblehead Pottery (Marblehead, Massachusetts) in 1907, with the intent to provide

therapy (in pottery, weaving, metals, and woodwork) for patients suffering from nervousness, but the technical requirements proved too difficult, and the pottery commercialized just a year later, in 1908. See Marilee Boyd Meyer, ed., "Marblehead Pottery," in *Inspiring Reform: Boston's Arts and Crafts Movement* (New York: Abrams, 1997), 222.

13. Kathleen Frydl, *The GI Bill* (Oxford: Oxford University Press, 2009), 121.

14. Wildenhain, *Pottery*, 57.

15. Otto Natzler, *Natzler Ceramics: Catalog of the Collection of Mrs. Leonard M. Sperry and a Monograph by Otto Natzler* (Los Angeles: Los Angeles County Museum of Art, 1968), 38.

16. Harold Rosenberg, "The American Action Painters," *Artnews* 51, no. 8 (December 1952): 23.

Chapter One

1. Jenelle Porter and Ingrid Schaffner, eds., *Dirt on Delight: Impulses That Form Clay* (Philadelphia: University of Pennsylvania Press, 2009). Held in 2009, the show was followed, two years later, by a series of interconnected exhibitions titled *Marvelous Mud* at the Denver Art Museum (DAM), held June 11 through September 18, 2011. DAM's shows spanned the ancient to the contemporary and functioned as a concerted museum-wide effort to coordinate a permanent collection and temporary exhibitions around the single thematic of clay.

2. Paul Mathieu, "The Dirt on *Dirt on Delight*," *Ceramics: Art and Perception* 86 (December 2011-February 2012), 74–77.

3. Anne Wilson, "Sloppy Craft: Origins of a Term," in *Sloppy Craft: Post-Disciplinarity and the Crafts*, ed. Elaine Cheasley Paterson and Susan Surette (London: Bloomsbury, 2013), 20.

4. See Garth Clark, *American Ceramics: 1876 to the Present* (New York: Abbeville Press, 1987); *Eccentric Teapot: Four Hundred Years of Invention* (New York: Abbeville Press, 1989); and *The Potter's Art: A Complete History of Pottery in Britain* (London: Phaidon, 1995). See also Elaine Levin, *History of American Ceramics 1607 to the Present: From Pipkins and Bean Pots to Contemporary Forms* (New York: Harry Abrams, 1988).

5. Cheryl Buckley, "Subject of History? Anna Wetherill Olmsted and the Ceramic National Exhibitions in 1930s USA," *Art History* 28, no. 4 (2005): 497.

6. Dr. Elizabeth Casson (1881–1954) founded the first school of occupational therapy at Dorset House in Bristol in 1929. See the British Association for Occupational Therapists/College of Occupational Therapists, http://www.cot.org/uk/public/aboutus/history.

7. Margery Williams Bianco, *The Velveteen Rabbit, or How Toys Become Real* (New York: Knopf, 1998), 10.

8. Jane Stafford, "Work Cures," *Science Newsletter* 29, no. 775 (February 15, 1936): 107–8.

9. Eileen Boris, *Art and Labor: Ruskin, Morris, and the Craftsman Ideal in America* (Philadelphia: Temple University Press, 1983).

10. Beth Linker, "Strength and Science: Gender, Physiotherapy, and Medicine in Early-Twentieth-Century America," *Journal of Women's History* 17, no. 3 (Fall 2005): 105–32.

11. Susan Peterson, "Ceramics in the West: The Explosion of the 1950s," in *Color and Fire: Defining Moments in Studio Ceramics, 1950–2000*, ed. Jo Lauria (New York: Rizzoli, 2000), 101.

12. See Anne M. Boylan, *The Origins of Women's Activism: New York and Boston, 1797–1840* (Chapel Hill: University of North Carolina Press, 2002). Boylan has mapped the factionalism of such endeavors and the gendered ideologies that gave way to various female-identified causes such as temperance, abolition, and the suffrage movement.

13. Christine Alexander and Maude Robinson, "Green Glazed Ware: Three Hellenistic Vases," *Metropolitan Museum of Art Bulletin*, New Series 3, no. 5 (January 1945): 133–36.

14. Victoria Thorson, "Essay from the Jane Hartsook 25th Anniversary Exhibition Catalog," http://www.greenwichpottery.org/literature_39833/Pottery_History.

15. During the art pottery movement of the early twentieth century, the generations of artists prior to Rhodes, such as Adelaide Alsop Robineau (1865–1929) and Myrtle Merritt French (1886–1960), had also trained as artists at Alfred, trained by Charles Harder's mentor, the British Arts and Crafts potter Alfred Binns. However, the program was called applied arts, rather than fine arts. See Janet Koplos and Bruce Metcalf, *Makers: A History of American Studio Craft* (Chapel Hill: University of North Carolina Press), 58–60. For an excellent history of French, see Cheryl R. Ganz, "Shaping Clay, Shaping Lives: The Hull-House Kilns," in *Pots of Promise: Mexicans and Pottery at Hull-House, 1920–1940*, ed. Cheryl R. Ganz and Margaret Strobel (Urbana: University of Illinois Press, 2004), 55–88.

16. Victor Salvatore, director, "Greenwich House Annual Report" (1929), box 1, page 1, Greenwich House Pottery Archives, New York.

17. The School for American Craftsmen was established by Mrs. Vanderbilt Webb in 1944, as the educational arm of the New York-based organization she founded, the American Craftsmen Council (later the American Craft Council). In 1950, the school shortened its name to the School for American Crafts and moved from New Hampshire to the Rochester Institute of Technology (RIT) in Rochester, NY.

18. By 1947, d'Harnoncourt had become a trustee of the ACCC's Educational Council. The School of American Craftsmen was a short-lived endeavor, merging with the School of Fine Arts in 1960 at its final location, the Rochester Institute of Technology (RIT), Rochester, New York. Regarding the foundry, see Thomas Hoving, *Making the Mummies Dance: Inside the Metropolitan Museum of Art* (New York: Simon and Schuster, 1993), 46.

19. Horace H. F. Jayne, "The Museums and the Craftsmen in Wartime," *Craft Horizons* 1, no. 2 (May 1942): 7.

20. *F. Carlton Ball: Ceramic Works, 1940–1990* (Tacoma, WA: Tacoma Museum of Art, 1990), 5. Artist's file, American Craft Council Library, Minneapolis, MN.

21. "Ceramics: The Socio-Economic Outlook," in *First American Craftsman's Conference, June 12, 1957, Asilomar, California* (New York: American Craftsman's Council, 1957), 11–13, 21. This discussion extends over multiple topics across nearly twenty-three pages of transcription. In the case of Ball, I have excerpted and merged his commentaries on pages 12 and 21, when he returns to the same topic from the earlier moment in the discussion.

22. John Hollander, "The Mad Potter," in *Harp Lake* (New York: Knopf, 1988), 22.

23. Max W. Sullivan, "Introduction," in *22nd Ceramic National Exhibition: A Survey of Contemporary Ceramics of the United States and Canada* (Syracuse, NY: Everson Museum of Art, 1962), 7.

24. Buckley, "Subject of History?," 514.

25. Ezra Shales, "The Paradoxes of the Everson's Ceramic Nationals," in *Northwest Modern: Revisiting the Annual Ceramic Exhibitions of 1950–1964* (Museum of Contemporary Craft, Portland, 2011), 2.

26. *F. Carlton Ball: Ceramic Works*, 15.

27. Lillian Boschen was the first director. For a more detailed history, see Peter Held, ed., *Fifty Years of the Archie Bray Influence: A Ceramic Continuum* (Seattle and Helena, Montana: University of Washington Press and the Holter Musum of Art, 2001).

28. Peter Voulkos, Black Mountain College Faculty Files, Series 2: "Research Files," box 25, North Carolina State Archives, Raleigh. See also Peter Voulkos's biography by the Frank Lloyd Gallery, http://www.franklloyd.com/dynamic/artist_bio.asp?ArtistID=34.

29. *Contemporary American Ceramics: Selected from the 17th Ceramic National* (Syracuse, NY: Syracuse Museum of Fine Arts, 1952), 7.

30. For a further analysis of Voulkos's architectonic impulses, and his oeuvre in relation to media hierarchies of the 1950s, see Andrew Perchuk's excellent chapter "Peter Voulkos: Out of Clay," in "From Otis to Ferus: Robert Irwin, Ed Ruscha, and Peter Voulkos in Los Angeles, 1954–1975" (PhD diss., Yale University, 2006), 20–92.

31. "Foreword," in *18th Ceramic National* (Syracuse, NY: The Syracuse Museum of Fine Arts, 1954), unpaginated.

32. Max W. Sullivan, "Introduction," in *25th Annual Ceramic National Exhibition* (Syracuse, NY: Everson Museum of Art, 1968), 7.

33. Jenelle Porter, "On Delight," in *Dirt on Delight: Impulses That Form Clay* (Philadelphia, PA: Institute of Contemporary Art, 2009), 51.

34. Ronald A. Kutcha, "Foreword," in *A Century of Ceramics in the United States, 1878–1978*, ed. Garth Clark and Margie Hughto (New York: E. P. Dutton, 1979), 8.

35. Margie Hughto, telephone conversation with the author, February 9, 2015.

36. Ibid.

37. Ronald A. Kuchta, "Foreword," in *New Works in Clay by Contemporary Painters and Sculptors* (Syracuse, NY: Everson Museum, 1976), 7.

38. Margie Hughto, "Notes on the Project," in *New Works in Clay by Contemporary Painters and Sculptors*, 87.

39. Ibid., 88–89.

40. For longer histories of women's labor and the alternative space movement, see my essays "The Feminist Nomad: The All-Women Group Exhibition," in *WACK! Art and the Feminist Revolution*, ed. Connie Butler and Lisa Mark (Los Angeles and Cambridge: MOCA and MIT Press, 2007), 458–71; and "Learning from Los Angeles: Gendered Pedagogy and Its Predecessors at the Woman's Building, 1973–1991," in *Doin' It in Public: Art and Feminism at the Woman's Building.* (Los Angeles: Otis College of Art and Design, 2011), 36–64.

41. Susan Peterson, oral history interview by Paul J. Smith, March 10, 2004, Archives of American Art, Smithsonian Institution, Washington, DC, hereafter Archives of American Art.

42. Margie Hughto, telephone conversation with the author, February 9, 2015.

43. Garth Clark, "Shifting Paradigms: A Memoir," in *Shifting Paradigms in Contemporary Ceramics: The Garth Clark & Mark Del Vecchio Collection* (New Haven and London: Yale University Press and the Museum of Fine Arts, Houston, 2012), 14.

44. A selection of these include *American Ceramics: 1876 to the Present*, a revision of *A Century of Ceramics*, and in its second printing (Abbevile Press, 1990); *American Potters: The Work of Twenty Modern Masters* (New York: Watson-Guptill, 1981); Garth Clark, ed., *Ceramic Echoes: Historical References in Contemporary Ceramics* (Kansas City, MO: Contemporary Art Society, 1983); *The Potter's Art: A Complete History of Pottery in Britain* (London: Phaidon Press, 1995); and Garth Clark, ed., *Ceramic Millennium: Critical Writings on Ceramic History, Theory and Art* (Halifax, NS: Press of the Nova Scotia College of Art and Design, 2005).

45. Glenn Adamson, "Implications: The Modern Pot," in *Shifting Paradigms in Contemporary Ceramics: The Garth Clark & Mark Del Vecchio Collection* (New Haven and London: Yale University Press and the Museum of Fine Arts, Houston, 2012), 44–45.

46. Garth Clark, "1940," in *American Ceramics: 1876 to the Present*, 2nd ed. (New York: Abbeville Publishers, 1987), 88.

47. Ibid., 63.

48. Garth Clark, *Beatrice Wood Retrospective* (Fullerton, CA: CSU Fullerton Art Gallery, 1983) and *It's All Part of the Clay: Viola Frey* (Philadelphia: Moore College of Art Gallery, 1984).

49. Garth Clark, "Foreword," in *Ceramic Echoes: Historical References in Contemporary Ceramics*, ed. Garth Clark (Kansas City, MO: Contemporary Art Society, 1983), 11.

50. Garth Clark, "How Envy Killed the Crafts Movement: An Autopsy in Two Parts" (Portland: Museum of Contemporary Craft and Pacific Northwest College of Art, 2008). For a condensed version, see Glenn Adamson, ed., *The Craft Reader* (Oxford: Bloomsbury, 2010), 515–20.

51. Jules Prown, "Mind in Matter: An Introduction to Material Culture Theory and Method," *Winterthur Portfolio* 17, no. 1 (Spring 1982): 1–19.

52. Rose Slivka, "The New Ceramic Presence," *Craft Horizons* 21, no. 4 (July/August 1961): 30–37.

53. Glenn Adamson, *Thinking Through Craft* (Oxford: Berg Publishers, 2007), 46.

54. John Coplans, "Abstract Expressionist Ceramics," *Artforum* (November 1966): 34–41. The same essay reappears in the exhibition catalog by the same name, curated by Coplans and held at UC Irvine's Art Gallery, also in 1966.

55. Perchuk, "From Otis to Ferus."

56. Rosalind E. Krauss, "Sculpture in the Expanded Field" (1978), in *The Originality of the Avant-Garde and Other Modernist Myths* (Cambridge: MIT Press, 1985), 276–90.

57. Rose Slivka, *Peter Voulkos: A Dialogue with Clay* (New York: Museum of Contemporary Crafts, 1977).

58. See Edmund de Waal, "Peter Voulkos and Otis," in *20th Century Ceramics* (London: Thames & Hudson World of Art, 2003), 156–63. Meant for a popular audience, there are no footnotes or endnotes for this volume, but de Waal's bibliography for this chapter refers back to the catalogs edited by Mary McNaughton and Jo Lauria (see next endnote).

59. Mary Davis McNaughton, ed., *Revolution in Clay: The Marer Collection of Contemporary Ceramics* (Claremont, CA, and Seattle: Scripps College and University of Washington Press, 1994); Jo Lauria, ed., *Color and Fire: Defining Moments in Studio Ceramics, 1950–2000* (Los Angeles and New York: LACMA and Rizzoli, Inc., 2000); Christy Johnson, ed., *Common Ground: Ceramics in Southern California, 1945–1975* (Pomona, CA: American Museum of Ceramic Art, 2012). See also Mary McNaughton, ed., *Clay's Tectonic Shift: John Mason, Ken Price, and Peter Voulkos, 1956–1968* (Claremont, CA: Scripps College, 2012), for which Clark was an advisor.

60. Richard Petterson, "California: A Climate for Craft Art," *Craft Horizons: A Special Issue on California* 16, no. 5 (October 1956): 11.

61. Ibid.

62. Ibid.

63. Beatrice Wood, *I Shock Myself: The Autobiography of Beatrice Wood* (San Francisco: Chronicle Books, 1985), 172.

64. Staci Steinberger, "Glen Lukens," in *A Handbook of California Design 1930–1965: Craftspeople, Designers, Manufacturers*, ed. Bobbye Tigerman (Cambridge: MIT Press, 2013), 172. See also Elaine Levin, *Glen Lukens: Pioneer of the Vessel Aesthetic* (Los Angeles: California State University, 1982).

65. Susan Peterson, "Ceramics in the West: The Explosion of the 1950s," in *Color and Fire*, ed. Lauria, 101.

66. Chouinard Art Institute, Accreditation History, subseries 2.7, box 9, folder 8, Chouinard Archives, California Institute of the Arts, Valencia, hereafter Chouinard Archives.

67. Ralph Bacerra, interview by Frank Lloyd, April 4–19, 2004, Archives of American Art.

68. For brief synopses and founding histories of each organization, see the webpages of the American Ceramics Society (http://ceramics.org/about/learn-about-acers/history/) and NCECA (http://nceca.net/static/about.php).

69. Chouinard Art Institute, contracts, subseries 2.7, box 9, folder 2, Chouinard Archives.

70. Otto and Vivika Heino, interview by Elaine Levin, March 4, 1981, Archives of American Art.

71. "Advertising Design Curriculum Minutes, 1959," subseries 2.7, box 8, folder 8, Chouinard Archives.

Chapter Two

1. Marguerite Wildenhain, *The Invisible Core: A Potter's Life and Thoughts* (Palo Alto, CA: Pacific Books, 1973), 28–29.

2. See Bernard Leach, "American Impressions," *Craft Horizons* 10, no. 4 (Winter 1950): 18–20; and Marguerite Wildenhain's rejoinder, "Potters Dissent: An Open Letter to Bernard Leach from Marguerite Wildenhain," *Craft Horizons* 13, no. 3 (May/June 1953): 43–44.

3. MW to Sam Maloof, May 16, 1977, fellows' files, American Craft Council Archives, Minneapolis, MN.

4. Wildenhain, *The Invisible Core*, 145.

5. The quote regarding Albers's commentary on clay comes from Trude Guermonprez Elsesser, interview with Mary Emma Harris, December 8, 1971, spool 83a, page 24, Black Mountain College Research Project, North Carolina State Archives, Raleigh.

6. Marianne Moore, "Review: "'If I Am Worthy, There is No Danger' Murder in the Cathedral by T.S. Eliot," *Poetry* 47 no. 5 (February 1936): 281.

7. T. S. Eliot to MW, January 2, 1953, reel 5048, Marguerite Wildenhain Papers, Archives of American Art.

8. Roy R. Behrens, *Recalling Pond Farm: My Memory Shards of a Summer with Bauhaus Potter Marguerite Wildenhain* (Dysart, Iowa: Bobolink Books, 2005), 5.

9. Wildenhain, *The Invisible Core*, 55.

10. Nicholas Fox Weber, *Josef and Anni Albers: Designs for Living* (New York: Cooper-Hewitt Museum, 2004), 98.

11. Billie Sessions. *Ripples: Marguerite Wildenhain and Her Pond Farm Students* (San Bernardino, CA: Robert V. Fullerton Art Museum, 2002), 84.

12. Wildenhain, *The Invisible Core*, 143.

13. Nathan McMahon, "Madam was Indisputably French," in *Marguerite Wildenhain and the Bauhaus: An Eyewitness Anthology*, ed. Dean and Geraldine Schwarz (Decorah, IA: South Bear Press, 2007), 483.

14. Philip Rieff, *Fellow Teachers* (New York: Harper & Row, 1973), 137. Best known for his work on Freud, Rieff was something of a celebrity-scholar, best known for his brief and scandalous marriage to his former student, Susan Sontag, while teaching in the sociology department at the University of Chicago. Rieff and M. C. Richards's former husband Bill Levi overlapped at the University of Chicago's philosophy department for a period of years, when Rieff was a student and Levi a newly-minted professor.

15. I am indebted to Sigrid Wetge-Wortmann's volume *Bauhaus Textiles: Women Artists and the Weaving Workshop* (London: Thames and Hudson, 1993), which details this phenomenon. See also Magdalena Droste, *Bauhaus, 1919–1933* (Berlin: Taschen Verlag, 1990), 38–40; and Hans Maria Wingler, *Bauhaus: Weimar, Dessau, Berlin, and Chicago*, trans. Wolfgang Jabs and Basil Gilbert (Cambridge: MIT Press, 1969).

16. Walter Gropius, "The Theory and Organization of the Bauhaus," in *Bauhaus: 1919–1928* (New York: Museum of Modern Art, 1986), 25.

17. This information has been gleaned from multiple sources: Wildenhain, *The Invisible Core*, 29; and the chronology found at the Bauhaus-Dessau Archive, www.bauhaus-dessau.de/index. See "Bauhausbuilding Chronologie: 1924 and 1925."

18. An alternative viewpoint exists. The German scholar Magdalena Droste has contested the validity of the Gropius-Marcks friendship, painting Marcks as a nationalist with anti-Semitic leanings. See Droste, "The Pottery Workshop," in *Bauhaus*, 68–72.

19. Still in existence today, the school is known as Burg Gibchenstein Hochschule für Kunst und Design in Halle-Saale. The cynical viewpoint of Wildenhain's appointment is that it indicated his desire for her: it is highly likely that Wildenhain and Marcks, both married to others, were lovers.

20. Diary entry, March 2, 1974, box 1, Frans Wildenhain Papers, Archives of American Art.

21. For a comprehensive overview of Frans Wildenhain's career, see Bruce A. Austin, *Frans Wildenhain, 1950–75: Creative and Commercial American Ceramics at Mid-Century* (Rochester, NY: Printing Applications Lab, 2012).

22. See Peter M. Rutkoff and William B. Scott, "The Politics of Disillusionment, 1933–1945" and "A Return to Social Theory: The Graduate Faculty, 1945–1960," in *New School: A History of the New School for Social Research* (New York: Free Press, 1986), 107–27, 196–217.

23. Arguably, the closest person accorded this honor would be the German language poet Paul Celan (1920–1970), whose reclamation of the German mother tongue, as a Jew and Holocaust survivor, was described by Shoshana Felman as "to annihilate his own annihilation in it, to reappropriate the language that had marked his own exclusion . . ." Shoshana Felman, "Education and Crisis, or the Vicissitudes of Teaching," in *Testimony: Crises of Witnessing in Literature, Psychoanalysis, and History*, ed. Shoshana Felman and Dori Laub (New York: Routledge, 1992), 35. Yet Celan is a poet renowned for his searing and intensely hermetic lyricism—hardly a universal gesture.

24. Hannah Arendt, "Approaches to the German Problem," in *Essays in Understanding: 1930–1954*, ed. Jerome Kohn (New York: Harcourt Brace and Co., 1994), 108–9. The essay was originally published in the *Partisan Review* 12, no. 1 (Winter 1945).

25. Philip Johnson, "Foreword to the 1995 Edition," in *The International Style: Architecture Since 1922*, ed. Henry Russell Hitchcock (New York: W. W. Norton & Co., 1997), 13.

26. Wildenhain, *The Invisible Core*, 30–31.

27. *Hallesche Nachrichten* and *Saale-Zeitung* (both January 20, 1930), reprinted in Dean and Geraldine Schwarz, eds., *Eyewitness*, 174.

28. Gerhard Masur, *Imperial Berlin* (London: Routledge and Kegan Paul, 1971), 22.

29. Richard Bessel, "The Nazi Capture of Power," *Journal of Contemporary History* 32, no. 2 (2004): 176.

30. Wildenhain, *The Invisible Core*, 35.

31. Ibid.

32. K.v.d. Mijl-Dekker to MW, July 26, 1945, box 1, Frans Wildenhain Papers, Archives of American Art.

33. See "Forms and Decorations, 1918–1933 Modern," at www.meissen.de/index.

34. Peter Guenther, "Gerhard Marcks," in *"Degenerate Art": The Fate of the Avant-Garde in Nazi Germany*, ed. Stephanie Barron (Los Angeles: Los Angeles County Museum of Art, 1991), 296–97.

35. Mario-Andreas von Lüttichau, "*Entartete Kunst*, Munich 1937: A Reconstruction," in *"Degenerate Art,"* 62.

36. Guenther, "Gerhard Marcks," 296.

37. Ibid.

38. Wildenhain, *The Invisible Core*, 36.

39. Marguerite Friedlaender Wildenhain, interview by Hazel Bray, March 14, 1982, reel 3199, Archives of American Art.

40. David J. Sorkin, *The Transformation of German Jewry, 1780–1840* (Oxford: Oxford University Press, 1987), 90.

41. Ibid., 140.

42. See Theodor Herzel's *The Jewish State* (New York: Dover, 1988). Written in German, *Der Judenstaat* was published in Vienna in 1896. In 1946, in the immediate postwar era, the book was translated into English and widely disseminated, published by the American Zionist Emergency Council.

43. Rudolf Stahl, "Vocational Retraining of Jews in Nazi Germany 1933–1938," *Jewish Social Studies* 1, no. 2 (April 1939): 169–94.

44. Ibid., 190.

45. Wingler, "Results of the Investigation Concerning Staatliche Bauhaus in Weimar," in *Bauhaus*, 39–41.

46. In their volume on Wildenhain, Dean and Geraldine Schwarz make the claim that Krehan and Wildenhain were lovers, based upon the discovery of an unsubstantiated diary.

47. See Bruce Saposnik, *Becoming Hebrew: The Creating of a Jewish National Culture in Ottoman Palestine* (Oxford: Oxford University Press, 2008), 67.

48. While this date has been published in the Dean Schwarz volume, it cannot be definitively confirmed.

49. Wildenhain, *The Invisible Core*, 36.

50. A. Gordon Herr to Jane Herr, April 28, 1939, in Dean and Geraldine Schwarz, eds., *Eyewitness*, 269.

51. Ibid., 53.

52. MW to Walter Gropius, March 21, 1940, Walter Gropius Archive, Berlin. A copy of this letter is reprinted in Dean and Geraldine Schwarz, eds., *Eyewitness*, 261.

53. Walter Gropius to MW, March 27, 1940, series III, folder 1730, Walter Gropius Papers (MS Ger 208), Houghton Library, Harvard College Library, Harvard University.

54. Walter Gropius to MW, May 23, 1940. Reprinted in Dean and Geraldine Schwarz, eds., *Eyewitness*, 263.

55. Wildenhain, *The Invisible Core*, 56–57.

56. WG to MW, May 13, 1940. Reprinted in Dean and Geraldine Schwarz, eds., *Eyewitness*, 262.

57. Michael Boylen, "Frans Wildenhain: Master of Form," *Studio Potter* 19, no. 2 (June 1991): 1, 5–16.

58. Victor Ries, "Religious Artistic Expression in Metal Sculpture," in *Renaissance of Religious Art and Architecture in the San Francisco Bay Area, 1946–1968*, transcription of an oral history conducted 1983 by Suzanne B. Riess, volume 2, Regional Oral History Office, Bancroft Library, University of California, Berkeley, 1985. Accessible online at www.bancroft.berkeley.edy/ROHO,

page 1083. See also "Homosexuals in War and Resistance," www.bevrjdinginterculterel.nl/eng/homoseksuelen.html.

59. In a postwar letter to Gropius, Marguerite describes Frans's and Henri's ordeal: "I must be grateful that in spite of all Rudolf [M's nickname for Frans] and my brother Henri, who was hidden away in a small attic for more than three years without ever getting out to the air, are still alive." MW to Walter Gropius, September 10, 1945, Walter Gropius Archive, Berlin. Reprinted in Dean and Geraldine Schwarz, eds., *Eyewitness*, 265.

60. "Maria Bruhn-Friedlander, 1997," Martyrs' and Heroes' Remembrance Authority, Virtual Wall of Honor, www1.yadvashem.org/heb_site/righteous/pdf/virtual_wall_of_honor/NETHERLANDS.pdf.

61. "Copy of A Letter of Mr. G.J. Haering, Chief Visa Division, Department of State, Washington D.C to Mr. Richard Gump, San Francisco," Marguerite Wildenhain Papers, Special Collections, Luther College, Decorah, IA.

62. Henri Friedlaender, "Declarations of Political Security," December 4, 1945, box 1, folder: "Letters, January-August, 1945," Frans Wildenhain Papers, Archives of American Art.

63. Trude Jalowetz Guermonprez, "Declarations of Political Security," December 5, 1945, box 1, folder: "Letters, January-August, 1945," Frans Wildenhain Papers, Archives of American Art.

64. "Lijst van uit te vosren goederen," box 1, folder: "Letters, January-August, 1945," Frans Wildenhain Papers, Archives of American Art.

65. Wildenhain, *The Invisible Core*, 60.

66. Josef Albers, *Interaction of Color* (New Haven: Yale University Press), 1963. Dedicated to his students, this book is the culmination of Albers's pedagogical efforts. The text itself is interspersed with many colored plates of his own work, reflecting the theory-practice integration of his oeuvre.

67. Eva Díaz, "The Ethics of Perception: Josef Albers in the United States," *The Art Bulletin* 90, no. 2 (June 2008): 281.

68. See Joanathan Guthrie Herr's memoir "Love, Because Nothing Else Matters," in Dean and Geraldine Schwarz, eds., *Eyewitness*, 314–36.

69. Trude Guermonprez, interview by Mary Emma Harris, December 8, 1971, spool 83Aa page 14, Black Mountain College Research Project, North Carolina State Archives, Raleigh, hereafter Black Mountain College Research Project.

70. McMahon, "Madam Was Indisputably French," 476.

71. Victor Ries, Oral History, 536. Author's Note: in a 2009 interview I conducted with Ries, when he was at the advanced age of 101, he was much less articulate about Pond Farm, but he emphatically affirmed this remark about Wildenhain, unprompted: "She was a man. Marguerite was a man." "Victor Ries at 101," interview conducted by Jenni Sorkin, May 27, 2009, Reutlinger Center for Jewish Living, Danville, CA.

72. Gerhard Marcks, "A Biography of Marguerite Friedlaender Wildenhain," box 22, Marguerite Wildenhain Pond Farm Collection, Special Collections, Luther College, Decorah, IA.

73. Ibid, 467.

74. Charles Edward Rossbach, "Fiber Arts Series," an oral history conducted 1983 by Ann Nathan, Regional Oral History Office, the Bancroft Library, University of California, Berkeley, 1987, www.bancroft.berkeley.edy/ROHO, 55.

75. "Victor Ries at 101," an interview conducted by Jenni Sorkin, May 27, 2009, Reutlinger Center for Jewish Living, Danville, CA.

76. Lawrence Langer, *Admitting the Holocaust: Collected Essays* (London: Oxford University Press, 1995), 38.

77. MW to GM, September 29, 1980; GM to MW, April 8, 1981, in Ruth Kath, ed., *The Letters of Gerhard Marcks and Marguerite Wildenhain, 1970–1981: A Mingling of Souls* (Ames and Decorah, IA: Iowa State University Press and Luther College Press, 1991), 189.

78. Kath, *Letters*, 103.

79. Michel Foucault, *Discipline and Punish: The Birth of the Prison*, trans. Alan Sheridan (New York: Vintage Books, 1977), 138.

80. Gilles Deleuze and Félix Guattari, *Anti-Oedipus*, trans. Robert Hurley, Mark Seem, and Helen R. Lane (Minneapolis: University of Minnesota Press, 1983), 17.

81. Sessions, *Ripples*, 111.

82. Gail Stewart, in attendance 1973 to 1980. Sessions, *Ripples*, 111.

83. Wayne Reynolds, in attendance seven summers between 1963 and 1979. Sessions, *Ripples*, 109.

84. John Conners Sessions, *Ripples*, 91.

85. Felman and Laub, eds., *Testimony*, 13.

86. Giorgio Agamben, *Homo Sacer: Sovereign Power and Bare Life*, trans. Daniel Heller-Roazen (Palo Alto, CA: Stanford University Press, 1998), 188.

87. Felman and Laub, eds., *Testimony*, 55. Interestingly, in his memoir, David Stewart, Wildenhain's studio assistant for sixteen summers, pointedly recalled a young woman once asking: "How come everything at Pond Farm is a crisis?" Dean and Geraldine Schwarz, eds., *Eyewitness*, 450.

88. The full, famed Rauschenberg quote is: "Albers was a beautiful teacher and an impossible person . . . I consider Albers the most important teacher I've ever had, and I'm sure he considered me one of his poorest students." Mary Emma Harris, *The Arts at Black Mountain College* (Cambridge: MIT Press, 1987), 126. See also Jeffrey Saletnik, "Josef Albers, Eva Hesse, and the Imperative of Teaching," *Tate Papers*, 7 (Spring 2007): n.p, www.tate.org.uk/research.

89. Slivka, "The New Ceramic Presence," 121. Akin to the Clement Greenberg of her field, Slivka maintained both power and fearlessness in her nearly singlehanded promotion of Voulkos. She is the author of his only monograph to date, *Peter Voulkos: A Dialogue with Clay* (New York: Reinholdt and New York Graphic Society), 1978.

90. Wildenhain, *The Invisible Core*, 14.

91. Andrew Perchuk, "From Otis to Ferus," 9.

92. David Stewart, a potter based in Southern California. According to Dean Schwarz, her final studio assistant, Stewart lost his summer gig with Wildenhain when he chose to stay home one summer with his wife rather than come to Pond Farm. Author's interview with Dean and Geraldine Schwarz, June 24, 2009, Decorah, IA.

93. Dean Schwarz taught pottery at Luther College from 1964 until 1986. He sent twenty-seven students to Pond Farm, but also invited potters like Voulkos to campus. Ibid.

94. *Frances Senska: A Life in Art* (Helena: Holter Museum of Art, 2004), unpaginated.

95. Frances Senska, oral history by Donna Forbes, April 16, 2001, Archives of American Art.

96. Ibid.

97. A devoted Voulkos supporter, Rose Slivka interprets the clash between Voulkos and his superior, the painter Millard Sheets, as rife with professional jealousies, as well as Sheets's academic conservativism, given his profound distaste for the work Voulkos and his students produced. See Slivka, *Dialogue with Clay*, 41–42.

98. For substantial essays on the so-called regionalisms and their various artistic networks, exhibition catalogs produced by LA MOCA, the Hammer Museum, and the Berkeley Art Museum

were crucial for mapping the postwar terrain during the 1980s and 1990s. However in 2011, the Getty Foundation's Pacific Standard Time initiative expanded the scope and breadth of California art history substantially, sponsoring sixty-nine exhibitions and corresponding publications.

99. Charles Talley, "School for Life," *American Craft* 50 (April/May 1991), 36.

100. Peter Voulkos, "The Great Move West, Part II," *Revolutions of the Wheel: The Emergence of American Clay Art* (London: Queens Row Productions, 1994), video, color and sound, 28:35 min.

101. Frances Senska, quoted in Slivka, *Dialogue with Clay*, 12.

102. Dean and Geraldine Schwarz, eds., *Eyewitness*, 473.

103. Ron Nagle, "Peter Voulkos and the Otis Group, Part III," *Revolutions of the Wheel* (London: Queens Row Productions, 1994), video, color and sound, 27:50 min.

104. Hunt Prothro, "Love that Cannot be Mastered," in Dean and Geraldine Schwarz, eds., *Eyewitness*, 586.

105. Marguerite Wildenhain, quoted in Mary Davis MacNaughton,"Innovation in Clay: The Otis Era, 1954–60," in *Revolution in Clay: The Marer Collection of Contemporary Ceramics* (Claremont, CA: Ruth Chandler Williamson Gallery, Scripps College, 1994), 56.

106. "A Song, to the Tune of Gilbert & Sullivan 'Pirates of Pinzans' Sung By the Students at S.S.' 71 at Traditional Final Party," series 7, box 1, Marguerite Wildenhain Collection, Special Collections, Luther College, Decorah, IA.

107. GM to MW, April 8, 1981, in Kath, ed., *Letters*, 199.

108. Frederick M. Logan, "Kindergarten and Bauhaus," *College Art Journal* 10, no. 1 (Autumn 1950): 36–43.

109. Hannah Arendt, *The Human Condition* (Chicago: University of Chicago Press, 1958), 161.

110. FW to MW, March 3, 1974, carbon copy, box 2: "Correspondence," Frans Wildenhain Papers, Archives of American Art.

111. Jane Kemp, "Pond Farmers Roundtable" (1994), box 14, Marguerite Wildenhain Collection, Special Collections, Luther College, Decorah, IA.

112. Andrew L. Zolnay, Sr. Land Agent, Negotiations to MW, December 23, 1963, series V: "Papers of Marguerite Wildenhain, Property Records, 1956–1982," Marguerite Wildenhain Collection, Special Collections, Luther College, Decorah, IA.

113. MW to Edward F. Dolder, Chief Division of Beaches and Parks, July 31, 1963, carbon copy, series V: "Papers of Marguerite Wildenhain, Property Records, 1956–1982," Marguerite Wildenhain Collection, Special Collections, Luther College, Decorah, IA.

114. Richard B. Petterson to Governor Pat Brown, July 10, 1963, series V: "Papers of Marguerite Wildenhain, Property Records, 1956–1982," Marguerite Wildenhain Collection, Special Collections, Luther College, Decorah, IA.

115. Ray Varley, Secretary to the Governor's Office, MW, August 6, 1963, series V: "Papers of Marguerite Wildenhain, Property Records, 1956–1982," Marguerite Wildenhain Collection, Special Collections, Luther College, Decorah, IA.

116. Arendt, *Human*, 7.

Chapter Three

1. Emmanuel Cooper, *Bernard Leach: Life and Work* (London and New Haven: Paul Mellon Centre and Yale University Press, 2003), 250.

2. John W. Dower, *Embracing Defeat: Japan in the Wake of World War II* (New York: W. W. Norton, 1999), 144.

3. The weaver Else Regensteiner, a peer of Anni's, classified Marilyn Bauer as Anni's "star pupil." See Else Regensteiner, interview by Mary Emma Harris, November 3, 1971, series IV: "Released Interviews," box 36, p. 7, Black Mountain College Research Project. For the Bauer quote, see: Marilyn Bauer Greenwald, interview by Mary Emma Harris, October 18, 1971, series IV: "Released Interviews," box 28, p. 10, Black Mountain College Research Project.

4. Mary Emma Harris, *The Arts at Black Mountain College* (Cambridge: MIT Press, 1987), 188.

5. Robert Chapman Turner (1913–2005) was a functional potter who earned an MFA from New York State College of Ceramics at Alfred University in 1949 and then established the ceramics program at Black Mountain College, which ran until 1953.

6. Fielding Dawson, *An Emotional Memoir of Franz Kline* (New York: Pantheon Books, 1967), 6, 22.

7. Clement Greenberg, "American-Type Painting," in *Art and Culture* (Boston: Beacon Press, 1961), 220.

8. Bert Winther-Tamaki, "Japanese Margins of Abstract Expressionism," in *Art in the Encounter of Nations: Japanese and American Artists in the Early Postwar Years* (Honolulu: University of Hawai'i Press, 2001), 56–57.

9. Marilyn Kushner, "Exhibiting Art at the American National Exhibition in Moscow, 1959: Domestic Politics and Cultural Diplomacy," *Journal of Cold War Studies* 4, no. 1 (Winter 2002): 6–26.

10. Charles Olson to Marguerite Wildenhain, May 24, 1952, series II: "Research Files," box 17, Black Mountain College Research Project.

11. This is funny in that native New Yorkers never use "downtown" as an adjective; it is utilized, instead, as a noun, where "downtown" is a location unto itself. See Fielding Dawson, *The Black Mountain Book* (Croton, NY: Croton Press Books, Ltd., 1970), 104.

12. Karen Karnes, Interviewed by Mark Schapiro, August 9–10, 2005, Archives of American Art.

13. Wildenhain agreed to serve as host-potter having previously declined the post of the resident potter. Marguerite Wildenhain to Charles Olson, May 26, 1952, series II: "Research Files," box 17, Black Mountain College Research Project.

14. Mary Emma Harris, interview with Karnes, April 9, 1972, spool 165, series III: "Research Files," box 25, Black Mountain College Research Project.

15. For the anecdote about students, see Garth Clark, *American Ceramics: 1876 to the Present* (New York: Abbeville, 1987), 114. For Albers reference, see M. C. Richards, interview by Martin Duberman, February 20, 1967, P. C. 1678.14, Martin Duberman Papers, North Carolina State Archives, Raleigh.

16. Robert Turner, interview with Mary Emma Harris, October 20, 1971, series IV, box 38, spool 52, p. 7, Black Mountain College Research Project.

17. M. C. Richards to James Leo Herlihy, December 6, 1948, series I, box 3, "Correspondence, 1948–1952," Mary Caroline Richards Papers, Getty Research Institute, Los Angeles, hereafter M. C. Richards Papers

18. M. C. Richards, "Letter to the Faculty (Summer, 1951)," series I, box 2, folder 4, M.C. Richards Papers.

19. M. C. Richards to Charles Olson, August 2, 1954, Charles Olson Research Collection, series II: "Correspondence,"box 208, folder "1954–1956," Special Collections, University of Connecticut, Storrs. Hereafter Charles Olson Papers.

20. Dawson, *The Black Mountain Book*; Michael Rumaker, *Black Mountain Days* (Asheville, NC: Black Mountain Press, 1996).

21. Dawson, *The Black Mountain Book*, 22–23; Rumaker, *Black Mountain Days*, 251.

22. Betty Baker, interview with Mary Emma Harris and Geraldine Berg, October 29, 1970, series II, box 30, pp. 2–3, Black Mountain College Research Project.

23. Ibid.

24. M. C. Richards, "A Summer Session in the Arts, July 9-August 31, 1951." This is the flipside to the aforementioned "Letter to the Faculty (Summer, 1951)," series I, box 2, folder 4, M. C. Richards Papers.

25. Ibid.

26. Harry Callahan, interview by Mary Emma Harris, March 16, 1971, spool 22, Black Mountain College Research Project.

27. The remaining correspondence between Olson and Bernard Leach is entirely one-sided: there are no incoming letters from Leach, who was making arrangements for himself, Shoji Hamada, and Soetsu Yanagi. See Black Mountain College Archives, May 1952, series II: "Research Files," box 17, Black Mountain College Project.

28. Charles Olson to Bernard Leach, March 19, 1952, series II, Box 25, Black Mountain College Project.

29. Meeting minutes of the Board of Fellows, October 14, 1952, series I, box 17, Black Mountain College Project.

30. Leach's first visit to the United States, a ten-week lecture and demonstration tour arranged by Robert Richman, Director of the Institute for Contemporary Art in Washington, DC, occurred in 1950. This tour took Leach to the following cities: Alfred, New York, Toronto, Wichita, Detroit, Ann Arbor, Madison, Minneapolis, San Francisco, and finally, Seattle, to stay with Tobey. After the 1952 Black Mountain Pottery Seminar, Leach, Hamada, and Yanagi continued on to the Midwest, where they conducted a separate two-week seminar, the Midwest Craftsmen's Seminar in St. Paul at the Minnesota School of Art, November 17–29, 1952, at the invitation of Warren and Alix MacKenzie. Leach, Hamada, and Yanagi encountered nearly every famed potter in the United States, making stops either by invitation or out of personal interest: Archie Bray Foundation, Helena, Montana (Peter Voulkos); San Ildefonso, New Mexico (Maria Martinez); Chinouard Art Institute, Los Angeles (Susan Peterson); Scripps College, Claremont, (Richard Petterson); and Mills College, Oakland (Antonio Prieto). See Cooper, *Bernard Leach*, 240–41.

31. Robert Creeley, "Introduction," *Black Mountain Review* 1, nos. 1–4, 1954 (New York: AMS Press, 1969): iii.

32. As Creeley was not in residence at BMC until the summer of 1954, he conflated the October 1952 Pottery Seminar with the summer session of 1953. Voulkos taught in 1953 and was not present for the pottery seminar. Voulkos first encountered Leach, Hamada, and Yanagi in December of 1952 in Helena, Montana, at Archie Bray, a fledgling pottery in a former brickyard that he, along with Rudy Autio, worked to establish as an artists' residency with the backing of businessman and namesake Archie Bray.

33. Mary Emma Harris, notes from an untaped interview with Marguerite Wildenhain, Guerneville, CA, December, 1971, box 40, Black Mountain College Research Project.

34. R[obert] D[iffendal], "Black Mountain Seminar covers art potters' role; U.S. results," *Ceramic Industry* (December 1952): 103–6.

35. Toshio Wantabe, ed., *Ruskin in Japan 1890–1940: Nature for Art, Art for Life* (Sheffield, UK, and Kamakura, Japan: Cogito, Inc., 1997), 314.

36. The Hunter Library of Western Carolina University, Cullowhee, NC, has created an excellent website from their extensive archives. See "Craft Revival: Shaping Western North Carolina Past and Present," www.wcu.edu/craftrevival/index.html. See also Jane S. Becker, *Selling Tradition:*

Appalachia and the Construction of an American Folk, 1930–1940 (Chapel Hill: University of North Carolina Press, 1998).

37. Western Carolina University has made a portion of its archive available online, through the Hunter Library website (http://wcudigitalcollection.edu). I have utilized its excellent finding aid and cataloging to cull biographical information related to Frances Goodrich. The fifth generation to attend Yale, Goodrich was the first woman in her family to do so, at a time when the university did not grant degrees to women. She attended from 1879–1882 and received a certificate, rather than a diploma.

38. Allen H. Eaton, *Handicrafts of the Southern Highlands* (New York: Russell Sage Foundation, 1937), 211.

39. Nancy Sweezy, "Jugtown Pottery," in *Raised in Clay: The Southern Pottery Tradition* (Washington, DC: Smithsonian Institution Press, Office of Folklife Programs, 1984), 211. I am indebted to Sweezy for the in-depth background her book provided in the preceding few paragraphs.

40. Mrs. Juliana Busbee to Black Mountain College, September 5, 1952, series II, box 25, Black Mountain College Research Project (hereafter cited as box 25).

41. Louise L. Pitman, Director, Southern Highland Handicraft Guild to Hazel Larsen, October 21, 1952, box 25.

42. Eaton, *Handicrafts*, 143.

43. Glenn Adamson has wrongly compared Ulmann to 1930s documentarian Walker Evans, misconstruing Ulmann as Evans's colleague in the Works Progress Administration (WPA) corps of photographers, and comments upon the "carefully staged" quality of her work, as well as her use of props and costumes. See Glenn Adamson, *Thinking Through Craft* (Oxford and New York: Berg Publishers, 2007), 112–13. In fact, Ulmann was not a WPA photographer. She was a former student of Clarence White and published often in *Pictorial Photography in America*, the 1920s-era journal of the Pictorial Photographers of America, of which she was a member.

44. See www.southernhighlandcraftguild.org for more information.

45. Edward DuPuy and Emma Weaver, "Louise Pitman," *Artisans of the Appalachians* (Asheville, NC: Miller Printing Company, 1967), 119.

46. Two years after her arrival in the United States, Wildenhain held a brief six-month appointment as the pottery director of the Appalachian Institute for the Arts and Crafts, North Carolina, an organization that opened and closed the same year, in 1942, dying along with its patron.

47. Edward L. Dupuy, interview by Mary Emma Harris, 1971, box 30, pp. 17–21, Black Mountain College Research Project.

48. Nicholas Fox Weber, *Josef and Anni Albers: Designs for Living* (New York: Merrell Publishers and Cooper Hewitt National Design Museum, 2004), 82–105.

49. M. C. Richards to Charles Olson, August 2, 1954, Charles Olson Papers. Richards resigned from Black Mountain in 1951, and while she had planned to attend the Pottery Seminar, she suffered from "female trouble," likely a miscarriage, around that time.

50. Victor C. Petchul, Editor, *Ceramic Industry*, to Robert L. Diffendal, September 30, 1952, box 25.

51. Hazel-Frieda Larsen to W. K. Burriss, Technical Editor, October 28, 1952, box 25.

52. Harris, *The Arts at Black Mountain College*, 175.

53. *Black Mountain College Bulletin-Newsletter* 2, no. 8 (August 1944): 2, State Library of North Carolina Digital Repository, http://digital.ncdr.gov. The GI Bill, passed in June, also reimbursed the college for costs associated with advertising to veterans. See Frydl, *The GI Bill*, 190.

54. This initiative began even earlier than 1944. See "Education in Wartime," *Black Mountain*

College Announcements, 1942–1943, 2–5, State Library of North Carolina Digital Repository, http://digital.ncdcr.gov.

55. Eric Martz to the author, January 22, 2008, e-mail correspondence.

56. Lilian Bochen to Constance Olson, September 27, 1952, box 25.

57. Rick Newby and Chere Jiusto, "'A Beautiful Spirit': Origins of the Archie Bray Foundation for the Ceramic Arts," in *A Ceramic Continuum: Fifty Years of the Archie Bray Influence*, ed. Peter Held (Seattle: University of Washington Press, 2001), 25.

58. On its website, the Archie Bray Foundation lists Lillian Boschen Tidball as a former resident artist, rather than its first director, further diminishing her reputation. See http://www.archiebray.org/artists.

59. Emmanuel Cooper, *Janet Leach: A Potter's Life* (London: Ceramic Review Publishing, Ltd., 2006), 26–27. Darnell became Bernard Leach's third wife in 1956.

60. Janet Darnell to Constance Olson, August 15, 1952, box 25.

61. A. J. Spencer to Constance Olson, September 24, 1952, box 25.

62. Cooper, *Bernard Leach*, 147. For an excellent and detailed account of Leach and Hamada in 1920s England, see chapter 7, "An Artistic Pottery: England, 1920–1926," 138–68.

63. Shoji Hamada, "Introduction: Yanagi and Leach," in *The Unknown Craftsman: A Japanese Insight Into Beauty*, ed. Soetsu Yanagi (Tokyo: Kodansha International, 1972), 9.

64. Karen Karnes, interview by Mark Schapiro, August 9–10, 2005, Archives of American Art.

65. Ibid.

66. Janet Leach, "Preface," in *Hamada: Potter* (Tokyo: Kodansha International, 1975), 10.

67. Rudy Autio, interview by Lamar Harrington, October 10 and 12, 1983, Archives of American Art.

68. Diffendal, "Black Mountain Seminar," 106.

69. "Program for the Pottery Seminar," typescript, box 25.

70. Val Cushing, interview by Margaret Carney, April 16, 2001, Archives of American Art.

71. Soetsu Yanagi, *The Unknown Craftsman: A Japanese Insight Into Beauty* (Tokyo: Kodansha International, 1972), 223.

72. Like any art school, the trail goes cold for large numbers of students and alumni. I was able to confirm that at least seven of the men in attendance were veterans, and two of the women had served the war effort, but specific biographical information is largely unknown for the majority of the attendees.

73. Otto and Vivika Heino, interview by Elaine Levin, March 4, 1981, Archives of American Art.

74. Donald S. Lopez, Jr., *The Story of Buddhism: A Concise Guide to Its History and Teachings* (New York: Harper Collins, 2001), 24.

75. Brian Moeran, *Folk Art Potters of Japan: Beyond an Anthropology of Aesthetics* (Surrey, England: Curzon Press, 1997), 24–25.

76. Susan Peterson, *Shoji Hamada: A Potter's Way and Work* (Tokyo: Kodansha International, 1974), 26.

77. Mark Schapiro, "Shoji Hamada," in *Innovation and Change: Ceramics from the Arizona State University Museum*, ed. Peter Held (Tempe: Arizona State University, 2009), 178.

78. Janet Leach, "Impressions of Mr. Hamada, His Pottery and Life at Mashiko, 1954," in *Shoji Hamada: Master Potter*, ed. Timothy Wilcox (London and Sussex: Lund Humphries and Ditchling Museum, 1998), 30.

79. Yanagi, *The Unknown Craftsman*, 121.

80. Leach, "Impressions," 32.

81. Peterson, *Hamada*, 39.

82. Ibid., 41.

83. According to Susan Peterson, Hamada spoke eight languages fluently, including English, French, and Korean. Hamada published a short text in English as an introduction to Yanagi's English-language volume of *The Unknown Craftsman* in the 1972 edition. See above, note 63.

84. Rudy Autio, interview by Lamar Harrington, October 10 and 12, 1983, Archives of American Art.

85. Yanagi, *The Unknown Craftsman*, 114.

86. See Branden Joseph, "John Cage and the Architecture of Silence," *October* 81 (Summer 1997): 80–104.

87. See Caroline A. Jones, "Finishing School: John Cage and the Abstract Expressionist Ego," *Critical Inquiry* 19, no. 4 (Summer 1993): 628–65.

88. Yuko Kikuchi, "The Myth of Yanagi's Originality: The Formation of 'Mingei' Theory in its Social and Historical Context," *Journal of Design History* 7, no. 4 (1994): 249–50.

89. Helen Westgeest, *Zen in the Fifties: Interaction Between East and West* (Zwolle, Netherlands: Wanders Uitgevers, 1996), 52.

90. Yuko Kikuchi, "Hybridity and the Orientalism of 'Mingei' Theory," *Journal of Design History* 10, no. 4 (1997): 344.

91. Wantabe, ed., *Ruskin in Japan*, 320–25. I am indebted to this volume for background context, both in Japan and Europe.

92. Robert H. Scharf, "The Zen of Japanese Nationalism," in *Curators of the Buddha: The Study of Buddhism under Colonialism*, ed. Donald S. Lopez, Jr. (Chicago: University of Chicago Press, 1995), 108.

Chapter Four

1. See Jay David Bolter and Richard Grusin, *Remediation: Understanding New Media* (Cambridge: MIT Press, 1999), 5.

2. For an in-depth understanding of the primary and subsequent receptions of Edelson's poster (which is in its third printing), see Linda S. Aleci, "In a Pig's Eye: The Offence of Some Living American Women Artists," in *The Art of Mary Beth Edelson* (New York: DAP, 2002), 32–33, 32b-32e.

3. Wagner, *Three Artists*, 24–25.

4. Mary Beth Edelson to the author, January 20, 2014, e-mail correspondence.

5. Bolter and Grusin, *Remediation*, 24.

6. Ann Laura Stoler, "Intimations of Empire: Predicaments of the Tactile and Unseen," in *Haunted by Empire: Geographies of Intimacy in North American History*, ed. Ann Laura Stoler (Durham: Duke University Press, 2006), 22.

7. Poster for *Clay Things to Touch, to Plant in, to Hang Up, to Cook in, to Look at, to Put Ashes in, to Wear, and for Celebration, Made by M. C. Richards*, 1958, box 1, M. C. Richards Papers. The poster is part of an addition to the original acquisition that completed the GRI's M. C. Richards archive.

8. Charles Mingus Jazz Workshop, 16 January 1959, box 48, folder 7, Charles Mingus Collection, Music Division, Library of Congress, Washington, DC.

9. MCR to CO, September 10, 1957, series II, box 208, folder: "1957–1959," Charles Olson Papers.

10. Paul Williams to MCR, September 30, 1953, box 2, folder 7, John Cage Collection, Special Collections, Music Library, Northwestern University, Evanston, IL, hereafter John Cage Papers.

11. M. C. Richards to Charles Olson, 10 September 1957, series II, box 208, folder: "1957–1959," Charles Olson Papers.

12. Allan Kaprow's *18 Happenings in 6 Parts* at the Reuben Gallery in New York occurred on October 4 and October 6–10, 1959, beginning at 8:30 p.m. each evening. The flier is reproduced in Eva Meyer-Hermann, Andrew Perchuk, and Stephanie Rosenthal, eds., *Allan Kaprow: Art as Life* (Los Angeles: Getty Research Institute, 2008), 120–21. For the original poster, see Getty Research Institute, flat folder, file 1, Allan Kaprow Papers, ca. 1940–1997 (980063). Claes Oldenburg's *The Store* was held December 1–31, 1961, at the Green Gallery in New York; its hours were 1 to 6 p.m., Fridays, Saturdays, and Sundays. The letterpress poster is in the print collection of the Museum of Modern Art, New York.

13. Back of poster for *Clay Things to Touch, to Plant in, to Hang Up, to Cook in, to Look at, to Put Ashes in, to Wear, and for Celebration, Made by M. C. Richards*, 1958, box 1, M. C. Richards Papers.

14. M. C. Richards, *Centering: On Pottery, Poetry, and the Person* (Middletown, CT: Wesleyan University Press, 1989), 3.

15. Numerous descriptions of the event abound, though the most comprehensive is William Fetterman, *John Cage's Theatre Pieces: Notations and Performances* (New York: Routledge, 1995). See also Martin Duberman, *Black Mountain: An Exploration in Community* (New York: Dutton, 1972), 347–54; Harris, *The Arts at Black Mountain College*, 226–28; and Branden W. Joseph, *Random Order: Robert Rauschenberg and the Neo-Avant-Garde* (Cambridge: MIT Press, 2003), 258.

16. Box 1, M. C. Richards Papers.

17. Judith F. Rodenbeck, *Radical Prototypes: Allan Kaprow and the Invention of Happenings* (Cambridge: MIT Press, 2011), 16.

18. The exhibition catalog had a slightly different title: *Living As Form: Socially Engaged Art from 1991–2011*, ed. Nato Thompson (Cambridge and New York: MIT Press and Creative Time), 2011.

19. Ibid., 25.

20. CO to MCR, June 21, 1957, box 22, folder 1, M. C. Richards Papers.

21. Richards, *Centering*, 111.

22. Harris, *The Arts at Black Mountain College*, 188–89.

23. Richards, *Centering*, 22–23.

24. Ibid., 29.

25. M. C. Richards, "Black Mountain College: A Golden Seed," *Craft Horizons* 27, no. 3 (June 1977): 70.

26. Paul Reps and Nyogen Senzaki, *Zen Flesh Zen Bones: A Collection of Zen and Pre-Zen Writings* (Boston: Tuttle, 1957, rpt. 1985), 121.

27. Paul Goodman, who, along with Frederick Perls, coauthored the influential text *Gestalt Therapy* (1951), taught literature and writing the summer of 1950 at Black Mountain. His contract was not renewed, as he was the source of much controversy, partly for his sexual pursuit of male students and partly for his clash with the more conservative Quaker members of the faculty. See Harris, *The Arts at Black Mountain College*, 214–16. This is detailed by Duberman, *Black Mountain*, 260.

28. These were second books for both Marcuse and Brown, close friends whose first books, Marcuse's *Eros and Civilization* (1955) and Brown's *Life Against Death: The Psychoanalytic Meaning Against History* (1959), are the beginning of their ongoing and spirited debates in print. Previously, the two émigrés served together in the Office of Strategic Services (forerunner of the CIA) during the war years, 1943 to 1946.

29. Richards, *Centering*, 38–39.

30. Marshall McLuhan, *Understanding Media: The Extensions of Man* (Cambridge: MIT Press, 1994), 98–99.

31. Richards, *Centering*, 49.

32. Ibid., 15.

33. Ong's "The Presence of the Word: Some Prolegomena for Cultural and Religious History" was initially delivered in three parts as the Terry Lectures in September of 1967 at Yale University.

34. MCR to JC, February 1962, box 4, folder 3, John Cage Papers. While the archive has labeled this letter February 1962, this seems impossible, since *Understanding Media* was not published until 1964. I believe the letter is from 1966, or, conversely, the book to which Richards refers is actually McLuhan's *Gutenberg Galaxy: The Making of Typographic Man*, which was published in 1962 and was about print culture. However, this would not explain how Richards would have an easy familiarity with the title of a book that was still two years away from publication.

35. Commencing at Vanderbilt University during the 1930s, The Fugitives was a group of Southern scholar-poets, including John Crowe Ransom, Andrew Lytle, Allen Tate, and Robert Penn Warren, that formed the basis for latter New Critics.

36. Mary Phelan Bowles, interview by Mary Emma Harris, December 21, 1971, series IV: "Released Interviews," box 35, p. 8, Black Mountain College Research Project.

37. Paul Williams, interview with Mary Emma Harris, 1971, series IV: "Released Interviews," box 40, p. 9, Black Mountain College Research Project.

38. Duberman, *Black Mountain*, 292.

39. It is my belief that Richards's letter was the actual catalyst for the Dreiers' and Alberses' resignations. In her book, Mary Emma Harris simply describes this period as: "The crisis that precipitated the resignation of the Alberses and others began in the summer of 1948" (*The Arts at Black Mountain College*, 164). This is because she had no access to this letter, which is not included among the college's papers. Duberman was given a carbon copy of this letter by Richards, acknowledged in his footnotes. He did not leave a copy in his own research files. The letter in its entirety can be found in Richards's papers, which were not acquired by the Getty until 1994. MCR to Ted and Bobbie Dreier, July 25, 1948, series I, box 1, folder 4, M. C. Richards Papers.

40. The child's ashes were buried at the base of the tree in front of the building. Alex Reed, an art student who worked closely with Anni Albers, designed and executed Quiet House in 1942, gathering stones, weaving the curtains himself, and building the benches out of wood with the guidance of Molly Gregory, the college's highly skilled carpenter and woodworker who was also a sculptor.

41. Norman O. Brown to MCR, September 31, 1962, series II, box 6, M. C. Richards Papers.

42. Stephen de Staebler, interview with Mary Emma Harris, December 13, 1971, box 30, pp. 21–22, Black Mountain College Research Project.

43. In February of 1949, more in-fighting broke out between remaining faculty, and three contracts were terminated, then immediately reinstated: Richards, Levi, and dramatist Joe Fiore.

44. M. C. Richards, interview by Martin Duberman, February 20, 1967, p. 23, P. C. 1678.14, Martin Duberman Collection, North Carolina State Archives, Raleigh. The previous three pages, 20–22, offer many more details.

45. Anni Albers to MCR, September 26, 1961, series II, box 1, folder 3, M. C. Richards Papers.

46. Theodore Dreier to MCR, September 12, 1961, series II, box 1, folder 3, M. C. Richards Papers.

47. Box 64, folder 2, M. C. Richards Papers.

48. Mary Emma Harris to Tram Combs, June 3, 1987, series III, box 2, folder 12, M. C. Richards Papers.

49. Paul K. Conkin, "Review," *Journal of American History* 60, no. 2 (September 1973): 512.

50. Ibid., 510.

51. William M. Ham, "Review," *History: Reviews of New Books* 1, no. 2 (Nov./Dec. 1972): 26.

52. Reed Whittemore, "Up in the Hills with Art," November 4, 1972, publication unknown, box 34: "Reviews 1972," Martin Duberman Collection, North Carolina State Archives, Raleigh.

53. Duberman, "Notes to Entries and Exits: Pages 288–294," *Black Mountain*, 472.

54. She was also extremely close to, but did not live with, Remy Charlip, a Merce Cunningham company dancer.

55. Diary entry, Sunday, May 20, 1951, series VI, box 56, folder 1, M. C. Richards Papers.

56. Whitney Davis, *Queer Beauty: Sexuality and Aesthetics from Winckelmann to Freud and Beyond* (New York: Columbia University Press), 246.

57. Carolyn Brown, *Chance and Circumstance: Twenty Years with Cage and Cunningham* (New York: Alfred A. Knopf, 2007), 88.

58. Numerous descriptions of the event abound. See Duberman, *Black Mountain*, 347–54; Harris, *The Arts at Black Mountain College*, 226–28; Joseph, *Random Order*, 258.

59. Harris, *The Arts at Black Mountain College*, 228.

60. Joseph, *Random Order*, 258. In this passage, Joseph's own endnote refers back to the Harris page from which I just quoted.

61. Robert Wunsch resigned in 1945, and Wesley Huss was appointed in 1950. Elizabeth and Peter Jennerjahn were the students who initiated the Light Sound Movement Workshop. According to Mary Emma Harris, other participants included Nick Cernovich, Mark Hedden, Dorothea Rockburne, and Vera Williams. See Harris, *The Arts at Black Mountain College*, 206–11.

62. Richards's translation of Cocteau's *Knights of the Round Table* appears in the chronology of her life in the exhibition catalog, *Imagine Inventing Yellow: The Life and Work of M. C. Richards* (Worcester, MA: Worcester Center for Crafts, 1999). She also directed W. B. Yeats's *The Death of Cuchulain* (1939), as in a June 4, 1950 playbill, Black Mountain College Research Project.

63. Richards, *Centering*, 50.

64. Antonin Artaud, *The Theatre and Its Double*, trans. Mary Caroline Richards (New York: Grove Press, 1958), 59.

65. James Leo Herlihy, interview with Mary Emma Harris, November 13, 1972, p 16, series IV: "Interviews," box 31, Black Mountain College Research Project.

66. *All Fall Down* (dir. John Frankenheimer, 1962) starred a young Warren Beatty and Eva Marie Saint. The playwright William Inge adapted the screenplay.

67. JLH to MCR, August 6, 1962, series III, box 14, M. C. Richards Papers.

68. In addition to Duberman, also in existence are numerous secondhand descriptions; see Harris, *The Arts at Black Mountain College*, 151.

69. Merce Cunningham Company, "Complete Chronology," www.merce.org/archive_chronology.html.

70. Daniel Charles, "Sixth Interview," in *For the Birds: John Cage in Conversation* (Boston: Marion Boyars, Inc., 1981), 164–67.

71. Duberman, *Black Mountain*, 351. A later version of *Theater Piece* was reprised in New York, March 7, 1960, at the Square Theater.

72. Fetterman, *John Cage*, 142.

73. Naseem Khan, "How We Met: Merce Cunningham and M. C. Richards," *Independent* (London), August 21, 1994, 56.

74. Brown, *Chance*, 18.

75. Molly Gregory, interview with Mary Emma Harris, n.d., 1971, p 14, Black Mountain College Research Project.

76. I am indebted to the following essays for excellent discussions of their respective oeuvres: Robert Orledge, "Satie & America," *American Music* 18, no. 1 (Spring 2000): 78–102; and Annette Shandler Levitt, "Jean Cocteau's Theatre: Idea and Enactment," *Theater Journal* 45 (1993): 363–72.

77. "January 9, 1949," series II: "Notebooks and Diaries," box 56, folder 1: "Diary 1948–1949," unpaginated, M. C. Richards Papers.

78. John Cage, "The Future of Music: Credo," in *Silence: Lectures and Writings* (Middletown, CT: Wesleyan University Press, 1961), 6.

79. Kenneth E. Silver, *Esprit de Corps: The Art of the Parisian Avant-garde and the First World War, 1914–1925* (Princeton: Princeton University Press, 1989), 116–39.

80. "Entry," series II: "Notebooks and Diaries," box 56, folder 1: "Diary 1947–1948," unpaginated, M. C. Richards Papers.

81. Deborah Menaker Rothschild, *Picasso's Parade: From Street to Stage* (London and New York: Sotheby's Publications and The Drawing Center, 1991), 236–37.

82. Ibid., 38–39.

83. Bolter and Grusin, *Remediation*, 47.

84. Michael Kirby and Richard Schechner, "An Interview with John Cage," *Tulane Drama Review* 10, no. 2 (Winter 1965): 52.

85. Rothschild, *Picasso's Parade*, 38–39.

86. MCR to Paul and Vera Williams, May 2, 1949, box 3: "Correspondence 1948–1952," M. C. Richards Papers.

87. Brown, *Chance*, 106.

88. Allan Kozinn, "David Tudor, 70, Electronic Composer, Dies," *New York Times*, August 15, 1996.

89. Charles, *For the Birds*, 124.

90. Both Karen Karnes (as of 1956) and the Williamses had young children they were raising, hence their domestic situations were profoundly different from Richards, who was childless.

91. MCR to CO, August 2, 1954, Charles Olson Papers.

92. Karen Karnes, interview with Mark Schapiro, August 9–10, 2005, Archives of American Art.

93. MCR to CO, August 2, 1954, Charles Olson Papers.

94. Brown, *Chance*, 107.

95. MCR to John Cage, November 1954, box 52, folder 3: "Correspondence B-C," David Tudor Papers (980039), Special Collections, Getty Research Institute, Los Angeles, hereafter David Tudor Papers

96. Williams to MCR, August 1962, M. C. Richards Papers.

97. An American composer, Seymour Barab (b. 1921), collaborated often with other writers, including Kurt Vonnegut (www.seymourbarab.com). Chanticleer is a rooster from various fables, the earliest of which appears in Chaucer's *Canterbury Tales*. A surviving copy of the libretto exists (M. C. Richards Papers, box 70). A vocal score remains in distribution by Boosey & Hawkes, Inc. (Cast: Soprano, Mezzo, Tenor, Baritone; Set: A farmyard. The present; Instrumentation: Flute, Oboe, 2 Clarinets, 2 Horns, Trumpet, Trombone, Timp-Bass, Percussion, Harp, Strings.)

98. MCR to CO, December 6, 1957, Charles Olson Papers.

99. MCR to CO, August 2, 1954, Charles Olson Papers.

100. MCR to CO, June 14, 1956, Charles Olson Papers.

101. Gavin Butt, *Between You and Me: Queer Disclosures in the New York Art World, 1948–1963* (Durham: Duke University Press, 2005), 13.

102. M. C. Richards, *Imagine Inventing Yellow: New and Selected Poems* (Barrytown, NY: Station Hill Arts, 1991), 73–74.

103. CO to MCR, June 7, 1951, M. C. Richards Papers.

104. Dawson, *The Black Mountain Book*, 21–22.

105. James Leo Herlihy, interview with Mary Emma Harris, November 13, 1972, p. 24, series IV: "Released Interviews," Black Mountain Research Project.

106. Molly Gregory, interview with Mary Emma Harris, March 31, 1971, p. 4, series IV: "Released Interviews," box 31, Black Mountain Research Project.

107. See Ann Eden Gibson, *Abstract Expressionism: Other Politics* (New Haven: Yale University Press, 1997); Michael Leja, *Reframing Abstract Expressionism: Subjectivity and Painting in the 1940s* (New Haven: Yale University Press, 1993).

108. Conniff's description is offered as a critique of the way he believes Martin Duberman disregarded Morley in his book, mentioning her only once. What Conniff fails to remember is that Morley's own acclaimed first book of poetry, *A Blessing Outside Us*, was not published until 1976—four years after Duberman's book. Brian Conniff, "Reconsidering Black Mountain: The Poetry of Hilda Morley," *American Literature* 65, no. 1 (March 1993): 120.

109. Ibid., 121–122.

110. MCR to CO, November 8, 1958, Charles Olson Papers.

111. Denise Levertov to Robert Duncan, October 7, 1958, in *The Letters of Robert Duncan and Denise Levertov*, ed. Robert J. Berthlolf and Albert Gelpi (Stanford: Stanford University Press, 2004), 141.

112. MCR to CO, April 21, 1964, series II, box 208, folder: "1961–1966," Charles Olson Papers.

113. Brown, *Chance*, 422.

114. Richards's first marriage, while she was a graduate student, lasted only a year.

115. MCR to John Cage, box 4, folder 3, John Cage Papers.

116. MCR to David Tudor, June 23 and August 10, 1960, box 58, folder 7, David Tudor Papers.

117. Norman O. Brown to MCR, July 27, 1960, series II: "Correspondence," box 6, M. C. Richards Papers.

118. Norman O. Brown to MCR, July 14, 1960, series II: "Correspondence," box 6, M.C. Richards Papers.

119. Wesleyan University Press pamphlet, box 26, folder 7, M. C. Richards Papers.

120. MCR to CO, April 21, 1964, Charles Olson Papers.

121. For a window into the repetition of players and spaces, see *Critical Mass: Happenings, Fluxus, Performance, Intermedia, and Rutgers University, 1958–1972*, ed. Geoffrey Hendricks (Amherst, MA: Mead Art Museum, 2003).

122. Sally Banes, *Greenwich Village 1963:Avant-Garde Performance and the Effervescent Body* (Durham: Duke University Press, 1993), 29.

123. Richards, *Centering*, 48–49.

124. Charles Olson to MCR, November 3, 1965, section II, box 22, folder 1, M. C. Richards Papers.

125. T. J. Clark, "On the Social History of Art," *Image of the People: Gustave Courbet and the Second French Republic* (Princeton: Princeton University Press, 1973), 14.

126. Richards, *Centering* Drafts, series III, box 18, M. C. Richards Papers.

127. Daniel Rhodes, "Review: Centering," *Craft Horizons* 25, no. 1 (January/February 1965):

52. Rhodes was not a stranger to Zen practices; he had spent a great deal of time making pottery in Japan, which would culminate in his book *Tamba Pottery* (Tokyo: Kodansha International, 1970).

128. MCR to W. Howard Adams, June 18, 1966, M. C. Richards Papers. Adams was an editor at Harper & Row who was encouraging Richards to write a "Black Mountain book"—the same one, incidentally, she had tried, unsuccessfully, to elicit institutional, foundation, and personal support for just five years prior. Only a partial letter can be found in MCR's papers. The complete letter—with its postal stamp of June 18, indicating it was also sent to Duberman—exists in reproduction in the M. C. Richards file, box 3, Martin Duberman Collection, North Carolina State Archives, Raleigh.

129. Christian Wolff and David Patterson, "Cage and Beyond: An Annotated Interview with Christian Wolff," *Perspectives of New Music* 32, no. 2 (Summer, 1994): 79.

130. JC to MCR, May 27, 1963, M. C. Richards Papers.

131. Box T1, M. C. Richards Papers, uncataloged.

132. Richards, *Centering*, 15.

133. JC to Jake Rose, Doubleday, October 11, 1968, John Cage Papers.

Chapter Five

1. McLuhan, *Understanding Media*, 181.

2. My argument is indebted to and enriched by Caroline Jones's notion of "post-studio production," articulated throughout *Machine in the Studio: Constructing the Post-War Artist* (Chicago: University of Chicago Press, 1996). The other touchstone for this chapter is David Joselit, *Feedback: Democracy After Television* (Cambridge: MIT Press, 2007).

3. Lauria, ed., *Color and Fire*, 104.

4. Robert Smithson, "Donald Judd," in *The Writings of Robert Smithson: Essays and Illustrations*, ed. Nancy Holt (New York: NYU Press, 1979), 21.

5. Susan Peterson, interview by Paul J. Smith, March 10, 2004, Archives of American Art.

6. "Upfront: The Grande Dames of Ceramics," *Ceramics Monthly* 52, no. 7 (September 2004), unpaginated section.

7. Susan Peterson, "Faux or Real Folk Art: The Story of Mingei," transcript of Dorothy Wilson Perkins Lecture, International Museum of Ceramic Art, Alfred University, October 26, 1999. Fellows' Files, American Craft Council, Minneapolis.

8. For an extended history of Chouinard and its influence on CalArts, and the gender divides in Los Angeles art pedagogy, see my essay "Learning from Los Angeles: Pedagogical Predecessors at the Woman's Building," in *Doin' It In Public: Feminism and Art at the Woman's Building* (Los Angeles: Otis College of Art and Design, 2011), 36–64. For a complete history of the school, see Robert Perine's *Chouinard, An Art Vision Betrayed: The Story of the Chouinard Art Institute, 1921–1972* (Encinitas, CA: Artra Publishing, 1985).

9. Susan Peterson, "Ceramics in the West: The Explosion of the 1950s," in Lauria, ed., *Color and Fire*, 101. According to Mary Emma Harris, it was only due to the GI Bill monies that BMC was able to pay the faculty a regular salary during the postwar period. See Harris, *The Arts at Black Mountain College*, 175.

10. "Registration Instructions, Spring 1958," subseries 2.3: "Admissions and Registration," box 5, folder 1, Chouinard Archives. National Defense Loans were a predecessor of the Perkins loan program established by the National Defense Education Act in 1958. See "History of Loan Forgiveness," www.finaid.org/loans.

11. Chouinard Art Institute course catalog, 1947–1948, subseries 2.3, box 11, folder 11, Chouinard Archives.

12. Chouinard Art Institute course catalog, 1954–1955, Chouinard Archives. The other majors were: fine arts, illustration, motion picture arts (four-year degree programs); advertising design, costume design, interior design, fashion illustration, animation and cartooning (three-year programs).

13. Vivika and Otto Heino, interview by Elaine Levin, March 4, 1981, Archives of American Art.

14. Ibid.

15. Biographical information was found on the following websites: Albert Henry King Papers, Smithsonian Institution, http://siris-archives.si.edu; Ed Traynor, in the Luther College/Pond Farm Collection, http://finearts.luther.edu/artists/traynor.html; and Lennox Tierney Interviews Collection Finding Aid, Special Collections, University of Utah, http://db3-sql.staff.library.utah.edu/lucene.

16. Richard Petterson, "California: A Climate for Craft Art," *Craft Horizons: A Special Issue on California* 16, no. 5 (October 1956), 11.

17. Author's interview with Susan Peterson, December 10, 2007.

18. See my essay, "A Prolonged Silence: John Cage and Still After," in *Silence* (New Haven: Yale University Press/The Menil Collection, 2012), 83–87.

19. Susan Peterson, interview by Paul J. Smith, March 10, 2004, Archives of American Art.

20. Martin Mayer, "Whatever Happened to Educational Television?" *Change* 4, no. 2 (March 1972): 48. Mayer's excellent article, published in the first decade of public television, is one of the few critiques and chronologies of the medium.

21. Ibid.

22. Ibid., 51–52.

23. David Joselit, *Feedback*, 24. For an excellent point-by-point discussion drawn from period literature, see the entirety of section 2, chapter 1, beginning on page 15.

24. Art Seidenbaum to SHP, July 28, 1967, box 11, folder 11, Susan Harnly Peterson Archives, Ceramics Research Center, Arizona State University Art Museum, Tempe, AZ, hereafter SHP Archives.

25. It is unclear if Peterson coined the phrase herself, but the slogan originates in a 1967 press release for the series's second run on CBS, during 1967–68. Box 11, folder 10, SHP Archives.

26. John Mason, interview by Paul J. Smith, August 28, 2006, Archives of American Art.

27. Susan Peterson, "Episode 1: Introduction: Overview of Ceramics," *Wheels, Kilns, and Clay*, 1964, two-inch tape transferred to VHS, 1997, SHP Archives.

28. John Mason, interview by Paul J. Smith, August 28, 2006.

29. Happy Price to the author, e-mail, January 26, 2010. Price was extremely ill in 2010, and passed away in 2012.

30. Ibid.

31. Susan Peterson, "Throwing Clay on the Potter's Wheel," in *Wheels, Kilns, and Clay: A Ceramics Text, 2nd Ed.* (Los Angeles: University of Southern California, 1972), 17. Box 24, SHP Archives.

32. Susan Peterson, *Wheels, Kilns, and Clay*, draft, box 11, folder 12, SHP Archives.

33. Janet Leach, "Lucie Rie," in *Lucie Rie: A Survey of Her Life and Work*, ed. John Houston (London: Crafts Council, 1981), 30.

34. Moira Vincentelli, *Women Potters: Transforming Traditions* (London: A & C Black, 2003), 10.

35. Royden K. Loewen, "Household, Coffee Klatsch, and Office: The Evolving Worlds of Mid-Twentieth-Century Mennonite Women," in *Strangers at Home: Amish and Mennonite Women in History*, eds. Kimberly D. Schmidt, Diane Zimmerman Umble, and Steven D. Reschly (Baltimore: Johns Hopkins University Press, 2002), 271.

36. Ibid.

37. I am indebted to an excellent anthology of essays on Amish and Mennonite women, Kimberly D. Schmidt, Diane Zimmerman Umble, and Steven D. Reschly, eds., *Strangers At Home: Amish and Mennonite Women in History* (Baltimore: Johns Hopkins University Press, 2002). Regarding Kansas in particular as a key site of Anabaptist modernization, see in this volume Royden K. Loewen, "Household," 259–83.

38. Paul W. Harnly, "The Life and Times of Paul Witmore Harnly" (1988), unpublished manuscript, 61, box 24, folder 8, SHP Archives.

39. Margaret C. Reynolds, "River Brethren Breadmaking Ritual," in *Strangers at Home*, 78–101. While the ceremony that Reynolds examines is particular to Old Order Brethren communities in Lancaster, Pennsylvania, surely Susan Peterson, also raised in a Brethren sect, would have experienced a version of this female ritual.

40. Reynolds, "River Brethren Breadmaking Ritual," 85.

41. Susan Peterson, *The Living Tradition of Maria Martinez* (Tokyo: Kodansha International, 1977), 192.

42. Susan Peterson, *Lucy M. Lewis: An American Potter* (Tokyo: Kodansha International, 1984), 24.

43. Ernest Welbeck, "As We Were Saying," *Idyllwild Town Crier* 21, no. 46 (December 15, 1967), unpaginated. Articles binder, box 11, SHP Archives.

44. See Elissa Auther, *String, Felt, Thread: The Hierarchy of Art and Craft in American Art* (Minneapolis: University of Minnesota Press, 2010), 1. Auther's book is the first scholarly analysis of 1960s-era American fiber art, but her ideas around classification systems are applicable to all craft media.

45. Susan Peterson, interview with Paul J. Smith, March 10, 2004.

46. Peterson, "Throwing Clay on the Potter's Wheel,", 14.

47. Peterson, *Wheels, Kilns, and Clay*, draft, box 11, folder 12, SHP Archives

48. WKC Correspondence, box 11, folder 13, SHP Archives.

49. Press release, University of Southern California News Bureau, October 22, 1967, box 11, folder 13, SHP Archives.

50. Episode titles are as follows: "Englobe Decorating on Bowl Forms" (24); "How to Mix an Englobe" (25); "Glaze Application on Bisqued Ware" (27); "Variety of Glaze Effects" (36); "Combining Oxides to Make Glazes" (37); "Mathematical Calculation of Glaze Formulas" (38); "Unusual Glazes: Ash, Lustre, Salt" (39); "Raku Glaze Firing" (40); and "Glaze Faults, Problems, Difficulties" (41).

51. For a succinct summary of Judd's career and oeuvre, see Ann Goldstein, "Donald Judd," in *A Minimal Future? Art as Object, 1958–1968*, ed. Ann Goldstein (Los Angeles and Cambridge: MOCA and MIT Press, 2004), 254–62.

52. "Specific Objects" originally appeared in *Arts Yearbook* 8, 1965. Reprinted in Donald Judd, *Complete Writings, 1975–1986/Donald Judd* (Eindhoven, Netherlands: Van Abbemuseum, 1987), 122. See also *Complete Writings, 1959–1975* (Halifax, NS, and New York: Press of the Nova Scotia College of Art and Design and NYU Press), 1975.

53. Judd, *Complete Writings, 1959–1975*, 98.

54. Peter Schjeldahl, "Light in Juddland: Flavin at Marfa," in *Let's See: Writings on Art from The New Yorker* (New York: Thames and Hudson, 2008), 39–40.

55. Smithson, *Writings*, 21.

56. Laura Schapiro, *Something from the Oven: The Rise of Packaged Food-Cuisine in 1950s America* (New York: Viking Press, 2009), 29.

57. See my previous description of this event in chapter 3.

58. Dan Aykroyd, "The French Chef," Saturday Night Live Transcripts, season 4, episode 8 (December 9, 1978), http://snltranscripts.jt.org/78/78hchef.phtml.

59. Susan Peterson, interview by Paul J. Smith, March 10, 2004, Archives of American Art.

60. Invented in 1956 by Ampex, a Southern California-based company, two-inch tape was lauded as an important invention that was a vast improvement (and replacement) for a film process called kinescoping, in which a film was made of a live broadcast via a television monitor, and the resulting print was duplicated and then transmitted widely. See the CBS and FCC websites, previously cited.

61. See Deirdre Boyle, *Subject to Change: Guerilla Television Revisited* (London: Oxford University Press, 1997). Boyle documents this phenomenon in her pioneering study of early video collectives such as TVTV and their forays into made-for-television documentaries.

62. Ibid., 5.

63. Miriam Schapiro, interview by Ruth Bowman, September 10, 1989, Archives of American Art.

64. Lucy Lippard, "Changing Since *Changing* (1976)," in *The Pink Glass Swan: Selected Essays on Feminist Art: 1970–1993* (New York: The New Press, 1995), 33.

65. For a fuller description of Womanhouse and its many other installations and performances, see Arlene Raven, "Womanhouse," in *The Power of Feminist Art: The American Movement of the 1970s, History and Impact*, eds. Norma Broude and Mary Garrard (New York: Abrams, 1995), 48–65.

66. Rebecca Morse, "Martha Rosler," in *WACK! Art and the Feminist Revolution*, eds. Connie Butler and Lisa Mark (Los Angeles and Cambridge: MOCA and MIT Press, 2007), 290–91.

67. Jennifer Way, "'Gold Mine in Southeast Asia': Russel Wright, Vietnamese Handicraft, and Transnational Consumption," October 2, 2009, symposium presentation at *A Long and Tumultuous Relationship: East-West Interchanges in American Art*, Smithsonian American Art Museum, Washington, DC.

68. For a discussion of Rosler in relation to other feminist colleagues working outside the space of the museum, see Helen Molesworth, "House Work and Art Work," *October* 92 (Spring 2000): 71–97.

69. Martha Rosler, interview by Jenni Sorkin, February 7, 2010, Mexico City.

70. Though, like Judd, she also maintains a treasured personal collection: her home contains many examples of early twentieth-century art pottery, including Gruby and Rookwood.

71. Lisa Bloom, "The California Work of Martha Rosler," in *Jewish Identities in American Feminist Art: Ghosts of Ethnicity* (New York: Routledge, 2006), 87.

72. Martha Rosler, February 7, 2010. I was fortunate enough to see this piece installed at Rosler's New Museum retrospective, New York, 2001.

73. Martha Rosler, *The East is Red, The West is Bending* (1977), video, color and sound, 19:57 minutes. Distributed by Electronic Arts Intermix, New York.

74. Martha Rosler, "For an Art Against the Mythology of Everyday Life," in *Decoys and Disruptions: Selected Writings, 1975–2001* (Cambridge: MIT Press, 2004), 8.

75. Susan Peterson, 1999 transcript, American Craft Council Fellow Archives.

76. "Eleanor Antin," in *Video Art: An Anthology*, eds. Ira Schneider and Beryl Korot (New York: Harcourt, Brace, and Jovanovich, 1976), 14.

77. Glenn Phillips, ed., *California Video: Artists and Histories* (Los Angeles: Getty Research Institute and Getty Publications, 2008), 200.

78. Vincentelli, *Women Potters*, 211.

79. For a comprehensive chronology of all-women exhibitions and spaces throughout the post-war period, see Jenni Sorkin and Linda Theung, "Selected Chronology of All-Women Group Exhibitions,

1943–1983," in *WACK! Art and the Feminist Revolution*, eds. Connie Butler and Lisa Mark (Los Angeles and Cambridge: MOCA and MIT Press, 2007), 473–99.

80. See Judy Chicago, *Through the Flower: My Struggle as a Woman Artist* (New York: Doubleday, 1975). There are many subsequent histories of the Feminist Art Program, which began at Fresno State College (known now as CSU Fresno) in 1971. For a fuller picture of the period, see Laura Meyer's excellent essay, "From Finish Fetish to Feminism: Judy Chicago's Dinner Party in California Art History," in *Sexual Politics: Judy Chicago's Dinner Party in Feminist Art History*, ed. Amelia Jones (Los Angeles and Berkeley: Hammer Museum and University of California Press, 1996).

81. Judy Chicago to the author, July 23, 2009, e-mail correspondence.

82. Helen Molesworth, "House Work and Art Work," *October* 92 (Spring 2000): 83.

83. I am indebted to my colleagues Michelle Moravec, Alex Juhasz, Vivien Fryd, and Jennie Klein for numerous conversations regarding the role of history at the Woman's Building.

84. Dena Kleiman, "Mrs. Mondale Shapes Clay as Aid to Arts," *New York Times*, April 6, 1978, page 1A.

Epilogue

1. Richards, "Introduction to the Second Edition," *Centering*, xix.

2. Karen Archey, "The Road to Somewhere," *Art in America*, August 20, 2010, http://www.artinamericamagazine.com/news-features/news/mildreds-lane/.

INDEX

An "f" following a page number indicates a figure. Color plates appear as "pl" followed by the plate number.